MAGNETIC STRATIGRAPHY

This is Volume 64 in the
INTERNATIONAL GEOPHYSICS SERIES
A series of monographs and textbooks
Edited by RENATA DMOWSKA and JAMES R. HOLTON

A complete list of books in this series appears at the end of this volume.

MAGNETIC

STRATIGRAPHY

NEIL D. OPDYKE

AND

JAMES E. T. CHANNELL
Department of Geology
University of Florida
Gainesville, Florida

ACADEMIC PRESS

San Diego   London   Boston
New York   Sydney   Tokyo   Toronto

Copyright © 1996 by ACADEMIC PRESS

All Rights Reserved.
No part of this publication may be reproduced or transmitted in any form or by any means, electronic or mechanical, including photocopy, recording, or any information storage and retrieval system, without permission in writing from the publisher.

Academic Press, Inc.
525 B Street, Suite 1900, San Diego, California 92101-4495, USA
http://www.apnet.com

Academic Press Limited
24-28 Oval Road, London NW1 7DX, UK
http://www.hbuk.co.uk/ap/

Library of Congress Cataloging-in-Publication Data

Opdyke, N. D.
    Magnetic stratigraphy / by N.D. Opdyke, J.E.T. Channell.
        p.   cm.
    Includes index.
    ISBN 0-12-527470-X (alk. paper)
    1. Paleomagnetism.   2. Stratigraphic correlation.   I. Channell,
J. E. T.   II. Title.
QE501.4.P35063   1996
551.7'01--dc20   96-5925
                                                            CIP

PRINTED IN THE UNITED STATES OF AMERICA
96 97 98 99 00 01 BC 9 8 7 6 5 4 3 2 1

In memory of S. K. Runcorn
1922–1995

Contents

1
Introduction and History

2
The Earth's Magnetic Field

3
Magnetization Processes and Magnetic Properties of Sediments

4
Laboratory Techniques

5
Fundamentals of Magnetic Stratigraphy

6
The Pliocene–Pleistocene Polarity Record

7
Late Cretaceous–Cenozoic GPTS

8
Paleogene and Miocene Marine Magnetic Stratigraphy

9
Cenozoic Terrestrial Magnetic Stratigraphy

10
Jurassic–Early Cretaceous GPTS

11
Jurassic and Cretaceous Magnetic Stratigraphy

12
Triassic and Paleozoic Magnetic Stratigraphy

13
Secular Variation and Brunhes Chron Excursions

14
Rock Magnetic Stratigraphy and Paleointensities

Preface

Magnetic polarity stratigraphy, the stratigraphic record of polarity reversals in rocks and sediments, is now thoroughly integrated into biostratigraphy and chemostratigraphy. For Late Mesozoic to Quaternary times, the geomagnetic polarity record is central to the construction of geologic time scales, linking biostratigraphies, isotope stratigraphies, and absolute ages.

The application of magnetic stratigraphy in geologic investigations is now commonplace; however, the use of magnetic stratigraphy as a correlation tool in sediments and lava flows has developed only in the past 35 years. As recently as the mid-1960s, magnetic polarity stratigraphy dealt with only the last 3 My of Earth history; today, a fairly well-resolved magnetic polarity sequence has been documented back to 300 My before present. An increasing number of scientists in allied fields now use and apply magnetic stratigraphy. This book is aimed at this expanding practitioner base, providing information about the principles of magnetostratigraphy and the present state of our knowledge concerning correlations among the various (biostratigraphic, chemostratigraphic, magnetostratigraphic, and numerical) facets of geologic time.

Magnetic Stratigraphy documents the historical development of magnetostratigraphy (Chapter 1), the principal characteristics of the Earth's magnetic field (Chapter 2), the magnetization process and the magnetic properties of sediments (Chapter 3), and practical laboratory techniques (Chapter 4). Chapter 5 describes the essential elements that should be incorporated into any magnetostratigraphic investigation.

The middle of the book is devoted to a survey of the status of magnetic stratigraphy in Jurassic through Quaternary marine (Chapters 6–8, 10, and 11) and terrestrial (Chapter 9) sedimentary rocks and describes the integration of the biostratigraphic and chemostratigraphic record with the geomagnetic polarity time scale (GPTS). Recent developments in extending the magnetic polarity record to pre-Jurassic sediments and the emergence of a magnetic polarity record for the Triassic, Permian, and Carboniferous periods are chronicled in Chapter 12.

The record of secular variation of the geomagnetic field has been important in attempting correlations in sediments that span the last 10,000 years (Chapter 13). Magnetic stratigraphy is expanding further through the use of nondirectional magnetic properties of sediments, such as magnetic susceptibility and a host of other magnetic parameters sensitive to magnetic mineral concentration and grain size (Chapter 14). It is now becoming apparent that geomagnetic intensity variations, as recorded in sediments, may well be the next important magnetostratigraphic tool, promising to provide high-resolution global correlation within polarity chrons.

The authors thank the following people, who made important suggestions for improving the manuscript: B. Clement, D. Hodell, K. Huang, E. Irving, D. V. Kent, E. Lindsay, B. MacFadden, M. McElhinny, S. K. Runcorn, and J. S. Stoner. Parts of this book were written while one of us (J.E.T.C.) was on sabbatical in Kyoto (Japan) and Gif-sur-Yvette (France). M. Torii, C. Laj, and C. Kissel made these sabbaticals possible. Finally, we express special appreciation to Marjorie Opdyke, who patiently and skillfully produced the original figures for this book with loving attention to detail.

The authors of *Magnetic Stratigraphy* received their doctoral training from the University of Newcastle-upon-Tyne, where Professor S. K. Runcorn inspired their life-long interest in paleomagnetism. We are saddened by his murder in December of 1995 and dedicate this volume to Dr. Runcorn as an expression of our gratitude to him as our friend and mentor.

Neil D. Opdyke
James E. T. Channell

Table Key

Throughout this book, compilations of important magnetostratigraphic studies are presented as black/white (normal/reverse) interpretive polarity columns. These are accompanied by tables that include the following information for each study.

Rock unit	A designation of the rock unit, usually as a formation name, locality, or core/hole/site number
Age range	Age range of the stratigraphic section(s)
Region	Country or region of study
λ	Site latitude (positive: north, negative: south)
Φ	Site longitude (positive: east, negative: west)
NSE	Number of sections
NSI	Number of sites or site interval (m). P (includes pass-through magnetometer measurements)
NSA	Number of samples per site
M	Section thickness (m)
D	Associated absolute age control: K (K-Ar), A (^{40}Ar/^{39}Ar), F (fission track), R (Rb/Sr), S (strontium isotopic ratios), O (δ^{18}O), As (astrochronology)
RM	Rock magnetic measurements/observations: K (susceptibility), J (Js/T), I (IRM), V (VRM), A (ARM), H (hysteresis measurements), TdI (thermal demagnetization of IRM), T (electron microscopy), O (optical microscopy), X (X-ray)
DM	Demagnetization procedure: A (alternating field), T (thermal), C (acid/chemical)
AD	How was the magnetization direction determined? B (blanket demagnetization), Z (Zijderveld/orthogonal

	projections), V (vector analysis), P ("principal component" or 3D least squares analysis)
A	Data presentation: F (Fisher statistics presented), D (declination plotted), I (inclination plotted), V (VGP latitudes plotted)
NMZ	Number of magnetozones (polarity zones)
NCh	Number of chrons
%R	Percentage of reverse polarity
RT	Reversal test: R+ [positive in the A, B, C rating of McFadden and McElhinny (1990) or according to McElhinny, (1964)], R− (negative)
F.C.T.	F (fold test), C (conglomerate test), Co (baked contact test); indicated as positive or negative
Q	Reliability index (see Chapter 5.7)

It should be noted that the quality of the data is sometimes not reflected in the reliability index. For example, some marine sediments are such good magnetic recorders that it is possible to get excellent quality magnetostratigraphic data with only basic techniques such as blanket demagnetization with alternating fields, as can be seen in some whole core data from the Ocean Drilling Program.

1

Introduction and History

1.1 Introduction

Paleomagnetism has had profound effects on the development of Earth sciences in the last 25 years. In the early days, paleomagnetic studies of the different continental blocks contributed to the rejuvenation of the continental drift hypothesis and to the formation of the theory of plate tectonics.

Paleomagnetism has led to a new type of stratigraphy based on the aperiodic reversal of polarity of the geomagnetic field, which is now known as *magnetic polarity stratigraphy*. Magnetic polarity stratigraphy is the ordering of sedimentary or igneous rock strata into intervals characterized by the direction of magnetization of the rocks, being either in the direction of the present Earth's field (normal polarity) or 180° from the present field (reverse polarity). This new stratigraphy greatly influenced the subject of plate tectonics by providing the chronology for interpretation of oceanic magnetic anomalies (Vine and Matthews, 1963). In the last 25 years, the geomagnetic polarity time scale (GPTS) has become central to the calibration of geologic time. The bridge between biozonations and absolute ages and the interpolation between absolute ages are best accomplished, particularly for Cenozoic and Late Mesozoic time, through the GPTS. In most Late Jurassic–Quaternary time scales, magnetic anomaly profiles from ocean basins with more or less constant spreading rates provide the template for the GPTS. Magnetic polarity stratigraphy on land or in deep sea cores provides the link between the GPTS and biozonations/bioevents and hence geologic stage boundaries. Radiometric absolute ages are correlated either directly to the GPTS in magnetostratigraphic section or indirectly through biozonations, and absolute ages are then interpolated using the GPTS (oceanic magnetic anomaly) template. In view of the paramount importance of the calibration of geologic time to understanding the rates of geologic processes, the contribution of magnetic polarity stratigraphy to the Earth sciences becomes self-evident.

The dipole nature of the main geomagnetic field means that polarity reversals are globally synchronous, with the process of reversal taking 10^3–10^4 yr. Magnetic polarity stratigraphy can therefore provide global stratigraphic time lines with this level of time resolution. Three other techniques in magnetic stratigraphy, not involving the record of geomagnetic polarity reversals, have become increasingly important. *Rock magnetic stratigraphy* refers to the use of nondirectional magnetic properties (such as magnetic susceptibility and laboratory-induced remanence intensities) as a means of stratigraphic correlation. *Paleointensity magnetic stratigraphy,* the use of the record of geomagnetic paleointensity, and *secular variation magnetic stratigraphy,* the use of secular directional changes of the geomagnetic field, have been used as a means of stratigraphic correlation in Quaternary sediments.

Individual texts exist for paleomagnetism in general, such as those of Irving (1964), McElhinny (1973), Tarling (1983), Butler (1992), and Van der Voo (1993); however, none is available for the rapidly growing discipline of magnetic stratigraphy.

1.2 Early Developments

Directions of natural remanent magnetization (NRM) reverse with respect to the present ambient magnetic field of the Earth were known to early pioneers in paleomagnetic research. Brunhes (1906) reported directions of magnetization in Pliocene lavas from France that yielded north-seeking magnetization directions directed to the south and up, rather than to the north and down. He attributed this behavior to a local anomaly of the geomagnetic field. Brunhes demonstrated that the baked contacts of igneous rocks are magnetized with the same polarity as the igneous rock. This was the first use of what is now referred to as the baked contact test, which can be used to determine the relative age of magnetizations in the vicinity of an igneous contact. When lava flows are extruded, or dikes are intruded into a host rock, the temperature of the rock or sediment into which the magma is intruded is raised above the blocking temperature of the magnetic minerals. The magnetization is, therefore, reset in the direction of the field prevailing at the time of baking. Brunhes carried out this test for both normally and reversely magnetized igneous contacts.

This work was followed by that of Matuyama (1929) on volcanic rocks from Japan and north China. This study included many examples of lavas that yielded directions that were reverse with respect to the present geomagnetic field (Fig. 1.1). Matuyama correctly attributed the reverse directions of magnetization to a reversal of the geomagnetic field. Matuyama separated his samples into two groups: Group I, which were all Pleistocene in age, gave NRM directions which grouped around the present direction of the geomagnetic field. The group II directions were antipodal to those of group I and to the present directions of the geomagnetic field, and most of these lavas were older than Pleistocene in age. This was the first hint that the polarity of the geomagnetic field might be age dependent. These early studies were followed by those of Chevallier (1925), Mercanton (1926), and Koenigsberger (1938).

The modern era of studies of reversals of the geomagnetic field began with those of Hospers (1951, 1953–1954) in Iceland and Roche (1950, 1951, 1956) in the Massif Central of France, which elaborated on the early work of Brunhes. In both of these studies, the relative age and position of the rocks were fixed stratigraphically. As in Matuyama's studies, the results indicated that rocks and sediments designated as Upper Pleistocene and Quaternary in age possessed normal directions of magnetization, whereas

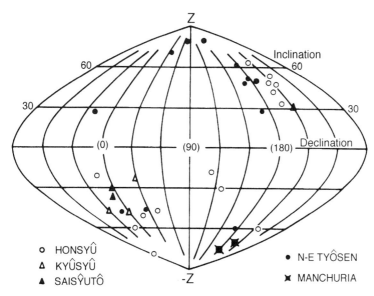

Figure 1.1 Directions of natural remanent magnetization in basalts from China and Japan (after Matuyama, 1929).

reverse directions of magnetization appeared in rocks of early Pleistocene or Pliocene age. Einarsson and Sigurgeirsson (1955) utilized a magnetic compass to detect reversals in Icelandic lavas and began to map the distribution of magnetic polarity zones in the rock sequences. They found about equal thicknesses of normal and reverse polarity strata, which implied that the magnetic field had little polarity bias in the Late Cenozoic. A study by Opdyke and Runcorn (1956) on lava flows from the San Francisco Peaks of northern Arizona showed that all young lavas studied were normally magnetized, whereas the stratigraphically older series of flows contained reversely magnetized lavas. Based on these studies, it was assumed that the last reversal of the field took place close to the Plio-Pleistocene boundary. Rutten and Wensink (1960) and Wensink (1964) built on Hosper's studies in Iceland and demonstrated that the youngest lavas were normally magnetized and that older underlying lavas were reverse, and these in turn were underlain by a normal sequence. They subdivided the Plio–Pleistocene lavas into three magnetozones N_1-R_1-N_2. They correlated N_2 to the Astien and R_1 to the Villafranchian and deduced that the earliest glaciations in Iceland were of Pliocene age.

Magnetometers capable of measuring the magnetization of igneous rocks had been available since before World War II. Johnson *et al.* (1948) developed a rock generator magnetometer which could measure natural remanence in sedimentary rocks. Blackett (1956) developed astatic magnetometers which were a further improvement in sensitivity. Paleomagnetic study of sedimentary rocks began with the classic study of Creer *et al.* (1954) which documented 16 zones of alternating polarity in a 3000-m section of the Torridonian sandstone (Scotland). These rocks passed the fold test, indicating a prefolding magnetization, implying that reversals were a long-term feature of the geomagnetic field. These authors also reported reversals from rocks of Devonian and Triassic age.

Simultaneous with the developments outlined above, which seemed to favor reversal of the dipole geomagnetic field as an explanation for the observations, a discovery was made in Japan which cast doubt on the field reversal hypothesis. Nagata (1952), Nagata *et al.* (1957), and Uyeda (1958) demonstrated that samples from the Haruna dacite, a hyperthermic dacite pumice from the flanks of an extinct volcano in the Kwa district of Japan, was self-reversing. The striking magnetic property of this rock is that, when heated to a temperature above 210°C and cooled in a weak magnetic field, the acquired thermal remanent magnetization (TRM) is in opposition to the applied field. Since self-reversal in rocks could be demonstrated, it became important to ascertain whether all reverse directions could be explained in this way, or whether both the self-reversal phenomenon and reversal of the main dipole field were taking place.

1.3 Evidence for Field Reversal

The baked contact test first employed by Brunhes played an important role is settling the self-reversal/field reversal controversy. If field reversal is the norm, one would expect baked contacts to be magnetized in the same polarity as the baking igneous rock. If self-reversal is the norm, one would expect the baked contacts to have different polarity from the igneous rock in a significant proportion of the cases, since it is unlikely that both igneous and country rock would have the same magnetic mineralogy. Wilson (1962) conducted a survey of such studies and found that in 97% of the contacts studied, the polarity of the baked contact was the same as that of the igneous body. These observations supported the hypothesis of geomagnetic field reversal.

By the late 1950s, several lines of evidence supported the hypothesis that the geomagnetic field reverse polarity. (1) All rocks of Late Pleistocene and recent age, both igneous and sedimentary, that had been studied possessed normal polarity magnetizations. (2) Reverse polarity magnetizations were observed in rocks of Lower Pleistocene or older age and in sedimentary and igneous rocks involving different magnetization processes. (3) Transitional (intermediate) directions were observed in Iceland in both reverse-to-normal and normal-to-reverse transitions, as well as in the Torridonian of Scotland. (4) Self-reversal had been observed in the laboratory in only one rock type (Haruna dacite) although many attempts had been made to replicate the experiment in other rock types. (5) As described above, in 97% of the contact tests, the baked contact and the baking rock possessed the same polarity. By the late 1950s, many paleomagnetists believed in geomagnetic field reversal although geophysicists in general were still skeptical.

In the early 1960s, the stage was set for the convincing demonstration of geomagnetic field reversal. A series of interdisciplinary studies were carried out in which K/Ar dating and the measurement of magnetization polarity were carried out on the same lavas (Cox *et al.*, 1963; McDougall and Tarling, 1963b). These studies established that rocks of the same age had the same polarity and led to the establishment of the first radiometrically dated polarity time scale. As information accumulated throughout the 1960s, the magnetic polarity sequence was progressively modified (Fig. 1.2), particularly with the discovery of short magnetic polarity intervals (Grommé and Hay, 1963). The long intervals of constant polarity of the geomagnetic field were designated by Doell and Dalrymple (1966) as magnetic epochs and named after the pioneers in the study of the geomagnetism (Brunhes, Matuyama, Gauss, and Gilbert); the shorter intervals (events)

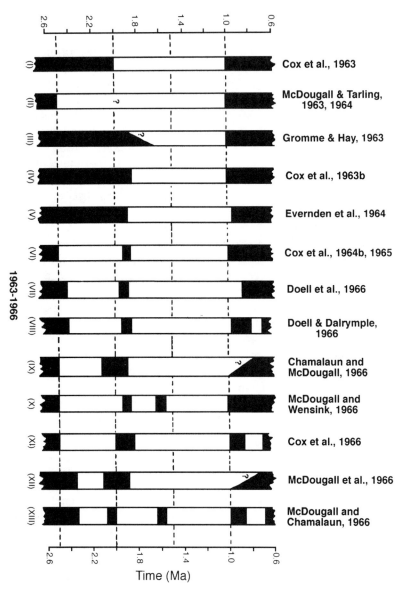

Figure 1.2 Evolution of the geomagnetic polarity time scale of the last 2.6 My (after Watkins, 1972).

were named after the locality of discovery such as Jaramillo Creek in New Mexico and Olduvai Gorge in Tanzania. It should be noted that the time scales were produced from lavas that were scattered over several continents and oceanic islands and were not in stratigraphic superposition, representing a major departure from standard stratigraphic procedures. The anomalous situation arose in which, unlike classical stratigraphy, no type sections for the magnetic epochs (polarity chrons) are available, yet type localities for magnetic events (polarity subchrons) are known.

The developments which led to the first dated polarity time scale were soon followed by studies of magnetic stratigraphy in sediments and investigations of oceanic magnetic anomalies. Khramov (1958) in the Soviet Union reported reverse and normal magnetizations in Plio–Pleistocene nonmarine sedimentary sequences from the Caucasus. These studies were followed by the discovery of reversals in marine sediments (Harrison and Funnel, 1964; Linkova, 1965). Opdyke *et al.* (1966) observed the same sequence of polarity

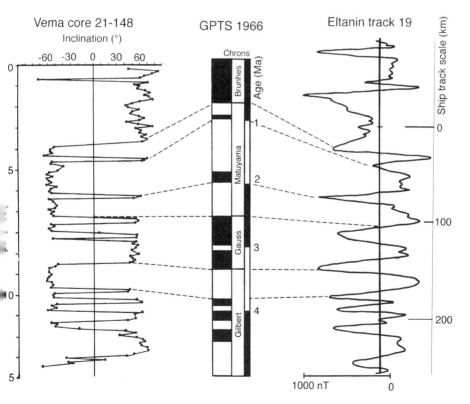

Figure 1.3 Comparison of oceanic magnetic anomalies from ship track Eltanin 19 (Pitman and Heirtzler, 1966) with magnetic polarity stratigraphy in core V21-148 (Opdyke, 1968) and the magnetic polarity time scale of Doell and Dalrymple (1966).

reversals in marine sediments that had been compiled by Cox *et al.* (1963) from dispersed volcanic outcrops. The same reversal sequence was then observed in oceanic magnetic anomalies by Vine and Wilson (1965) and Pitman and Heirtzler (1966). The fact that the same reversal sequence appeared to have been recorded by three different recording media (Fig. 1.3) closed the argument in favor of field reversal and relegated self-reversal to a subsidiary process. The resolution of this argument in favor of field reversal tipped the scales in favor of the hotly debated hypothesis of plate tectonics and sea floor spreading.

2

The Earth's Magnetic Field

2.1 Introduction

The magnetic field of the Earth has fascinated human beings for well over 2 millennia. The Chinese invented the magnetic compass in the second century B.C. (Needham, 1962) and knowledge of the magnetic compass reached Western Europe over a thousand years later in the twelfth century A.D. The first truly scientific paper on geomagnetism was written in 1262 by Petrus Peregrinus and entitled "Epistola de Magnete" (Smith, 1970). In a series of experiments on lodestone (magnetite) spheres, Peregrinus defined the concept of polarity and defined the dipolar nature of a magnet stating that like poles repel and unlike poles attract; however, the treatise was not published until 1558. The work by Peregrinus undoubtedly influenced the later work of William Gilbert, who published the book "De Magnete" in 1600. Gilbert studied the variation of the angle of inclination (or magnetic dip) over lodestone cut into the shape of a sphere, perhaps one of the greatest model experiments ever made in Earth science. His conclusion was that "the Earth itself is a great magnet." From the 14th century to the present, compasses were widely used as a navigation aid on both land and sea and were carried by explorers throughout the world. The angular distance between the polar star, which does not move in the heavens, and the direction of the north-seeking end of the compass needle was often faithfully recorded. This magnetic deviation from true north is called the magnetic declination (D). In midlatitudes, from 40°N to 40°S, the declination of the present geomagnetic field can vary up to 40° from true north. In polar regions, near the magnetic poles, declination anomalies of up to 180° are observed (Fig. 2.1).

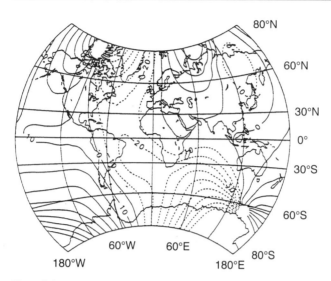

Figure 2.1 Declination for the International Geomagnetic Reference Field (IGRF) 1990 (after Baldwin and Langel, 1993).

If a magnetized needle is suspended on a fiber and allowed to swing freely, the north-seeking end of the needle will not only point north, it will also point down in the northern hemisphere and up in the southern hemisphere. This property of the earth's field, its inclination, was discovered by George Hartmann in 1544 (Smith, 1968). If a dip needle which measures the angle of magnetic inclination is carried from high northern latitudes to high southern latitudes, it will be seen that the inclination of the field varies systematically from vertical down at the north magnetic pole, to the horizontal at some point at low latitude (magnetic equator), and vertical once more (but with the north-seeking end of the needle pointing vertically upward) at the south magnetic pole (Fig. 2.2). It should be noted that the north and south magnetic poles are not 180° apart and that the magnetic equator is not equidistant from the two magnetic poles.

If the geomagnetic field is analogous to that of the magnetized spheres studied by William Gilbert, the intensity of the field, as well as the inclination and declination, would vary systematically over the Earth. After Gauss invented the deflection magnetometer, this was found to be the case, the magnetic field strength at the magnetic poles being almost twice that at the magnetic equator.

The field at any point on the Earth's surface is a vector (F) which possesses a component in the horizontal plane called the horizontal component (H) which makes an angle (D) with the geographical meridian (Fig.

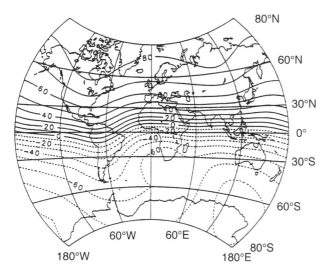

Figure 2.2 Inclination for the International Geomagnetic Reference Field (IGRF) 1990 (after Baldwin and Langel, 1993).

2.3). The declination (D) is an angle from north measured eastward ranging from $0°$ to $360°$. The inclination (I) is the angle made by the magnetic vector with the horizontal. By convention, it is positive if the north-seeking vector points below the horizontal or negative if it points above. The horizontal component (H) (Fig. 2.3) has two components, one to the north X and one to the east Y. The following equations relate the various quantities:

$$H = F \cos I \quad Z = F \sin I \quad \text{Tan } I = Z/H \tag{2.1}$$

$$X = H \cos D \quad Y = H \sin D \quad \text{Tan } D = Y/X \tag{2.2}$$

$$F^2 = H^2 + Z^2 = X^2 + Y^2 + Z^2. \tag{2.3}$$

The dipole nature of the geomagnetic field, originally suggested by Gilbert, was first put to a mathematical analysis by Gauss (1838), who used measurements of declination, inclination, and intensity from 84 widely spaced locations to derive the first 24 coefficients in the spherical harmonic expansion of the geomagnetic field. The results of this analysis indicated that, within the uncertainties of the data, the geomagnetic field had no sources external to the Earth and that it had a dominant dipole component. Since 1835, when Gauss first determined the intensity of the dipole moment, the calculations have been repeated many times, and this value has been decreasing steadily over the last century and a half. During this century,

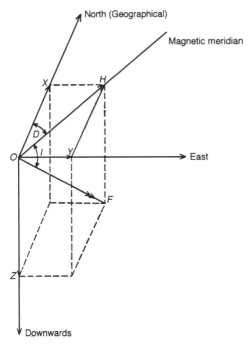

Figure 2.3 The direction and intensity of the total field vector (F) resolved into declination from geographical north (*D*) and inclination from horizontal (*I*). Equations (2.1), (2.2), and (2.3) relate the various quantities.

the decay rate has increased, and during the last 30 years has reached a rate of 5.8%/century (see Barton, 1989). If the trend were to continue, the main dipole field would disappear by the year 4000 A.D., a possible but unlikely event.

Modern spherical harmonic analyses confirm Gauss' findings and lead to the conclusion that: (1) the geomagnetic field is almost entirely of internal origin, (2) ~90% of the field observed on the surface can be explained by a dipole inclined to the Earth's axis of rotation by 11.5° (Fig. 2.4), (3) the magnitude of the dipole moment is 7.8×10^{22} Am^2, (4) the axis of the centered dipole (geomagnetic pole) emerges in the northern hemisphere between Greenland and Ellesmere Island at 79.0°N, 70.9°W. The great circle midway between the geomagnetic poles is called the geomagnetic equator (Fig. 2.4).

Magnetic observations of the elements of the geomagnetic field were begun as early as the 16th century in London and the 17th century in Paris. It was realized early on that the declination and inclination of the

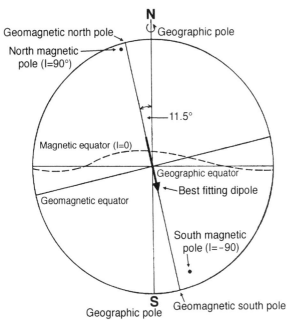

Figure 2.4 Graphical presentation of the magnetic, geomagnetic, and geographic poles and equators (after McElhinny, 1973).

geomagnetic field change with time (Fig. 2.5). This relatively rapid change of the magnetic field with time is called secular variation. The declination of the field at London has been changing at a rate of about 14° per century from 0° in 1650 to 24°W in 1800 and is now 5°W. The change in the magnetic elements is not constant over the Earth's surface and, for example, changes less rapidly over the Pacific Ocean than in, say, South America. Global secular variation is often displayed as "isoporic" charts, which are contour maps of equal annual change of a particular element of the field, such as the vertical component. From isoporic charts constructed from observations dating back about 200 years, isoporic centers have drifted westward at rates of up to about 0.28°/yr over this time interval. This observation is a manifestation of the westward drift of dipole, quadrupole, and octupole components of the field. Drift rates of these components can be calculated directly from the spherical harmonic coefficients and their changes with time. Changes in the rate of westward drift and the growth and decay of isoporic features occur on a decadal time scale. The westward drift is generally attributed to differential angular velocity of Earth's lithosphere and outer core, where the nondipole components of the field are thought

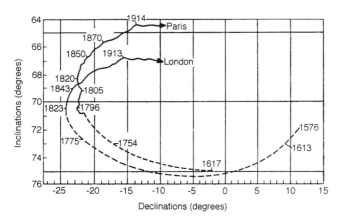

Figure 2.5 Change of declination and inclination at London and Paris from observatory records (after Gaiber-Puertas, 1953).

to originate. It is important to note that although westward drift is dominant, not all secular variation features drift westward; some are more or less stationary and some are moving slowly eastward.

The geomagnetic field varies over a wide range of time scales (Table 2.1). The longer term behaviors (4–7, Table 2.1) are produced in the Earth's interior, and the short-term behaviors (1–3, Table 2.1) are atmospheric or ionospheric in origin. Secular variations of the field, first observed in observatory records, can be monitored further back in time using paleomagnetic data from marine and lake sediments with high accumulation rates. The longer term variations (5–7, Table 2.1) have also been documented by paleomagnetic study of rocks and sediments. Since some lavas and archaeological artifacts can become permanently magnetized in a matter

Table 2.1
Scales of Geomagnetic Variability

Geomagnetic behavior	Duration
1. Pulsations or short-term fluctuation	minutes
2. Daily magnetic variations	hours
3. Magnetic storms	hours to days
4. Secular variations	10^2–10^3 yr
5. Magnetic excursions	10^3–10^4 yr
6. Reversal transition	10^3–10^4 yr
7. Interval between reversals	10^5–10^6 yr

of hours, it is theoretically possible to detect ancient field behavior on this time scale from paleomagnetic data.

2.2 The Dipole Hypothesis

The fact that the geomagnetic field can be modeled as a geocentric dipole is of fundamental importance in paleomagnetic studies. If we assume that the *time-averaged* dipole is aligned with the axis of rotation of the Earth and situated at the center of the Earth (an axial geocentric dipole), the geomagnetic and geographic axes will coincide, and the geomagnetic equator will coincide with the geographic equator. The geomagnetic inclination (I) will vary systematically with latitude (λ) according to the formula:

$$\text{Tan } (I) = 2 \text{ Tan } (\lambda).$$

If M is the moment of the geocentric dipole and the radius of the Earth is r, the horizontal (H) and vertical (Z) components at any latitude (λ) can be derived from the following geometric relationship (Fig. 2.6), where I is the inclination from the horizontal. Using the formulae used by Gauss for his deflection magnetometer,

$$H = (M \cos \lambda) / r^3$$
$$Z = (2M \sin \lambda) / r^3 \qquad\qquad (2.4)$$

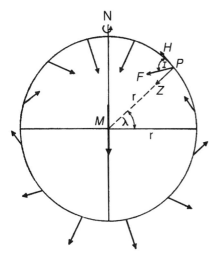

Figure 2.6 Field lines at the Earth's surface for the axial geocentric dipole (after McElhinny, 1973).

$$\text{Tan } I = Z/H = 2 \text{ Tan } (\lambda), \tag{2.5}$$

if $D = 0°$, colatitude (θ) is given by

$$\text{Tan } I = 2 \cos \theta \ (0 < \theta < 180°). \tag{2.6}$$

The relationship makes it possible to determine the paleolatitude of a site from mean paleomagnetic directions at any point on the Earth's surface. In the discussion given above, the declination (D) was presumed to be zero, although in most paleomagnetic investigations this is not so. The paleomagnetic pole represents the position on the Earth's surface where the dipole axis (oriented to give a field in agreement with the mean paleomagnetic direction) cuts the surface of the globe. Every paleomagnetic investigation, if successful, will yield a declination (D) which represents the angle between the paleomagnetic vector and the present geographic meridian. The declination defines a great circle that passes through the sampling site and the paleomagnetic pole. The mean paleomagnetic inclination gives the distance along the great circle (or paleomeridian) to the paleomagnetic pole, according to the dipole formula given above (2.5).

The pole position can be calculated in the present geographical coordinate system using the following relationship. If D_m, I_m are known at a sampling site with latitude λ and longitude Φ then the position of the paleomagnetic pole $P (\lambda', \Phi')$ can be calculated

$$\sin \lambda' = \sin \lambda \cos \theta + \cos \lambda \sin \theta \cos D_m \tag{2.7}$$

$$\Phi' = \Phi + \beta \quad \text{(when } \cos \theta > \sin \lambda \sin \lambda') \tag{2.8}$$

$$\Phi' = \Phi + (180 - \beta) \quad \text{(when } \cos \theta < \sin \lambda \sin \lambda'), \tag{2.9}$$

where $\sin \beta = \sin \theta \sin D/\cos \lambda'$.

By convention, latitudes are positive for the northern hemisphere and negative for the southern hemisphere; longitudes are measured eastward from the Greenwich meridian and lie between 0 and 360°. Other symbols are defined in Fig. 2.7. In paleomagnetic research, two types of paleomagnetic pole positions are calculated. A *virtual geomagnetic pole* (VGP) is a pole calculated from magnetic data which represent a very short period of time, such as the observatory record of D and I of the present field, or a spot paleomagnetic field reading from a single lava flow. A *paleomagnetic pole*, on the other hand, represents the field averaged over a considerable length of time (on the order of 50 ky) such that short-term (secular) variations are averaged out. A paleomagnetic pole position is essentially the mean vector sum of many virtual geomagnetic pole positions.

In the middle 1950s, it was known that the mean Plio–Pleistocene paleomagnetic poles more or less coincide with the pole of rotation of the Earth. A compendium of poles by Tarling (1983) yields an average

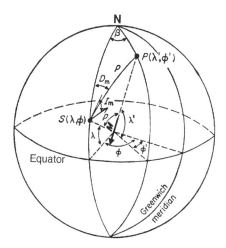

Figure 2.7 Calculation of a paleomagnetic pole from mean directions of magnetization (Dm, Im) at sampling site (S) generates a paleomagnetic pole (P) (after McElhinny, 1973).

paleomagnetic pole indistinguishable from the present pole of rotation (Fig. 2.8). These observations support one of the basic hypotheses in paleomagnetism, the dipole hypothesis, which states that the Earth's dipole field averages to an axial geocentric dipole over periods of time longer than the averaging time for secular variation (10–50 ky). The dipole hypothesis may be tested using widely spaced sampling sites for rocks of the same age, or by testing the relationship of inclination to site latitude in marine cores. Opdyke and Henry (1969) and Schneider and Kent (1990a) demonstrated that Pleistocene paleomagnetic inclinations in widely spaced marine cores are consistent with the dipole hypothesis (Fig. 2.9). Merrill *et al.* (1979) and Schneider and Kent (1988, 1990a) have documented asymmetry between normal and reverse intervals over the last few million years, which can be accounted for by increasing the quadrupole term for reverse polarity intervals relative to normal polarity intervals. This asymmetry is difficult to explain due to the symmetry of the induction equations (Gubbins, 1994). Outer core boundary conditions (in the mantle or inner core) can be invoked to break the symmetry of the dynamo equations (see Clement and Stixrude, 1995).

2.3 Models of Field Reversal

Although the geomagnetic field may be molded as if the Earth were a permanent magnet, it is clear that this is not the case as thermal disordering (Curie) temperatures of permanently magnetized materials are exceeded

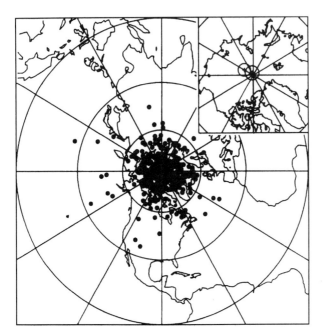

Figure 2.8 Paleomagnetic poles for igneous rocks younger than 20 Ma (after Tarling, 1983).

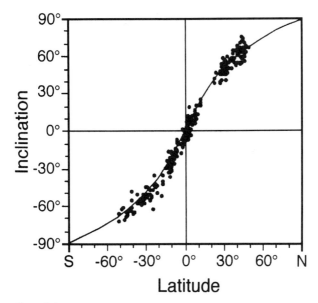

Figure 2.9 Inclination versus latitude in Pleistocene marine sediments. Line is expected inclination derived from the dipole formula (after Schneider and Kent, 1990a).

by burial greater than 10–20 km. The recognition of polarity reversals of the dipole field also provided a powerful argument against permanent magnetization. In 1947, Blackett proposed that planetary dipole fields are an intrinsic property of rotating bodies and subsequently designed experiments that refuted the claim (Blackett, 1952). Beginning with the work of Elsasser (1946) and Bullard (1949), it is now known that the geomagnetic field arises from convective and Coriolis motions in the fluid nickel–iron outer core. According to dynamo theory, outer core flow can amplify a small instantaneous axial field to observed dipole values, although the details of the process remain poorly constrained. No detailed theory predicting the form of geomagnetic secular variation or the mechanism of geomagnetic reversals is yet available. Gubbins (1994) and Jacobs (1994) have summarized recent developments on this topic.

Uncertainties in the details of the geomagnetic field generation process lead to little consensus on the process of geomagnetic polarity reversal. The models of Cox (1968, 1969), Parker (1969), and Levy (1972) involved interaction between the main dipole field and the nondipole field. According to Parker (1969) and Levy (1972), changes in distribution of cyclonic eddies, which account for the nondipole field, can produce regions of reverse torodial field which effect reversal of the poloidal (dipole) field. According to Cox (1968, 1969), the critical factor is the relative field strength of the high-frequency oscillations of the nondipole component relative to the lower frequency (cyclic) oscillations of the dipole component. A high-amplitude nondipole field coincident with a low in the main dipole field could, in this model, produce an instantaneous reverse total field, which is then reinforced by the dynamo process. According to Laj *et al.* (1979), the supposed coupling of the nondipole and dipole fields in the Cox model is unlikely in view of the different time constants of the two field components, and the mechanism (if feasible) would be expected to generate very frequent reversals.

Hillhouse and Cox (1976) supposed that the dipole component of the field dies in one direction and grows in the opposite direction during a polarity reversal and that the nondipole components do not change over times sufficient to span several reversals. The model predicts that sequential reversals, of opposite and same sense, recorded at a sampling site should yield similar transitional VGP paths. This has not been found to be the case and hence this model is no longer favored.

The models of Hoffman (1977) and Fuller *et al.* (1979) are based on the idea that the reversal process floods through the core from a localized zone. These models can be tested as they predict the VGP paths during a polarity transition for either quadrupole or octupole transition field geometries. The VGP paths are predicted to lie either along the site longitude (near-sided) or 180° away (far-sided) depending on the sense of the reversal, the hemisphere of observation, and where in the core the reversal process is initiated. These models are generally no longer favored due to the lack of evidence for near-sided or far-sided VGP paths.

Olson (1983) attributed reversal to a change in sign of the helicity (the correlation between turbulent velocity and vorticity), triggered by a change in the balance between heat loss at the core–mantle boundary and solidification at the inner-outer core boundary. In this model, the duration of a polarity chron will depend on the strength of the dipole field and the process of reversal would be rapid (~7500 yr), more or less consistent with paleomagnetic observations.

The solar dipole fluid reverses polarity approximately every 11 years (Eddy, 1976) and the observations of the solar reversal process provide clues to the reversal process on Earth (see Gubbins, 1994). The high frequency of solar field reversal allows the process to be observed directly. In addition, the flux distribution at the solar surface can be monitored, in contrast to the Earth, where the flux distribution at the core–mantle boundary must be approximated by downward continuation. Relatively low temperature sunspots are regions of intense magnetic field, considered as sites of expulsion of toroidal flux. The locations of sunspot pairs move from the solar poles toward the equator, leading up to a reversal of the solar dipole field.

Bloxham and Gubbins (1985) and Gubbins (1987a,b) interpret flux concentrations in maps of the radial component of the geomagnetic field as due to eddies at the core–mantle boundary, which they liken to sunspots. Two "core spots" have been identified beneath South America and southern Africa. The process at the Earth's core–mantle boundary leading to the reversal of the geomagnetic dipole may be similar to the solar reversal process, with flux spots beneath South America and southern Africa being analogous to sunspots. These core spots produce local fields opposed to the Earth's main dipole field. Bloxham and Gubbins (1985) have suggested that these areas of high magnetic flux account for the high rate of secular variation in the Atlantic region. The core spots are apparently strengthening as the main dipole field decays. They are apparently moving to the south and could possibly lead to a reversal of the main dipole field. As the core spots in these regions become larger, they might be observed at the surface as large regional deviations of the field (magnetic excursions).

2.4 Polarity Transition Records and VGP Paths

In the past decade, an increasing amount of high-quality paleomagnetic data have been acquired from polarity transitions in high deposition rate sedimentary sequences (e.g., Clement, 1991; Tric *et al.*, 1991a; Laj *et al.*, 1991) and in volcanic rocks (e.g., Prévot *et al.*, 1985; Hoffman, 1991, 1992). The records provide important insights into the configuration of the field during a polarity transition; however, the interpretations are controversial. Some data indicate that transition fields have a large dipolar component. For example, the records of the Cobb Mountain subchron from the western Pacific and North Atlantic have similar VGP (virtual geomagnetic pole) paths, implying dipolar transitional fields (Clement, 1992). The VGP paths of some spatially distributed Matuyama-Brunhes polarity transitions coincide (Clement, 1991) whereas others do not (Valet *et al.*, 1989). Another controversy has involved the apparent clustering of polarity transition VGPs along two longitudinally constrained paths, one at ~80°W longitude over North and South America and the other at ~100°E over eastern Asia and Australia (Fig. 2.10). Some studies emphasize the statistical significance of the preferred VGP paths (e.g., Clement, 1991; Laj *et al.*, 1991, 1992). Valet *et al.* (1992) have disputed both the claim that transition fields are dipolar and the statistical clustering of transitional VGPs to longitudinal paths. Quidelleur and Valet (1994) considered that the apparent clustering of VGP paths is biased by uneven site distribution. The highest concentration of sampling sites is in southern Europe, ~90° away from the VGP longitudinal path across the Americas. Egbert (1992) showed that, except in the case where the transition field is entirely dipolar, the VGPs are likely to cluster ~90° in longitude away from the sampling site. According to Constable (1992), the preferred VGP path across the Americas is apparent even after the sites located ~90° away from the path are removed from the record. It is important to realize that preferred longitudinal VGP paths do not necessarily mean that the geomagnetic field is dipolar during polarity transitions (Gubbins and Coe, 1993).

Prévot and Camps (1993) examined a large population of transitional records from volcanic rocks younger than middle Miocene and concluded that the transitional VGPs have no statistical bias in their distribution. On the other hand, more selective use of data from volcanic rocks leads to a different conclusion. Hoffman (1991, 1992) found that transitional VGPs from distributed sites of volcanic rocks younger than 10 Ma fall into two clusters in the southern hemisphere, one over Australia and the other over South America (Fig. 2.10). The clusters are located close to the Asian and American longitudinal VGP paths resolved from sedimentary transition records. These data lead to the conclusion that transitional field states are long-lived (i.e., characteristic of sequential reversals) and have a strong dipolar component, which is inclined to the Earth's rotation axis.

Bloxham and Gubbins (1987) have downward continued the field since 1700 A.D. to the core–mantle boundary, and it appears that magnetic flux

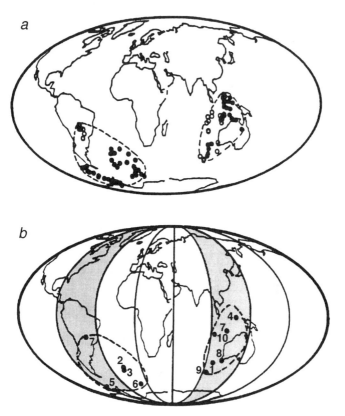

Figure 2.10 (a) Clustered VGPs concentrated over Southeast Asia and South America/Antarctica from polarity transition records from lava flows. (b) Means of VGPs from lava flows superimposed on shaded preferred longitudinal VGP paths from sedimentary reversal transition records (after Hoffman, 1992).

at the core–mantle boundary is also concentrated approximately along the east Asian and American longitudinal paths. Gubbins and Kelly (1993) have analyzed paleomagnetic data for the last 2.5 My from both lavas and sediments and suggested that the magnetic field at the core's surface has remained similar to that of today over this period of time. The persistence of the preferred paths through time and the fact that paths pass through regions of high modern flux activity at the core–mantle boundary suggest that these regions are pinned in their present position and have been for some time. The long time scales involved indicate that the position of these regions is controlled by mantle (as opposed to outer core) processes. Laj *et al.* (1991) and Hoffman (1992) have pointed out that radial flux centers associated with rapid secular change in today's field are closely associated with regions of fast-P-wave propagation in the lower mantle (Dziewonski and Woodhouse, 1987), indicating the presence of relatively cold mantle; however, this interpretation of mantle tomography in terms of temperature variations is not universally accepted.

The preferred longitudinal VGP paths lie either side of the Pacific Ocean, which is a region characterized by low-amplitude secular variation. At present, there are no foci of rapid secular change in the Pacific region and none have been known since measurements of the magnetic field in the region began several hundred years ago. Runcorn (1992) suggested that this low rate of secular change may be caused by a well-developed high conducting D'' shell under the Pacific. According to Runcorn (1992), as the dipole begins to reverse polarity, the magnetic field induces currents in the Pacific D'' shell, causing a torque that rotates the outer core relative to the mantle until the reversing dipole lies along the American or East Asian longitudinal paths.

Clement and Stixrude (1995) have invoked magnetic susceptibility anisotropy of the inner core to account for several problematic features of the paleomagnetic field. The Earth's inner core exhibits anisotropy in seismic velocity which can be explained in terms of the preferred orientation of hexagonal close-packed (hcp) iron. Room temperature analogs of hcp iron have a strong magnetic susceptibility anisotropy, which could produce an inner core field inclined to the rotation axis. This may help to explain not only the well-known far-sided effect (Wilson, 1972) but also the preferred VGP paths during polarity transitions. When the outer core field is weak during a polarity reversal, the persisting inner core field may bias the total field to produce the observed VGP paths and clusters observed in transition records. During polarity reversal, as the outer core field grows in the opposite direction, it may take a few thousand years to diffuse through the inner core and reset the inner core field.

Polarity reversals in the paleomagnetic record appear to be accompanied by geomagnetic intensity lows (e.g., Opdyke *et al.,* 1966), although in some sedimentary records, the apparent lows may be an artifact of the remanence acquisition process. Paleointensity studies by Valet and Meynadier (1993) imply reducing geomagnetic field intensity during Plio–Pleistocene polarity chrons, with intensity recovery immediately post reversal. The present axial dipole moment is 7.8×10^{22} Am^2. If it is assumed that the calibration of the relative paleointensity to absolute values is correct, then the dipole moment during reversals falls to about 1×10^{22} Am^2, or about a 10% of its present value. Valet and Meynadier (1993) pointed out that the lows in apparent paleointensity within the Brunhes Chron correspond to the age of documented field excursions (Ch. 14).

2.5 Statistical Structure of the Geomagnetic Polarity Pattern

As the GPTS is refined and our record of polarity reversals becomes more complete, estimates of the statistical structure of the polarity sequence become more realistic. From present to 80 Ma, since the Cretaceous long normal superchron, the polarity sequence approximates to a gamma process where the gamma function (k) equals ~1.72 (Phillips, 1977) or, in an updated estimate, ~1.5 (McFadden, 1989). The Poisson process is the special case of a gamma process where $k = 1$, corresponding to a random process. The observation that the gamma function is greater than 1 means that either the process has "memory," which reduces the likelihood of a reversal immediately post reversal, or that the record of reversals for the last 80 My is incomplete with many short-duration polarity chrons missing from the record. The oceanic magnetic anomaly record is characterized by numerous "tiny wiggles" referred to as cryptochrons in the time scale (Cande and Kent, 1992b). Inclusion of cryptochrons in the record can, depending on their duration, reduce the value of k to values very close to 1, indicative of a Poisson distribution (see Lowrie and Kent, 1983). Although the issue of tiny wiggles in the oceanic magnetic anomaly records is not settled, it is now popular to interpret them as due to paleointensity variations rather than short polarity chrons. If this is correct, it implies that polarity reversal of the geomagnetic field is not a Poisson process but that the probability of a reversal drops to zero immediately after a reversal and then rises to a steady value.

The rate of reversal has varied with time and the variation in rate is quite smooth, with a steady decrease from 165 Ma to 120 Ma and a steady increase from 80 Ma to the present. The change in reversal rate results in a change in the potential resolution of magnetic polarity stratigraphy as a correlation tool. From 120 Ma to 80 Ma, during the Cretaceous normal superchron, the reversal process was not functioning, which may imply that the geomagnetic field is more stable in the normal polarity state. McFadden and Merrill (1984) analyzed the relative stability of normal and reverse polarity states and concluded that there is no evidence for a stability bias between the two polarity states. The existence of a superchron of reverse polarity in the Late Carboniferous–Permian of comparable duration to the Cretaceous normal superchron tends to support the notion that the polarity of superchrons is not dictated by stability bias. There is evidence, however, for a difference in the relative proportions of zonal harmonics which make up the time-averaged field for the two polarity states. On the basis of data for the last 5 My, Merrill and McElhinny (1977) concluded that the reverse polarity field has a larger quadrupole and octupole content than the normal polarity field. Schneider and Kent (1988) reached a similar conclusion based on the analysis of deep-sea core inclination data. This apparent asymmetry for normal and reverse polarity states may be attributed to an axis of magnetic susceptibility anisotropy in the inner core inclined to the rotation axis (Clement and Stixrude, 1995).

3

Magnetization Processes and Magnetic Properties of Sediments

3.1 Basic Principle

All matter responds to an applied magnetic field, due to the effect of the field on electron motions in atoms, although the response is very weak for most materials. The ions of some transitional metals, notably Fe^{2+}, Fe^{3+}, and Mn^{2+}, carry an intrinsic spin magnetic moment and are referred to as *paramagnetic ions.* Materials which contain these cations exhibit an enhanced response to an external field.

The weakest response is called *diamagnetism.* Diamagnetic materials do not contain paramagnetic ions in appreciable concentrations and when placed in a magnetic field (H), a weak magnetization (M_i) is induced, by the effect of the magnetic field on orbiting electrons, which is antiparallel to the applied field (H). The *susceptibility* (k) (where $M_i = kH$) is therefore small and negative. No (remanent) magnetization (M_r) remains after the magnetizing field is removed. Many of the most common minerals (such as quartz, halite, kaolinite, and calcite) are diamagnetic, and rocks which are composed largely of these minerals will have negative susceptibility.

An enhanced response to a magnetic field is referred to as *paramagnetism.* Paramagnetic materials contain paramagnetic ions. The susceptibility is large (relative to diamagnetic susceptibility) and positive and due to the alignment of the intrinsic magnetic moments in the applied field. The susceptibility is proportional to the reciprocal of absolute temperature. When the applied field is removed, the weak alignment of spin moments is randomized by thermal vibrations and paramagnetic materials do not retain a remanent magnetization. Many clay minerals and common iron-bearing minerals such as siderite, ilmenite, biotite, and pyrite are paramagnetic.

A few minerals, some iron oxides, oxyhydroxides, and sulfides, exhibit a third type of magnetic behavior called *ferromagnetism (sensu lato)*. In these materials, the paramagnetic cations are closely juxtaposed in the crystal lattice such that spin moments are coupled either by direct exchange or by superexchange through an intermediate anion. Ferromagnetic susceptibility values vary widely but are high (relative to paramagnetic susceptibility values) and positive. Unlike diamagnetic and paramagnetic susceptibility, ferromagnetic susceptibility is strongly dependent on the strength of the applied field (H). At low H, the induced magnetization (M_i) is proportional to H, and therefore the *initial susceptibility* (k) is constant, and the magnetization is lost after removal of the applied field. However, at higher values of H, part of the magnetization is retained by the material after the removal of the applied field. This behavior can be summarized by use of a *hysteresis loop*, which is a plot of magnetization (M) against applied field (H) (Fig. 3.1). At low H, M increases proportionally (hence, k is constant for low H) and the magnetization process is reversible. At higher H, the magnetization is not proportional to H, the curve is not reversible, and a small proportion of the magnetization, referred to as the *remanent magnetization* (M_r), is retained by the material after field removal. Eventually, as

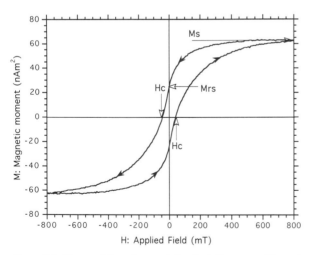

Figure 3.1 Hysteresis loop from a pink pelagic limestone, illustrating the saturation magnetization (Ms), saturation remanence (Mrs), and coercive force (Hc). The loop is constricted (wasp-waisted) due to presence of low-coercivity magnetite and high-coercivity hematite.

H increases further, the *saturation magnetization* (M_s) is reached. Removal of the applied field leaves the sample with a *saturation remanent magnetization* (M_{rs}). If the applied field is now increased in the opposite sense ($-H$), the total magnetization of the sample will fall to zero at a value of $-H$ referred to as the *coercive force* (H_c) for that sample. The back field which leaves the sample with zero remanent magnetization is referred to as the *remanent coercivity* or the *coercivity of remanence* (H_{cr}). The hysteresis loop can be completed by taking the sample to its saturation magnetization in the back field and imposing an increasing positive H until the saturation is achieved again. The hysteresis properties of ferromagnetic materials are destroyed at a characteristic temperature for the mineral referred to as the *Curie or Néel temperature,* above which the behavior is paramagnetic.

The term ferromagnetism (*senso lato*) is used to refer to materials which show hysteresis behavior. There are various types of ferromagnetism which differ in magnetic properties due to different arrangements of spin moments within the crystal lattice. In the case of *ferromagnetism (sensu stricto)*, the adjacent spin moments are parallel and of uniform magnitude (Fig. 3.2). This behavior is restricted to native iron, nickel, and cobalt where paramagnetic ions are sufficiently closely juxtaposed for direct exchange interactions. Minerals (on Earth) do not exhibit this type of behavior. In *ferrimagnetism* adjacent spin moments are antiparallel but a net moment results due to their unequal magnitude. In *antiferromagnetism,* adjacent spins are antiparallel and of uniform magnitude and hence no net moment results. If, however, the antiparallelism of the spin moments is imperfect due to canting, a weak net moment results. Ferrimagnetism and canted antiferromagnetism are the behaviors exhibited by minerals which are capa-

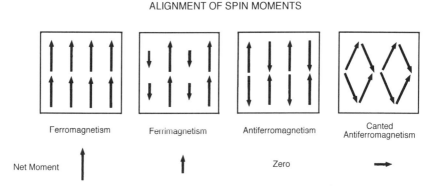

Figure 3.2 Arrangement of spin moments associated with different classes of magnetic behavior.

ble of carrying magnetic remanence, the *magnetic minerals*. The ferrimagnetic minerals generally have high susceptibilities, high magnetization intensities, and relatively low coercivities, whereas the converse is true for antiferromagnetic minerals.

3.2 Magnetic Minerals

a. Magnetite (Fe_3O_4) and the Titanomagnetites ($xFe_2TiO_4.[1 - x]Fe_3O_4$)

Magnetite is one of the end members of the ulvospinel-magnetite solid solution series (Fig. 3.3) and is the most common magnetic mineral on Earth. Magnetite is a cubic mineral with inverse spinel structure. The disordering (Curie) temperature varies more or less linearly with Ti content from $-150°C$ for ulvospinel to $580°C$ for magnetite (Fig. 3.4a). Hence, ulvospinel is paramagnetic at room temperature. Titanomagnetites are ferrimagnetic at room temperature for values of x below about 0.8. Complete

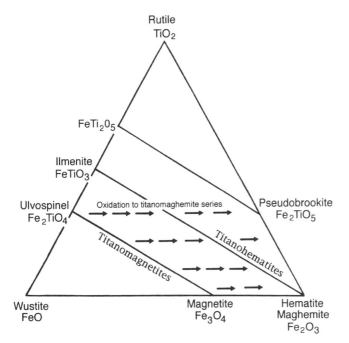

Figure 3.3 Ternary diagram indicating the important iron oxide magnetic minerals, the titanomagnetite and titanohematite solid solution series, and the titanomaghemite oxidation trend.

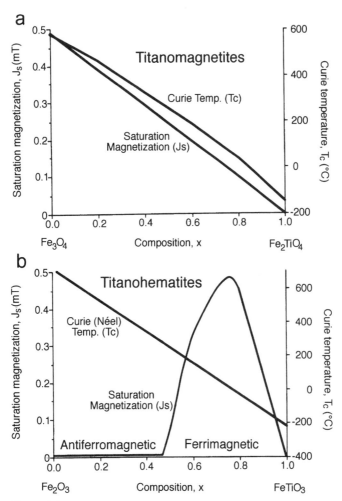

Figure 3.4 Saturation magnetization (J_s) and Curie temperature (T_c) as a function of (a) titanomagnetite composition and (b) titanohematite composition (after Nagata, 1961; Butler, 1992).

solid solution between ulvospinel and magnetite occurs only above 600°C, and exsolution lamellae are common in igneous rocks. The exsolution is usually not between ulvospinel and magnetite, but between ilmenite and magnetite due to the high-temperature (*deuteric*) oxidation of ulvospinel to ilmenite, which usually accompanies exsolution (Haggerty, 1976a,b). As the ilmenite lamellae develop, the intervening titanomagnetite composition becomes progressively depleted in titanium. Further high-temperature oxidation results in the replacement of ilmenite and magnetite by aggregates of hematite, rutile, and pseudobrookite. High deuteric oxidation states are most common in acidic plutonic rocks which cool slowly and have high water content. A wide variety of deuteric oxidation states can be observed within a single lava flow and the controls are not clearly defined. In general, unexsolved titanomagnetites are most common in basic igneous rocks which have cooled rapidly, such as newly erupted submarine basalts. The exsolution process involves a change in grain size, grain shape, and composition of the titanomagnetite. The changes will generally involve a decrease in x, a decrease in grain size, and an elongation of the grain shape, and will therefore tend to increase the coercivity and blocking temperature of the grain. Therefore, the exsolution of magnetite and ilmenite by deuteric oxidation may serve to increase the stability of the remanence acquired during the process. The ilmenite separating the exsolved titanomagnetite rods is paramagnetic at room temperature and behaves as a nonmagnetic matrix for the ferrimagnetic titanomagnetite grains. Exsolution lamellae are a characteristic of titanomagnetite grains affected by deuteric oxidation. Low-temperature oxidation of titanomagnetites by weathering processes and/or oxidizing diagenetic conditions results in maghematization without exsolution, and subsequent formation of hematite.

Titanomagnetites are a common constituent of igneous and metamorphic rocks and are therefore a prominent detrital component in sediments. Many sediments also contain low-titanium magnetite grains, which are uncommon in igneous rocks and are generally believed to be biogenic. Many organisms, including bacteria and molluscs, produce magnetite either by extracellular precipitation or as an integral part of their metabolism (e.g., Blakemore, 1975; Kirschvink and Lowenstam, 1979; Blakemore *et al.*, 1985; Frankel, 1987; Lovley *et al.*, 1987; Lovley, 1990; Bazylinski *et al.*, 1988; Chang and Kirschvink, 1989; Sparks *et al.*, 1990). Electron microscopy of magnetic extracts of lake sediments (Snowball, 1994) and a wide range of marine sediments (Vali *et al.*, 1987; McNeill, 1990; Yamazaki *et al.*, 1991) has shown that magnetite is often found in a restricted single domain (SD) grain size range, similar to the grain size of magnetite produced by living bacteria (Fig. 3.5). The grain size and low-titanium composition of this magnetite phase is such that it is an important carrier of stable magnetic remanence (see Moskowitz *et al.*, 1988, 1994).

b. Hematite (α-Fe$_2$O$_3$) and Titanohematite (xFeTiO$_3$.[1 − x]Fe$_2$O$_3$)

Mineral compositions intermediate between ilmenite and hematite are referred to as the titanohematites. Hematites have rhombohedral symmetry and corundum structure. As for the titanomagnetites, the disordering (Néel)

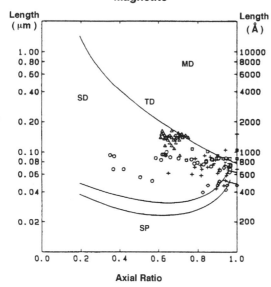

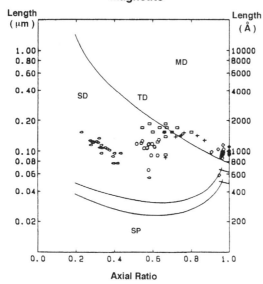

Figure 3.5 Comparison of size and shape of marine sedimentary magnetite and modern bacterial magnetite from electron microscopy (after Vali *et al.*, 1987). SP: superparamagnetic; SD: single domain; TD: two domain; MD: multidomain.

temperature varies with x, from $-218°C$ for ilmenite to $675°C$ for hematite (Fig. 3.4b). Compositions corresponding to values of x greater than about 0.8 are paramagnetic at room temperature. Other compositions are ferrimagnetic for values of x in the 0.5–0.8 range and antiferromagnetic for $0 < x < 0.5$. Complete solid solution occurs only above about 1000°C and exsolution of the two end members is common except where $x < 0.1$ or $x > 0.9$. Titanohematites are associated with some metamorphic and plutonic rocks but almost pure hematite is much more abundant in metamorphic, sedimentary, and igneous rocks. Hematite is a very important as a remanence carrier in both clastic and chemical sediments. The hematite may occur in sediments as detrital specularite or as an authigenic pigment which grows during diagenesis. Hematites are antiferromagnetic, the weak net moment being due to slight canting of the spin moments (Fig. 3.2). The saturation magnetization is much lower than that of magnetite, and the remanent coercivity is significantly higher.

c. Maghemite (γ-Fe$_2$O$_3$) and Titanomaghemite (xFeTiO$_3$.[1 − x]Fe$_2$O$_3$]

Maghemite has the spinel structure of magnetite and the chemical composition of hematite. Maghemite and titanomaghemite are metastable phases, with maghemite and titanomaghemite inverting to hematite and magnetite, respectively, above 250°C. The disordering (Curie) temperature, is therefore, difficult to determine but is believed to be about 640°C. The magnetic properties are similar to those of magnetite, with slightly lower saturation magnetization and comparable remanence coercivity. Maghemite is common as a low-temperature oxidation product of magnetite in both igneous and sedimentary rocks.

d. Goethite (α-FeOOH)

Goethite is the most common of the iron oxyhydroxides and is antiferromagnetic with a disordering (Néel) temperature of about 120°C (Hedley, 1971). Goethite has weak saturation magnetization and very high coercivity (Hedley, 1971; Rochette and Fillion, 1989). It dehydrates to hematite at about 300°C. Goethite is an important product in low-temperature oxidation, particularly of iron sulfides, and is a constituent of manganese nodules and of the Pacific red clay facies, where it may precipitate directly from seawater. Lepidocrosite (γ-FeOOH) is less common than goethite, has a Néel temperature of $-196°C$, and is therefore not an important remanence carrier, but it dehydrates to maghemite and may therefore indirectly contribute to magnetic remanence.

e. Pyrrhotite (FeS_{1+x}, where $x = 0-0.14$)

Fe_7S_8, the most magnetic pyrrhotite, is monoclinic and ferrimagnetic, with a Curie temperature of about 325°C. Fe_9S_{10} is less important magnetically, being ferrimagnetic in a very restricted temperature range ($\sim 100-200$°C). The saturation magnetization of Fe_7S_8 is about ten times that of hematite and about half that of magnetite and maghemite (Clark, 1984). Remanence coercivity is generally greater than that of magnetite and maghemite (Dekkers, 1988, 1989). Pyrrhotite occurs in basic igneous rocks (Soffel, 1977, 1981) and can grow during sediment diagenesis in certain reducing environments (Kligfield and Channell, 1981; Freeman, 1986). The most common iron sulfide, pyrite (FeS_2), is paramagnetic.

f. Greigite (Fe_3S_4)

Greigite is ferrimagnetic and carries a stable magnetization in some lake sediments (Snowball and Thompson, 1990; Snowball, 1991) and in rapidly deposited clastic marine sediments (Tric *et al.*, 1991a; Roberts and Turner, 1993; Reynolds *et al.*, 1994). Coercivities are similar to those of magnetite and therefore generally somewhat less than for pyrrhotite. As for pyrrhotite, the Curie temperature is about 320°C, and thermomagnetic curves are characterized by an increase in magnetic moment (and susceptibility) as the iron sulfide oxidizes to magnetite.

3.3 Magnetization Processes

a. Thermal Remanent Magnetization (TRM)

The remanent magnetization acquired by a grain during cooling through its *blocking temperature* is referred to as thermal remanent magnetization. The blocking temperature of a grain is the temperature above which the magnetic ordering in the grain is randomized by thermal energy during laboratory heating. The magnetic ordering of the grain will eventually become randomized at any temperature; the time that it is likely to take is referred to as the *relaxation time* (T_r), which is a function of coercive force (H_c), grain volume (v), spontaneous magnetization (J_s), and temperature (T).

$$T_r = T_o \exp \left(v H_c J_s / 2kT \right)$$

where k is Boltzmann's constant and T_o is the frequency factor ($\sim 10^{-9}$ s). The exponential form of this equation means that, for a particular grain, the relaxation time increases logarithmically with decreasing temperature.

The temperature at which the relaxation time becomes large relative to laboratory experimental time is referred to as the blocking temperature. For a rock or sediment sample, the grain population will have a *blocking temperature spectrum* reflecting the compositional and grain size range of the magnetic minerals. For weak magnetic fields, the TRM intensity is approximately proportional to the applied field, although lower cooling rates will enhance the TRM intensity. It is important to note that natural blocking temperature of a grain during igneous cooling will be lower than the laboratory-determined blocking temperature. The differences are greater at lower igneous cooling rates and become smaller when the blocking temperature is close to the Curie temperature (Dodson and McClelland-Brown, 1980).

b. Chemical Remanent Magnetization (CRM)

Chemical remanent magnetization is produced by an increase in grain volume at temperatures below the blocking temperature of the grain. The process is analogous to TRM acquisition but in the case of CRM the relaxation time [see Eq. (3.1)] is increased by change in grain volume (v) rather than ambient temperature (T). The logarithmic increase in relaxation time with grain volume leads to the concept of *blocking volume,* which is the critical grain volume at which the relaxation time becomes geologically significant. CRM is a very important acquisition process in sediments because magnetic minerals commonly grow authigenically during diagenesis. Magnetite, hematite, maghemite, goethite, and pyrrhotite can all grow authigenically in certain diagenetic conditions. Alteration during weathering can produce a CRM carried by goethite and/or hematite in a wide variety of rock types.

c. Detrital Remanent Magnetization (DRM)

This process is an important source of primary magnetization in sediments. Detrital or biogenic magnetic mineral grains with a preexisting TRM or CRM can align themselves with the ambient field during deposition. The alignment occurs largely within the upper few centimeters of soft sediment rather than in the water column (Irving and Major, 1964). Magnetite, titanomagnetite, hematite, and maghemite are common carriers of DRM in sediments. The lock-in depth of DRM has been estimated by laboratory experiments (Kent, 1973; Lovlie, 1974; Tauxe and Kent, 1984), by comparing the position of reversals with oxygen isotope data in cores with varying sedimentation rates (de Menocal *et al.,* 1990), and by observation of the damping of secular variation records (Hyodo, 1984; Lund and Keigwin, 1994). The estimates often lie in the 10–20 cm depth range below the sediment/water interface. The lock-in depth and width of the lock-in zone will depend on the grain size spectrum, sedimentation rate, and bioturbation depth.

d. Viscous Remanent Magnetization (VRM)

VRM is an important source of secondary magnetization in rocks and sediments. Grains with relatively low relaxation time will tend to become remagnetized in the ambient weak field. The intensity of the VRM has been found both theoretically and experimentally to be proportional to the logarithm of time, and the proportionality constant (referred to as the viscosity coefficient) increases with temperature (Dunlop, 1983). Grains with volumes close to the blocking volume (i.e., close to the threshold between single domain and superparamagnetic behavior) are likely to be the most important carriers of VRM, although multidomain grains can also carry appreciable VRM. For grains with high coercivity, such as fine-grained hematite, the VRM acquired during the Brunhes chron can have high coercivity and blocking temperature up to about 350°C. VRM components of the NRM will generally be aligned with the present Earth's field.

e. Isothermal Remanent Magnetization (IRM)

This is a high-field magnetization process and therefore does not generally occur in nature except in the case of lightning strikes. As the properties of the IRM are useful for determining the magnetic mineralogy and grain size, IRM is produced in the laboratory using a pulse magnetizer, electromagnet, or superconducting magnet. The IRM is the remanence measured after removal of the sample from the magnetizing field. The intensity of the IRM will increase as the strength of the magnetizing field is increased until the saturation IRM (SIRM) is acquired. The magnetizing field required to attain the SIRM is a function of the remanent coercivity (H_{cr}) of sample grains.

f. Anhysteretic Remanent Magnetization (ARM)

ARM is an artificial remanence produced in the laboratory by placing the sample in an alternating high field (typically 100 mT) with a low d.c. biasing field (typically 50–100 μT) and allowing the alternating field to ramp down smoothly to zero. ARM is very useful because it has similar characteristics to TRM, and the characteristics of ARM can be used to determine grain size and composition of magnetic minerals present in the sample. The anhysteretic susceptibility (k_{arm}) is the ARM divided by the biasing field.

g. Self-Reversal Mechanisms

Self-reversal is a phenomenon where the remanent magnetization of a sample is antiparallel to the direction of the ambient magnetic field at the time of remanence acquisition. It is now clear that self-reversal is rare and the phenomenon is restricted to a few well-documented cases, particularly in titanohematites. Self-reversal can occur as a result of magnetostatic interaction between adjacent magnetic phases with different blocking temperatures and usually results from negative exchange coupling between a strongly magnetic, low Curie point phase and a weaker high Curie point phase with moment aligned with the ambient magnetizing field. The phases may be distinct mineral compositions or compositional gradients within grains. If the lower blocking temperature phase has greater volume and spontaneous magnetization than the higher blocking temperature phase, the net remanent magnetization can be antiparallel to the ambient field. The best documented cases of self-reversal are in a dacite containing titanohematite in the $0.45 < x < 0.60$ compositional range (Uyeda, 1958) and in some oxidized titanomagnetite compositions (Schult, 1976; Heller and Petersen, 1982). Pyroclastics from the 1985 eruption of Nevado del Ruiz (Columbia) are reversely magnetized, and the self-reversal process has been replicated in laboratory-induced acquisition of TRM (Heller *et al.*, 1986).

3.4 Magnetic Properties of Marine Sediments

The magnetic properties of marine sediments depend not only on the nature of primary magnetic minerals contributed from detrital or biogenic sources, or by precipitation from seawater, but also on the diagenetic conditions which determine the alteration of primary magnetic phases as well as the authigenic growth of secondary magnetic minerals. The important remanence-carrying minerals in marine sediments are magnetite (Fe_3O_4), titanomagnetite, hematite (α-Fe_2O_3), maghemite (γ-Fe_2O_3), geothite (α-$FeOOH$), and iron sulfides such as pyrrhotite (FeS_{1+x}) and greigite (Fe_3S_4). Of the magnetic minerals listed above, magnetite, titanomagnetite, greigite, and goethite can be *primary* remanence carriers in marine sediments. Hematite, pyrrhotite, and greigite occur as authigenic secondary minerals formed during early diagenesis, and goethite can be a product of low-temperature oxidation (weathering), particularly of iron sulfides. The magnetic properties of sediments from the four marine environments for which there is most information are reviewed below.

a. Shallow Water Carbonates

The majority of shallow water platform limestones are unreliable recorders of the geomagnetic field. However, there are several notable exceptions where early magnetizations have been recorded in this facies. For example, Silurian reef-slope limestones from Indiana yield magnetite magnetizations which predate compaction-related tilting (McCabe et al., 1983). Well-defined Plio-Pleistocene magnetostratigraphies have been acquired from shallow water limestones and dolomites recovered by drilling in the Bahama Bank (McNeill et al., 1988) and from the Mururoa Atoll (Aissaoui et al., 1990). Single domain magnetite of probable biogenic origin is thought to be the carrier of remanence in the partially dolomitized carbonates from the Bahamas, and the dolomitization process does not appear to remagnetize the carbonates (McNeill, 1990). A similar conclusion was drawn from study of Carboniferous limestones in Wyoming (Beske-Diehl and Shive, 1978).

In the Mediterranean area, the "southern Tethyan" Mesozoic platform limestones generally carry very weak magnetizations, which are either too weak for precise measurement or are secondary and due to the weathering of pyrite to hematite and/or goethite. "Helvetic" limestones from the European Mesozoic continental margin carry a recent VRM with no resolvable primary magnetization components (Kligfield and Channell, 1981). The more marly shallow water Jurassic limestones from southern Germany carry a weak low-coercivity magnetization attributed to detrital magnetite; however, the purer limestone facies (bafflestones) carry a secondary magnetization due to both goethite and hematite produced by late oxidation of pyrite (Heller, 1978). A magnetite magnetization in marly shallow water Late Cretaceous limestones from the Munster Basin predates latest Cretaceous folding and therefore appears to be primary in origin (Heller and Channell, 1979).

In North America, Paleozoic neritic and lagoonal limestone exposed in New Jersey and New York such as the Ordovician Trenton limestone (McElhinny and Opdyke, 1973; McCabe et al., 1983), the Siluro-Devonian Helderberg Series (Scotese et al., 1982), and the overlying Onodaga limestone (Kent, 1979) carry a synfolding or postfolding remanent magnetization with a direction implying remagnetization during Late Carboniferous– Early Permian time (Scotese et al., 1982; McCabe et al., 1983). These limestones contain spheroidal/botryoidal magnetite which grows after pyrite during diagenesis (Suk et al., 1990, 1993), and the particular diagenetic conditions which bring about this transformation have been associated with the migration of hydrocarbons during Alleghenian thrusting (McCabe et al., 1983). Comparison of the coercivity of IRM and ARM suggests that the magnetite is mainly in the MD size range (McElhinny and Opdyke, 1973; Kent, 1979). Curie point determinations and energy-dispersive X-ray analysis indicate that the carrier is a low-Ti magnetite (Scotese et al., 1982; McCabe et al., 1983). Remagnetized Paleozoic limestones from North

America and Britain have characteristically "wasp-waisted" hysteresis loops (Fig. 3.6) (Jackson, 1990; McCabe and Channell, 1994). The origin of such wasp-waisted loops can be attributed to mixtures of grains with contrasting coercivities. In North American remagnetized limestones, the wasp-waisted characteristic of the loops is observed not only in the bulk rock but also in the "nonmagnetic" residue after magnetic extraction of the large (> few micron) sized spheroidal/botryoidal magnetite (Sun and Jackson, 1994). The large magnetite grains do not, therefore, appear to contribute to the characteristic wasp-waisted loop shape, which could be produced by mixing of SD magnetite with a volumetrically dominant SP magnetite phase (Jackson *et al.*, 1993; McCabe and Channell, 1994; Channell and McCabe, 1994).

b. Pelagic Limestones and Calcareous/Siliceous Ooze

Deep-sea (pelagic) limestones, and the unlithified analogues (calcareous ooze) recovered from ocean drilling, are the group of marine sediments that have been most thoroughly studied from a magnetic viewpoint. The interest in this sediment type stems from their efficiency as recorders of the ancient geomagnetic field and their usefulness as a means of establishing the correlation of polarity reversals to microfossil biozonations. There are many examples of the use of this type of sediment in magnetostratigraphic studies (e.g., Opdyke, 1972; Opdyke *et al.*, 1974; Alvarez *et al.*, 1977; Channell *et al.*, 1979; Lowrie *et al.*, 1982; Tauxe *et al.*, 1983c).

Many pelagic limestones and calcareous/siliceous oozes have rather simple magnetic properties, exhibiting a single magnetization component carried by ferrimagnetic magnetite or titanomagnetite, typically with hysteresis ratios lying in the pseudo-single domain (PSD) grain size range (Fig. 3.7). These sediments acquire their magnetization (DRM) by mechanical rotation of magnetite grains into line with the ambient geomagnetic field in the bioturbated upper few centimeters of soft sediment. As discussed above (§3.3c), the depth (below the sediment/water interface) and size of the "lock-in zone" in which the grains lose their mobility will depend on grain size, magnetic mineralogy, sedimentation rate, and the extent of bioturbation. Estimates of this depth for pelagic sediments are generally a few tens of centimeters, which correspond to a few times 10^4 years, at typical pelagic sedimentation rates. Sediments which carry a magnetite DRM are often excellent recorders of the direction of the geomagnetic field at a time close to the time of deposition.

Magnetite in pelagic sediments was thought to be entirely detrital (e.g., Lovlie, 1974); however, Curie temperatures of magnetic separates from deep-sea sediments often indicate the presence of almost pure magnetite

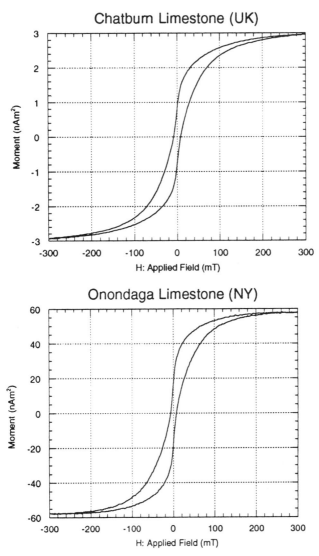

Figure 3.6 Hysteresis loops from limestones remagnetized during Late Carboniferous–Early Permian time: top, Carboniferous limestones from Britain; bottom, Ordovician limestones from New York State (after McCabe and Channell, 1994).

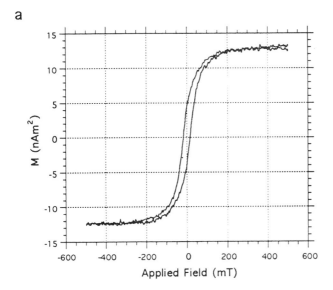

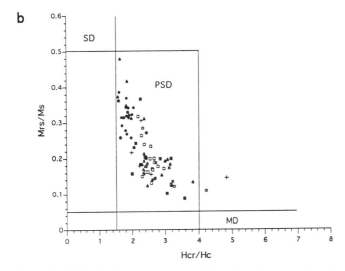

Figure 3.7 (a) Typical hysteresis loop for white pelagic limestone (Maiolica Formation, Italy). (b) Hysteresis ratios for pelagic limestones of the Maiolica Formation, single domain (SD), pseudo-single domain, and multidomain (MD) magnetite fields according to Day *et al.* (1977) (data after Channell and McCabe, 1994).

(Fe_3O_4). As pointed out by Henshaw and Merrill (1980), this is inconsistent with a totally detrital source as detrital source rocks (such as mid-ocean ridge basalts and igneous rocks exposed on land) contain titanomagnetite, rather than pure magnetite. The relatively recent discovery that several marine organisms, including bacteria and molluscs, precipitate nearly pure magnetite either extracellularly or as an integral part of their metabolism (Blakemore, 1975; Kirschvink and Lowenstam, 1979; Moskowitz *et al.*, 1988, 1994) may provide the answer to this dilemma. Magnetite crystals of probably bacterial origin have been observed in pelagic sediments using transmission electron microscopy (Petersen *et al.*, 1986; Vali *et al.*, 1987; Yamazaki *et al.*, 1991). These typically octahedral or parallelepiped-shaped magnetite grains are usually 0.05 to 0.1 μm across and within the single domain grain size range, which is optimal for retention of magnetic remanence (Fig. 3.5). Magnetite of bacterial origin may form at a few centimeters below the sediment-water interface at the transition from iron-oxidizing to iron-reducing conditions (Karlin *et al.*, 1987). It appears that nearly pure magnetite of biogenic origin is a major carrier of magnetic remanence in deep-sea sediments. Peterson *et al.* (1986), distinguished two populations of magnetite grains in pelagic sediment from the South Atlantic: (1) a poorly sorted, irregular-shaped titanomagnetite of probable detrital origin and (2) a well-sorted, euhedral SD low-Ti magnetite of probable bacterial origin.

Although the magnetites are the dominant remanence-carrying minerals in pelagic limestones and calcareous oozes, there is ample documentation that diagenetic alteration of primary magnetic minerals can occur in this environment. At typical pelagic sedimentation rates of the order of 10 m/My, organic matter tends to get oxidized at the sediment/water interface. Hence, redox conditions in the soft sediment approach those in seawater, and under these conditions magnetite may be slowly but progressively oxidized. Tucker and Tauxe (1984) show a progressive oxidation of primary magnetite downcore at DSDP Leg 73 sites. Maghemite is the probable product of this oxidation, and the oxidation may occur by leaching of iron from the magnetic spinel phases rather than by incorporation of extra oxygen into the lattice. Maghemite has been adduced to occur in pelagic limestones cropping out on land (Channell *et al.*, 1982a; Lowrie and Heller, 1982).

The magnetic properties of Tethyan pelagic limestones exposed on land and in the Mediterranean area are strongly color dependent (Lowrie and Alvarez, 1975, 1977a; Channell *et al.*, 1982a; Lowrie and Heller, 1982; Channell and McCabe, 1994). Red and pink pelagic limestones contain maghemite and hematite, in addition to the primary magnetite. The contrasting coecivities of magnetite and hematite produce wasp-waisted hysteresis loops (e.g., Fig. 3.1). White pelagic limestones are magnetite bearing

(Fig. 3.7), do not appear to contain maghemite or hematite, but often contain optically visible pyrite, which readily weathers to goethite and fine-grained hematite. The mineral goethite can also play a more primary role in the magnetization of pelagic sediments. It is a constituent of manganese nodules and can precipitate directly from seawater and, together with iron-bearing clay minerals may be a precursor to pigmentary hematite (in the red limestones) and to pyrite (in the white limestones). Neither maghematization of primary magnetite nor authigenic growth of pigmentary hematite appears to occur in the white pelagic limestones. The remanence carried by pigmentary hematite in the red limestones appears, in some cases, to be due to growth of hematite in the uppermost few meters of soft sediment (Channell *et al.*, 1982a). The dominant control on limestone color and on magnetic properties is probably organic content, the burial of organic matter being largely controlled by sedimentation rate.

c. Pacific "Red Clay" Facies

Not all deep-sea sediments are efficient recorders of the geomagnetic field at the time of deposition. The lack of a primary magnetization may be due to the magnetic mineralogy or grain size of the detritus, to adverse diagenetic conditions which result in the alteration of primary magnetite, or to the growth of authigenic magnetic mineral phases. The so-called red clay facies occurs over a large part of the midlatitude Pacific (Davies and Gorsline, 1976). It is devoid of calcareous and siliceous microfossils and accumulated below the CCD at rates of about 25 cm/My. In this facies, the primary magnetization generally degrades at a few meters depth below the sediment-water interface. The boundary between the primary and stable magnetic records often occurs in the later part of the Gauss chron and coincides closely with the late Pliocene onset of northern hemisphere glaciation (Opdyke and Foster, 1970; Kent and Lowrie, 1974; Prince *et al.*, 1980). A number of different explanations have been offered to account for the degradation of the primary magnetization. Kent and Lowrie (1974) considered that oxidation of primary magnetite to maghemite accounts for the observed increase in VRM downcore. The SP-SD grain size threshold for maghemite is greater than that for magnetite, and therefore the oxidation process might be expected to increase the VRM contribution to remanence. In this scenario, the increased sedimentation rates in the upper part of the core characterized by stable primary remanence retard the oxidation (due to increased burial of organic matter). Johnson *et al.* (1975) also attributed the instability of the magnetization to the diagenetic growth of maghemite downcore, but considered the secondary magnetization to be a CRM rather than a VRM. Henshaw and Merrill (1980) have suggested that secondary ferromanganese oxides and oxyhydroxides [such as the jacobsite ($MnFe_2O_4$) solid solution series] are important authigenic magnetic phases and can carry a CRM which masks the primary DRM. Yamazaki and Katsura (1990) used a suspension method to determine the magnetic moment distribution and ferrimagnetic grain size in redeposited red clay. The mean grain diameter decreases below 1 m depth in the sediment, and this is consistent with changes in the frequency dependence of susceptibility, which indicates an increase in SP grains below this depth. These data imply that the increased importance of VRM downcore may be due to reduced grain size of eolian detrital magnetite in preglacial time (due to less vigorous atmospheric circulation), rather than to downcore maghematization.

d. Hemipelagic Sediments and Turbiditic "Flysch" Deposits

In areas of high sedimentation rate and/or restricted bottom-water circulation, the burial of organic matter can result in reducing diagenetic conditions and the formation of iron sulfides. Paramagnetic pyrite is by far the most common iron sulfide in sediments, and it can grow by reduction of primary magnetite. The metastable iron sulfides mackinawite (FeS) and greigite (Fe_3S_4) are, unlike pyrite, capable of carrying magnetic remanence, as are some compositions of the more stable pyrrhotite (FeS_{1+x}). Pyrrhotite coexists with magnetite and pyrite in young siliciclastic sediments from the Sea of Japan (Kobayashi and Nomura, 1972) and in Mesozoic "helvetic" limestones (Kligfield and Channell, 1981). Greigite in siliciclastic marine sediments from reducing depositional/diagenetic environments has been recognized in northern Italy (Tric *et al.*, 1991a), New Zealand (Roberts and Turner, 1993), and the North Slope of Alaska (Reynolds *et al.*, 1994).

Karlin and Levi (1985) have documented the progressive dissolution of primary magnetite, due to reducing diagenetic conditions, in hemipelagic silts and clays from the Oregon coast and in laminated diatomaceous oozes from the Gulf of California. Sedimentation rates were estimated to be 121 and 135 cm/ky, respectively. A decrease in NRM, SIRM, and ARM intensity occurs in the topmost meter as dissolution reduces the population of fine-grained magnetite. Increases in coercivity and relative stability of ARM relative to IRM, from about 1 m depth in the cores to the base at 4 m, are interpreted in terms of a reduction in mean effective magnetite grain size as dissolution proceeds. A similar sequence of events has been observed at ODP Site 653, in a high sedimentation rate core from the Tyrrhenian Sea (Channell and Hawthorne, 1990).

Turbiditic "flysch" deposits can carry a primary magnetization; however, the principal remanence carrier is usually multidomain (MD) titanomagnetite, hence the remanent coercivity is low and the sediments tend to readily acquire VRM. In the Late Cretaceous Gurnigel Flysch, the VRM acquired in the Brunhes Chron dominates the NRM such that the primary magnetization cannot be resolved (Channell *et al.*, 1979). In turbiditic sequences, it is generally good practice to sample the finest upper parts of individual flows; however, even in the finer silt horizons, MD titanomagnetite dominates the magnetite fraction. Turbiditic deposits are attractive for magnetostratigraphic study as their high sedimentation rates give high potential magnetostratigraphic resolution. The high sedimentation rates may, however, result in a high rate of burial of organic matter, sulfate reduction to sulfide, and the alteration of primary detrital titanomagnetite to iron sulfide minerals.

3.5 Magnetic Properties of Terrestrial Sediments

a. Lake Sediments

In early studies of lake sediments, detrital hematite was considered to be the major carrier of remanence on the basis of the apparent observation of the Morin transition at $-10°C$ (Creer *et al.*, 1972). Later studies showed that the reduction in magnetization intensity at this temperature was due to the reorientation of grains by growth of ice crystals and that the dominant carrier is fine-grained single domain magnetite (Stober and Thompson, 1977; Turner and Thompson, 1979). The grain size of the magnetite ranges from multidomain to superparamagnetic in some North American lake sediments (King *et al.*, 1982), detrital sources and bacterial activity being the controlling factors. Bacterial magnetite has also been documented in European lakes (e.g., Snowball, 1994).

Sediments acquire a detrital remanent magnetization (DRM) by mechanical rotation of the magnetite grains in response to the ambient geomagnetic field. Although the DRM may be acquired within a few days in redeposited lake sediment (Barton *et al.*, 1980), observations of magnetic coercivity of lake sediment from close to the sediment/water interface gave an estimate of lock-in time for a moderately stable DRM of about 100 years (Stober and Thompson, 1977). Lock-in times of a few hundred years may significantly distort the secular variation record particularly if the lock-in process occurs over an appreciable depth range in the sediment. If the lock-in or moment fixing function can be estimated (by experiment) for a particular sediment, the secular variation record can be deconvolved by numerical methods (Hyodo, 1984; Lund and Keigwin, 1994). Comparison of the magnetic properties of natural and redeposited lake sediments has shown that the grains are probably not fixed in place by dewatering of the sediment to a critical level, but possibly by gels that form in the pores of the wet sediment and are particularly prevalent in organic-rich sediment (Stober and Thompson, 1979). The process of sediment drying often disrupts the fidelity of the record; however, this may be due to alteration of remanence-carrying iron sulfides. Greigite is an important carrier of magnetic remanence in lake sediments (Giovanoli, 1979; Snowball and Thompson, 1988; Snowball, 1991). Greigite has coercivity similar to that of magnetite, however it can be recognized by thermomagnetic analysis and Mossbauer spectroscopy. Although greigite is an authigenic mineral in lake and marine sediments, it grows during early diagenesis and can carry a high fidelity record of the geomagnetic field (e.g., Tric *et al.*, 1991b).

b. Loess Deposits

The first magnetic stratigraphies of Chinese loess deposits resulted in drastic revision of the duration of Chinese loess deposition, from ~1.2 My to ~2.5 My (Heller and Liu, 1982, 1984). In addition, magnetic susceptibility records define lithologic loess/paleosol cycles which correlate with glacial-interglacial marine d^{18}O record (Kukla *et al.*, 1988; Liu *et al.*, 1985). At first, it was thought that the susceptibility fluctuations were controlled by variations in the concentration of windblown detrital magnetite, due to increased loess deposition during glacials superimposed on a more constant background deposition of windblown magnetite (Kukla *et al.*, 1988). It is now generally accepted that the enhanced susceptibility in the paleosols (during interglacials) is a result of authigenic "magnetic enhancement" as a result of elevated rainfall/temperature. The magnetic enhancement is generally attributed to bacterial production of fine-grained SP and SD magnetite during pedogenesis (Zhou *et al.*, 1990; Heller *et al.*, 1991; Maher and Thompson, 1991, 1992; Banerjee *et al.*, 1993; Evans and Heller, 1994), although some authors consider maghemite to be important (Eyre and Shaw, 1994). The loess sediment has higher coercivities than intervening paleosols (Fig. 3.8) due to the greater contribution of hematite in the loess (Heller and Liu, 1984; Maher and Thompson, 1992).

c. Continental Red Beds

Continental nonmarine clastic "red beds" often have stable magnetization carried by two distinct phases of hematite, one detrital and one authigenic (Collinson, 1965, 1974; Purucker *et al.*, 1980; Tauxe *et al.*, 1980). The magnetization of coarser grained (often drab-colored) sandstone units is often dominated by "specular" hematite, of presumed detrital origin, whereas finer grained units (often more reddened) tend to have CRMs dominated

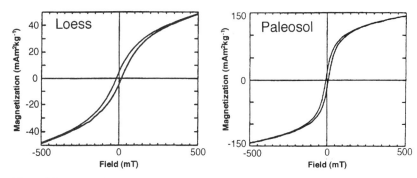

Figure 3.8 Hysteresis loops from Chinese loess and intervening paleosol (after Heller and Evans, 1995).

by fine-grained authigenic "pigmentary" hematite. Continental red beds have been important in magnetostratigraphic studies of the Paleozoic (Di-Venere and Opdyke, 1990, 1991a,b), Mesozoic (e.g., Helsley and Steiner, 1974), and Cenozoic (Tauxe and Opdyke, 1982); however, the age of the hematite (and its magnetization) has been the subject of considerable debate (Purucker *et al.*, 1980; Walker *et al.*, 1981; E. E. Larson *et al.*, 1982; Steiner, 1983; Lovlie *et al.*, 1984). Redeposition experiments of naturally disintegrated hematite-bearing sediments indicate that the hematite grains can carry a detrital remanence (DRM) closely aligned to the ambient field (Tauxe and Kent, 1984; Lovlie and Torsvik, 1984). Characteristic magnetization components in rip up clasts from Triassic and Carboniferous red beds are randomly directed, indicating an early origin for the magnetization (Molina-Garza *et al.*, 1991; Magnus and Opdyke, 1991). Probably the most convincing evidence for the penecontemporaneous acquisition of red-bed remanence is the extensive studies of the red beds of the Siwalik system of Pakistan by Johnson *et al.* (1982, 1985), Opdyke *et al.* (1982), Tauxe and Opdyke (1982), and many others. The Siwalik sediments contain bright red hematite-bearing units which range in age from Middle Miocene (15 Ma) to Late Pliocene (2.5 Ma). The hematite magnetizations predate Pliocene-Pleistocene folding, and the Miocene sediments contain no evidence of detrital magnetite. Tauxe *et al.* (1980) carried out a conglomerate test and demonstrated an early magnetization in red siltstone clasts formed in an overbank environment and subsequently incorporated in the gray sandstone of a river channel. Upon thermal demagnetization, the sedimentary clasts were randomly, but stably magnetized, showing that the CRM directions were acquired early in the history of the sediment. Further evidence that the characteristic magnetization of these beds was acquired early in their history is the fact that the magnetic stratigraphy from these sediments can be correlated to the GPTS derived from the oceanic magnetic anomaly record. Independent fission track dates on tuffs in the sedimentary sequence confirm the correlation. Tauxe *et al.* (1980) suggest that the CRM of these sediments was acquired within 20 ky or less of deposition. This will not always be the case in reddened sediments, where authigenic pigmentary hematite can significantly postdate deposition.

4

Laboratory Techniques

4.1 Introduction

Sampling in indurated rocks is usually conducted using a hand-held drill. A sample 5 to 10 cm in length is cored and left standing in the hole. The sample is then oriented using a slotted tube which fits over the core. Orientation devices usually have a plate on which a sun or magnetic compass is mounted and aligned with the slotted tube. The compass is leveled and the azimuth and dip of the sample are recorded. A mark is made on the sample along the slot, and the downward direction is marked on the sample. In volcanic areas, it is usually necessary to take sun compass readings, as local magnetic anomalies may distort the Earth's magnetic field. Most modern orientation devices are equipped for sun compass readings. The direction of a sundial shadow is recorded, along with the precise time of observation. The direction and dip of the bedding at the site are recorded to enable the bedding plane to be returned to the horizontal. Oriented hand-sized samples are often taken from outcrop and in this case the dip and strike of a flat surface are marked and recorded. The samples are then returned to the laboratory, where they are fashioned into a standard size, either cores with diameter 2.5 cm and length 2.3 cm or, alternatively, cubes of 8–15 cm³.

The direction and intensity of the natural remnant magnetization (NRM) are measured using a magnetometer. Most laboratories in North America, Europe, and Japan are equipped with cryogenic magnetometers in which the pick-up coils and SQUID sensors operate at liquid helium temperature (Goree and Fuller, 1976; Weeks *et al.*, 1993). Other types of magnetometers such as the rock generator type (Johnson *et al.*, 1948) or fluxgate spinner magnetometers (Foster, 1966) are still used in some laboratories (see Collinson, 1983). The intensity of the sample magnetic field is measured in three orthogonal directions. This allows the calculation of the total moment of the sample as well as the declination and inclination of the magnetization relative to the orientation line on the core or cube. The magnetization direction of the sample is then rotated into geographic coordinates and rotated again, about the strike of the bedding, to account for bedding tilt.

The NRM of rock and sediment samples is often the resultant of more than one magnetization component. Individual components may be acquired at different stages in the history of the sample. Components acquired during deposition of the sediment or cooling of the igneous body are referred to as "primary." Those associated with later geological events such as sediment diagenesis, deformation, uplift, and weathering are referred to as "secondary." The objective of this chapter is to provide a practical guide to methods of determining (1) the component content of a magnetization and (2) magnetic minerals which carry the magnetization components. For detailed discussions of mineral magnetic properties and of instrumentation, see Collinson (1983), O'Reilly (1984), and Dunlop (1990).

Information concerning modes of occurrence of magnetic minerals in marine sediments has increased considerably in the last few years (Ch. 3) such that an insight into the magnetic mineralogy is an important step in determining the relative age of resolved magnetization components. In magnetostratigraphic studies, it is necessary to establish which, if any, of the resolved magnetization components are primary and can be associated with the depositional or cooling age of the sedimentary or igneous rocks. Estimating the age of magnetization components is a critical step in paleomagnetic and magnetostratigraphic studies. Determination of the mineralogy of magnetic carriers does not generally provide unequivocal evidence for (but is often a guide to) the relative age of magnetization components. Field tests (the tilt test and conglomerate test) (§ 5.3) are additional important means of constraining the age of magnetization components. Note that the normal and reverse directions recorded by a particular component do not necessarily indicate that the component is primary; on the other hand, the absence of antiparallel reverse directions may be an indication that the resolved components are secondary, or influenced by secondary components.

4.2 Resolving Magnetization Components

a. Alternating Field Demagnetization

The strength of the external magnetizing field which is required to reduce the remanent magnetization of a population of grains to zero is referred to as the remanent coercivity, or the coercivity of remanence (§ 3.1). The remanence direction of individual grains will be altered by different external field strengths depending on the "coercivity" of the individual grain, which is controlled by composition and grain size. A natural sample will have a "coercivity spectrum" reflecting the range of magnetic grain sizes and compositions present in the sample. This dependence of coercivity on composition and grain size is utilized in alternating field demagnetization to resolve the magnetization components carried by different grain populations. If the NRM of the sample is a composite of individual magnetization components which have distinct coercivity spectra, the alternating field method will be suitable for resolving the individual components. The normal procedure is to progressively demagnetize the sample by a stepwise increase in peak alternating field—the peak field is smoothly ramped down to zero—in field-free space in order to avoid acquisition of ARM (Fig. 4.1), and the remaining remanence measured after each step. In a natural sample, grain remanences with coercivity lower than the peak alternating field will be reoriented as the peak field is ramped down, and their contributions to the remaining remanence will be randomized (Fig. 4.1). The net effect is to "demagnetize" those grains. As the peak demagnetizing field is progressively raised, a larger proportion of the total magnetic grain population becomes demagnetized and ceases to contribute to the remanence of the sample.

There are two principal types of alternating field demagnetizers in current use. For both types, a Mu metal shielded solenoid generates an alternating field, with a frequency of a few hundred hertz, which is smoothly ramped down to zero from a designated peak alternating field. The sample is either placed in this field in three steps, such that the alternating field acts along the three orthogonal axes of the sample, or rotated in a specially designed tumbler about three (or in some cases two) orthogonal axes simultaneously, within the solenoid. The rotation of the sample within the coil system has the advantage of reducing the tendency for ARM acquisition; however, the observation that spurious remanent magnetization can be acquired along the axis about which the sample is rotated (Wilson and Lomax, 1972) has led to wide use of demagnetizers in which the sample is not rotated and is demagnetized in three individual steps along the orthogonal axes of the sample. The availability of highly efficient Mu metal shields has tended to negate the advantage of rotational methods in reducing the problem of ARM acquisition. For a full description of the design of alternating field demagnetizers, see Collinson (1983).

b. Thermal Demagnetization

The time required for magnetic remanence of a grain to decay in zero external field (the relaxation time) is a function of ambient temperature,

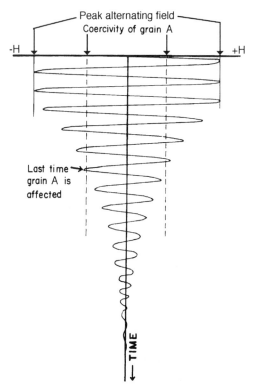

Figure 4.1 Decay of alternating field strength during AF demagnetization. As the experiment is conducted in Mu metal shields in the absence of the Earth's field, the net magnetic moment of a large population of grains with coercivities less than the peak field is zero (after Van der Voo, 1993).

grain size, and composition (Fig. 4.2). The blocking temperature (or unblocking temperature) is the temperature at which the relaxation time of the grain falls within the time range of laboratory experiments ($\sim 10^3$–10^4 s). As the function is exponential, the exact length of time of the experiment does not, within reason, critically affect the observed blocking temperature. Two populations of magnetic grains with different origin, and therefore different composition and/or grain size, might be expected to have distinct blocking temperature spectra. This principle is utilized in thermal demagnetization to separate magnetization components carried by different magnetic grain populations. Normal practice is to heat the sample in progressive temperature increments (until the magnetization falls below the magnetometer noise level), allowing the sample to cool in field-free

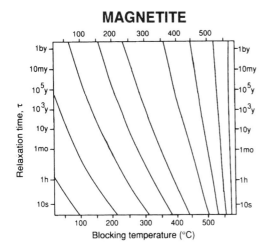

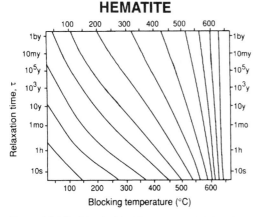

Figure 4.2 Blocking/unblocking contours joining time–temperature points for populations of SD magnetite and hematite (after Pullaiah *et al.*, 1975) illustrating that high temperatures in the laboratory time scale (say 1 hour) can unblock a thermoviscous magnetization acquired at lower temperatures on geologic time scales (say 10 My).

space after each heating increment. The remanence of all grains with blocking temperatures less than the maximum temperature will become randomized as the sample cools in field-free space. The magnetization measured after each heating step can be used to determine the directions of magnetization components with distinct blocking temperature spectra.

As for alternating field demagnetizers, the advent of highly efficient Mu metal magnetic shielding has greatly improved and simplified the shielding of thermal demagnetization equipment. Most thermal demagnetizers are built inside a long (~2 m) Mu metal shield and a cylindrical furnace capable of holding the samples lies adjacent to a cylindrical blown-air cooling chamber. The sample boat is pushed from the furnace into the cooling chamber inside the Mu metal shielding, and samples can be cooled while another set of samples is being heated. The furnace must be noninductively wound and the ambient field in the cooling chamber should not exceed a few nT, or a partial TRM may be acquired by the samples.

c. Chemical Demagnetization

Leaching with hydrochloric acid can provide a means of selective destruction of magnetic grains. The method is applicable to siliceous rocks, particularly red sandstones, which contain a secondary magnetization due to authigenic growth of fine-grained, pigmentary hematite. The dissolution rate of coarse-grained hematite (specularite) of detrital origin is lower than that of the finer-grained, pigmentary hematite of diagenetic origin, due to the higher ratio of surface area to grain volume for finer grains. Leaching in 3N to 10N hydrochloric acid, and remanence measurement at certain time intervals during the leaching, provides a means of resolving components with distinct "leaching spectra." Slots are cut in the sample to maximize the surface area of the sample in contact with the acid.

d. Orthogonal Projections

The most widely used method for analyzing changes in magnetization direction and intensity during demagnetization (using any of the above methods) is the orthogonal projection (Wilson, 1961), also referred to as the Zijderveld diagram (Zijderveld, 1967). This is a simple but extremely useful means of displaying the change in magnitude and direction of the magnetization vector during demagnetization. The end points of the magnetization vector are plotted both on the horizontal plane, where the axes are N, S, E, and W, and on the vertical plane where the axes are Up and Down (Fig. 4.3). Straight-line segments on these plots indicate that the magnetization vector being removed in this interval has constant direction. This direction can be computed and defined as a component of the magnetization. It is important to note that straight-line segments on orthogonal projections may not always represent single components. They may also represent the composites of two or more components with identical demagnetization (coercivity, blocking, or leaching) spectra. Invariably, in a particular projection, there

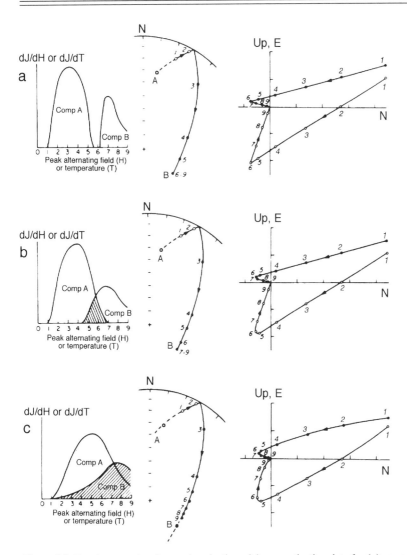

Figure 4.3 Equal area and orthogonal projection of demagnetization data for (a) nonoverlapping, (b) partially overlapping, and (c) overlapping coercivity or blocking temperature spectra (after Dunlop, 1979).

may be several straight-line segments indicating the presence of multiple components. Curvature of the lines at the junction between two straightline segments indicates overlap in demagnetization spectra (Fig. 4.3). The magnetization components cannot be resolved if the overlap in demagnetization spectra is too great. If the *blocking* temperature spectra overlap, this does not necessarily mean that the *coercivity* spectra will overlap, and the choice of demagnetization technique will be critical to whether the component content of the magnetization can be resolved.

The points on the orthogonal projection which define straight-line segments are commonly picked by eye. The component direction associated with a particular straight-line segment can then be computed by a "least-squares fitting" technique (e.g., Kirschvink, 1980). This method gives a maximum angular deviation (MAD) value which provides an estimate of the goodness of fit to the straight line to the points on the projection. MAD values greater than ~15° usually indicate that the component is poorly defined.

e. Hoffman–Day Method

In the special case in which three magnetization components with overlapping demagnetization spectra are present, the Hoffman–Day method (Hoffman and Day, 1978) may help to resolve the component directions. The method involves plotting subtracted vectors on an equal area stereographic projection, at each stage in the progressive demagnetization process (Fig. 4.4). In a demagnetization interval characterized by two overlapping demagnetization spectra, the subtracted vectors will plot on a great circle between the two component directions. If three overlapping demagnetization spectra are present (Fig. 4.4), the intersection of the two great circles can give an estimate of the direction associated with the component (B) with intermediate blocking temperature, even though this component direction is not directly resolved by the demagnetization process. Orthogonal projection in this case would not resolve the intermediate component.

f. Remagnetization Circles

If two magnetization components with partially overlapping demagnetization spectra are present in the sample, the resultant vector will trace a great circle in stereographic projection as the overlapping interval is demagnetized. If one of the two components has dispersed directions, the great circles from individual samples will converge to give the direction of the undispersed component (Fig. 4.5). The method is applicable in the special case where one of the components is dispersed. If both components have uniform directions, there will be no appreciable convergence of the great circles, and the method cannot be used on its own. The dispersion of one of the components can occur in a number of situations. For example, the

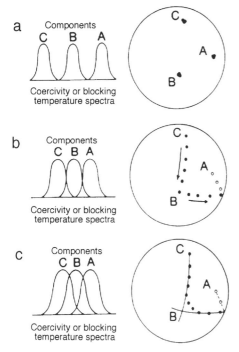

Figure 4.4 Difference vector paths corresponding to (a) nonoverlapping, (b) partially overlapping, and (c) overlapping coercivity or blocking temperature spectra (after Hoffman and Day, 1978).

primary component in conglomerate clasts will be dispersed relative to the secondary component, which postdates conglomerate deposition. In this case the secondary component could be resolved by remagnetization circle analysis. The method can also be applied in circumstances in which one of the two components postdates and the other predates the folding. In theory, both components can be resolved by applying the method both before and after tilt correction. Remagnetization circle analysis is most often used in conjunction with individual component directions resolved from orthogonal projections. If curvature of lines on orthogonal projections precludes the resolution of individual components for some but not all samples from a site, remagnetization circle analysis of these direction trends can augment and improve the estimate of the site mean direction derived from the orthogonal projections, using the procedure of McFadden and McElhinny (1988).

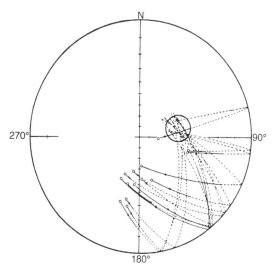

Figure 4.5 Remagnetization circles obtained by AF demagnetization of a two-component magnetization in which one component is dispersed. Dashed segments are great circle extrapolations of the data (after Halls, 1976).

4.3 Statistics

The techniques outlined above are designed to isolate individual components of magnetization. An individual component is often combined with others from the same horizon or from a particular sampling site. These vectors are represented as unit vectors and often displayed as points on an equal area (stereographic) projection, representing points on a sphere. Fisher (1953) developed a method for analyzing the distribution of points on a sphere. He predicted that these points would be distributed with the probability density P given by

$$P = \frac{\kappa}{4\pi \sinh(\kappa)} e^{\kappa \cos \psi}, \qquad (4.1)$$

where ψ is the angle between the direction of magnetization of a sample and the direction ($\psi = 0$) at which the density of the observations is a maximum. If the density of the points is axially symmetrical about the true mean direction, with the points being distributed through 360°, then the probability of finding a point between θ and $\theta + \delta\theta$ is:

$$P_\theta \, \delta\theta = (\kappa/2 \, \sinh(\kappa))e^{\kappa\cos\theta} \sin\theta \, \delta\theta. \tag{4.2}$$

The parameter κ determines the distribution of the points and is called the "precision parameter." If $\kappa = 0$ then the points are distributed evenly over the sphere and the directions are random. If the values of κ are large, the points are clustered tightly about the true mean direction.

If a group of points representing directions of magnetization are distributed on a sphere, the best estimate of the mean direction of magnetization (the center of these points) can be found by the vector addition of all directions. The direction of magnetization of a rock is given in terms of the declination D measured over 360° clockwise from true north and the inclination I measured positive (negative) downward (upward) from the horizontal. The unit vector representing this direction can be determined by its three direction cosines:

north component $\qquad l = \cos D \cos I$

east component $\qquad m = \sin D \cos I$

down (up) component $\qquad n = \sin I$.

The resultant direction of N such directions may be determined by summing the direction cosines and is given by:

$$X = 1/R \sum_{i=1}^{N} l_i$$

$$Y = 1/R \sum_{i=1}^{N} m_i \tag{4.3}$$

$$Z = 1/R \sum_{i=1}^{N} n_i.$$

The vector sum will have a length R which is equal to or less than the number of observation and is given by:

$$R^2 = \left(\sum_{i=1}^{N} l_i\right)^2 + \left(\sum_{i=1}^{N} m_i\right)^2 + \left(\sum_{i=1}^{N} n_i\right)^2. \tag{4.4}$$

The declination and inclination of the vector mean direction R are given by:

$$\text{Tan } D_R = \sum_{i=1}^{N} m_i / \sum_{i=1}^{N} n_i \tag{4.5}$$

$$\text{Sin } I_R = 1/R \sum_{i=1}^{N} n_i. \tag{4.6}$$

The best estimate k of the precision parameter κ for $k > 3$ is:

$$k = (N - 1)/(N - R). \qquad (4.7)$$

The angle from the true mean within which 95% of the directions will lie is given approximately by:

$$\psi_{95} = 140/(\sqrt{k}) \text{ degrees} \qquad (4.8)$$

This value represents the angle from the mean beyond which only 1/20 of the directions will lie.

Fisher (1953) has shown that the true mean direction of a group of (N) observations will lie within a circular cone about the resultant vector R with a semiangle α for a certain probability level $(1 - P)$ when $k > 3$. P is usually taken as 0.05 and if α is small, the circle of 95% confidence is given by:

$$\alpha_{95} = 140/(\sqrt{kN}). \qquad (4.9)$$

The value of α_{95} is often used to determine whether two sets of directions differ from one another. If the α_{95} values of two mean directions of magnetization do not overlap the mean directions of magnetization are usually believed to be significantly different at the 95% confidence level. A more rigorous test of whether two populations are different was proposed by Watson (1956a).

Fisher statistics are almost universally used in paleomagnetic analysis; however, the Fisher distribution is not ideally suited to paleomagnetic data for two reasons. Firstly, paleomagnetic directions are not always symmetrically distributed about the mean direction; elliptical distributed data sets are commonly observed. Secondly, paleomagnetic directions are often bimodal comprising both normal and reverse polarity directions.

The Bingham distribution (Bingham, 1974; Onstott, 1980) is better suited to paleomagnetic data as noncircular (elliptical) and bimodal data sets are allowed. In this treatment, the "orientation matrix" (computed as the normalized matrix of sums of squares and products) is similar to the moment of inertia matrix with unit mass assigned to each data point on the sphere. The largest (principal) eigenvector provides a very satisfactory estimate of the mean direction for the population; however, the estimation of confidence limits about the mean is more problematic, resulting in the low popularity of this method in paleomagnetic analyses.

Tauxe *et al.* (1991) advocated the use of "bootstrap" methods to calculate the confidence limits of paleomagnetic mean direction. The advantage of the method is that less stringent assumptions are made concerning the nature of the distribution (the probability density function). The disadvantage is that it is not suitable for data sets smaller than "a few dozen." The method is therefore inapplicable in cases where less than ~30 magnetization components are resolved (samples collected) at a paleomagnetic sampling site. The bootstrap method involves random sampling of the data set (population: n) to produce a large number $(\sim n^2)$ of pseudo-data sets each with the same population (n) as the original data set. Obviously, some data points may occur more than once in a given pseudo-data set. The parameter of interest, such as the mean direction, is calculated for each pseudo-data set, and the variability is used to calculate the confidence limits. For each pseudo-data set, the normal and reverse modes are separated about the normal to the principal eigenvector. The mean direction for each mode in the pseudo-data set is estimated as the sum of unit vectors, as in the Fisher procedure [Eq. (4.3)]. The mean of the bootstrapped means, the principal eigenvector of the orientation matrix derived from the bootstrap means, and the mean of the original data set should be very nearly the same. The confidence circle (α_{95}) associated with the mean is then estimated from the bootstrap means assuming that they have a Kent (1982) distribution, which is the elliptical analogue of the Fisher distribution.

4.4 Practical Guide to the Identification of Magnetic Minerals

Because magnetic mineral phases tend to have characteristic modes of occurrence, the identification of magnetic remanence carriers is a useful clue to the age of magnetization components. The following is designed to provide a guide to the identification of common remanence-carrying magnetic minerals. The focus is on convenient methods which are useful in sediments where the magnetic mineral grains are well dispersed and in low concentration. Most of the methods utilize instruments which are commonly available in paleomagnetic laboratories. In weakly magnetized sediments, the characteristics of artificially imposed remanences (such as ARM and IRM) are very useful for determining the magnetic mineralogy of the sample. However, in order to determine the carriers of NRM components, the characteristics of the NRM itself must be determined as fully as possible.

a. NRM Analysis

As described above, progressive demagnetization is the standard procedure for determining the component content of the NRM. Thermal and alternating field demagnetization characteristics indicate the blocking temperature and coercivity spectra, respectively, of the NRM components. Blocking temperature is a function of composition and grain size of the magnetic carrier. For SD grains of constant composition, the blocking temperature will be proportional to grain size [Eq. (3.1)]. Due to the large grain size of hematite at the SD/MD threshold (~ 2 mm), the SD range for hematite covers most of the remanence-carrying grain sizes encountered in sediments. Therefore, in the case of hematite, blocking temperature is a useful guide to grain size. For titanomagnetites ($x\mathrm{Fe_2TiO_4}.(1-x)\mathrm{Fe_3O_4}$) the blocking temperature is a function of composition and grain size. As x decreases, Curie point increases almost linearly (Fig. 3.4), and as blocking temperature cannot exceed the Curie point, the available blocking temperature range will increase as x decreases. For SD grains of uniform shape and composition, the blocking temperature will be a function of grain shape as well as grain volume. In the MD range, the rapid decrease in coercivity with increasing grain size might be expected to give rise to an accompanying decrease in blocking temperature. The blocking temperature range in itself is not generally diagnostic of magnetic mineralogy, other than the fact that the maximum blocking temperature cannot exceed the Curie or Néel point of the mineral composition. In the case of minerals with relatively low Curie or Néel points, such as geothite ($\sim$120°C), the blocking temperature range is rather diagnostic because it is so restricted. Mineralogical changes as a result of heating during thermal demagnetization can obscure the blocking temperature spectrum. For example, maghemite alters to hematite at about 300°C, iron sulfides will oxidize in air to form magnetite at about 300°C, and hematite tends to convert to magnetite at temperatures above 600°C. These phase changes, if identified as such, can be useful as a means of identification of magnetic minerals.

The coercivities of remanence of hematite and geothite are typically about two and three orders of magnitude greater, respectively, than that of SD magnetite. Coercivity of remanence in naturally occurring hematite and titanomagnetite is largely a function of grain volume in the case of hematite and grain shape and volume in the case of titanomagnetites. For magnetites, coercive force (H_c) increases with x and with decreasing grain size in a wide MD grain size range (0.1–1000 μm) where H_c is proportional to $d^{-0.6}$, with crushed grains having higher coercivity than hydrothermally grown or naturally occurring crystals (Heider *et al.*, 1987) (Fig. 4.6). Below about 0.05 μm, the coercivities will decrease sharply close to the SP-SD grain size threshold (see Maher, 1988).

Although demagnetization of remanence carriers from AF or thermal demagnetization data is rarely unequivocal, combinations of these data from the same samples can be fairly diagnostic. For example, in Fig. 4.7, three components, A, B, and C, are apparent in the orthogonal projection of thermal demagnetization data. The blocking temperature spectra of components B and C overlap. In the AF demagnetization diagram for a specimen from the same sample, only components A and C are apparently resolved. Component B has low blocking temperature (200–300°C) and high coercivity (>80 mT). In view of the discussion above, fine-grained hematite is a possible carrier for this component. This is consistent with the red pigment of the limestone sample and suggests that this component is a secondary component of diagenetic origin. Component C has a maximum

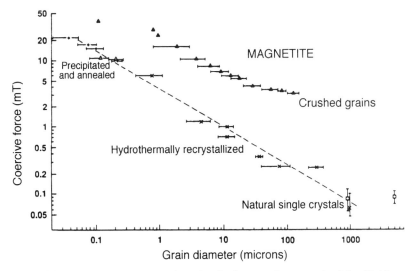

Figure 4.6 Coercive force as a function of grain diameter for magnetite (after Heider *et al.*, 1987). The apparent higher coercivities of crushed grains are attributed to their stressed state and adhering finer fragments.

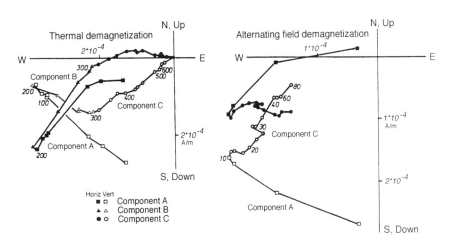

Figure 4.7 Orthogonal projection of a three-component magnetization (from a pink pelagic limestone) illustrating contrasting response to thermal (temperatures in °C) and alternating field (peak fields in mT) demagnetization (after Channell *et al.*, 1982a).

blocking temperature of over 600°C and is therefore carried, at least in part, by relatively coarse-grained hematite. The coercivity spectrum of component C suggests that a low-coercivity mineral (magnetite) contributes to this component. Component A has low blocking temperature and low coercivity, and the *in situ* direction is close to that of the present field; therefore the component is a recent (probably viscous) overprint. It is important to note that the more primary magnetization component is not necessarily the final component revealed by the demagnetization procedure. In the example mentioned above, the secondary component C has the highest coercivity of the components present in the sample.

b. J_s/T Curves

The change in saturation magnetization with temperature is an important indicator of magnetic mineralogy, because it is independent of grain size and shape. The temperature at which the saturation magnetization disappears is the Curie or Néel temperature, which is the temperature at which thermal energy precludes spin moment alignment. This temperature is characteristic of the magnetic mineralogy. Most Curie balance systems are not sufficiently sensitive to determine Curie or Néel points from weakly magnetized sediments without extraction and concentration of magnetic minerals. Magnetic extraction procedures are inevitably grain size and compositionally selective, and therefore the J_s/T curves may not be representative of the magnetic mineral content of the sample. Significant improvements in Curie balance sensitivity have, however, been achieved by sinusoidally cycling the applied field and making continuous drift corrections through Fourier analysis of the output signal (Mullender *et al.*, 1993). Horizontal balance systems are generally preferred to vertical systems, as they reduce weight loss problems arising from dehydration of clay minerals. For titanomagnetites and titanohematites, Curie and Néel temperatures indicate the composition (x) (Fig. 3.4), and in favorable circumstances Curie temperatures corresponding to several magnetic phases can be observed (Fig. 4.8). Alteration of magnetic minerals during heating may obscure the Curie or Néel point of the original mineral and produces irreversible curves. This irreversibility can be useful for detecting the presence of maghemite, which converts to hematite at about 300°C (Fig. 4.8), and for determining the presence of iron sulfides, which alter to magnetite when heated in air. Ideally, irreversible curves should be rerun *in vacuo* or in an inert gas atmosphere.

c. IRM Acquisition and Thermal Demagnetization of 3-Axis IRM

The shape of the curve of acquisition of IRM with increasing magnetizing field is a measure of the coercivity spectrum of magnetic minerals present

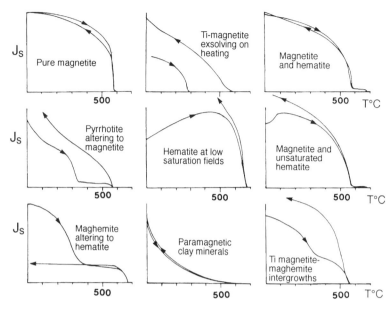

Figure 4.8 Typical saturation magnetization–temperature (Js/T) curves for various minerological ical compositions (after Piper, 1987).

in the sample (Dunlop, 1972). The common magnetic phases present in sediments, SD magnetite, MD magnetite, fine SD hematite, coarse SD hematite, and goethite have rather characteristic and distinct coercivity spectra and IRM acquisition curves (Fig. 4.9). The intensity of the SIRM and the magnetizing field at which the SIRM is acquired vary with mineralogy and grain size. Thermal demagnetization of a composite 3-axis IRM (Lowrie, 1990) is very useful for determining magnetic mineralogy. An IRM is applied sequentially along three orthogonal axes of the sample using decreasing magnetizing fields, typically 1 T, 0.4 T, and 0.12 T. Thermal demagnetization of the three orthogonal "soft," "medium," and "hard" IRMs can give insights into the magnetic mineralogy (Fig. 4.10).

d. Hysteresis Loops

The advent of the alternating gradient force magnetometer (μMAG) allows hysteresis loops to be generated for small (~50 mg) samples, even for weakly magnetized sediments. Saturation magnetization (M_s), saturation remanence (M_{rs}), and coercive force (H_c) can be determined from the loops, and back field experiments on the μMAG can yield values of remanent coercivity (H_{cr}) (Fig. 4.11). For samples containing magnetite only, hystere-

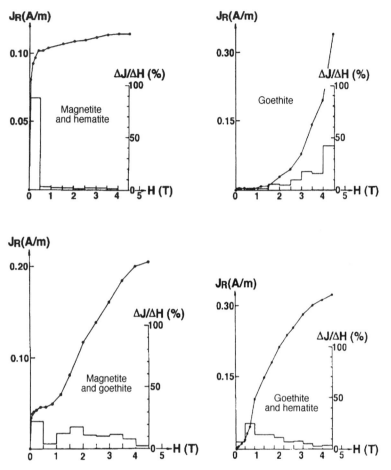

Figure 4.9 Acquisition of isothermal remanent magnetization (IRM) for various mineralogical compositions in limestones (after Heller, 1978).

sis parameters determined in this fashion can be plotted on a hysteresis ratio plot to estimate the domain state of the magnetite population (Day *et al.*, 1977) (Fig. 4.12).

e. AF Demagnetization of IRM and ARM

A modification of the Lowrie-Fuller (1971) test introduced by H. P. Johnson *et al.* (1975), provides a useful means of determining the relative importance of SD and MD grains in magnetite-bearing rocks and sediments. The original Lowrie-Fuller test is based on the observation and theory that the

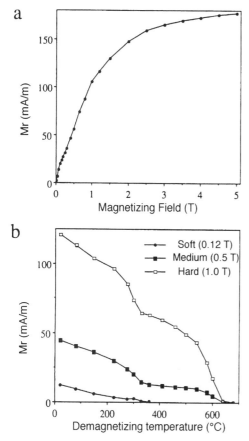

Figure 4.10 For a limestone sample containing pyrrhotite and hematite, (a) acquisition of isothermal remanent magnetization (IRM) and (b) thermal demagnetization of a 3-axis IRM (after Lowrie, 1990).

stability to AF demagnetization of thermal remanent magnetization (TRM) in SD magnetite increases with decreasing magnetizing field, whereas the converse is true for MD grains. This observation is not strictly a domain state delimiter; however, it is a useful grain size parameter (Dunlop, 1986). In the modified version of this test (H. P. Johnson *et al.* 1975), it is assumed that the stability of strong field TRM will be similar to that of SIRM and that the stability of low-field TRM will be similar to that of ARM. The modified test compares the stability of AF demagnetization of SIRM and ARM, is very simple to apply, and avoids the need to heat the sample, which may produce mineralogical change. According to the H. P. Johnson *et al.* (1975) test, if the ARM is more stable to AF demagnetization than

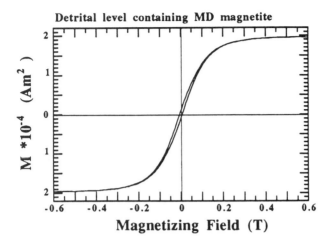

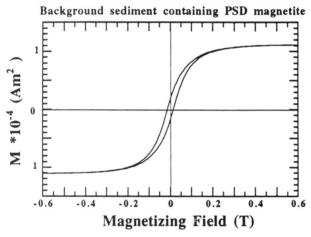

Figure 4.11 Hysteresis loops for Upper Pleistocene detrital layers from the Labrador Sea containing MD magnetite and intervening background sediment containing PSD magnetite (after Stoner *et al.*, 1995a).

SIRM, the magnetite is considered to be predominantly in the SD grain size range (Fig. 4.13). Cisowski (1981) has shown that strongly interacting SD grains respond to the test, but that noninteracting or weakly interacting SD grains do not give the predicted response. Cisowski (1981) proposed a test for the degree of interaction based on the point of intersection (on the

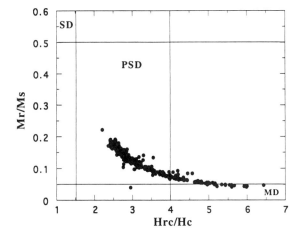

Figure 4.12 Hysteresis ratios plotted on a Day *et al.* (1977) diagram, illustrating a mixing line in sediments containing PSD and MD magnetite (after Stoner *et al.*, 1995a).

normalized moment axis) of the IRM acquisition and IRM demagnetization curves (Fig. 4.13).

f. Low-Temperature Treatment of IRM

Transitions in the magnetic properties of magnetite, pyrrhotite, and hematite occur at low temperatures and they provide a potential means of magnetic mineral identification. Magnetite exhibits a crystallographic phase transition from cubic to monoclinic at 110–120 K. Associated with this transition, the anisotropy constant goes through zero as the easy axis of magnetization changes from [100] to [111] (Nagata *et al.*, 1964). The so-called Verwey transition in magnetite can be recorded by a decrease in intensity of an IRM imposed at low temperature as it is allowed to warm through the transition temperature or alternatively as an IRM imposed at room temperature is allowed to cool through this temperature (Fig. 4.14). The transition has been clearly observed in the NRM iron ore samples (Fuller and Kobayashi, 1967) and the SIRM of European Mesozoic limestones (e.g., Heller, 1978; Mauritsch and Turner, 1975; Channell and McCabe, 1994) and deep-sea sediments (e.g., Stoner *et al.*, 1996, Fig. 4.14). The transition should be manifest as a relatively sharp change in magnetization intensity and should not be confused with the steady decrease in low-temperature IRM intensity due to unblocking of superparamagnetic grains. It was considered that the Verwey transition was restricted to MD grains (Fuller and Kobayashi, 1967); how-

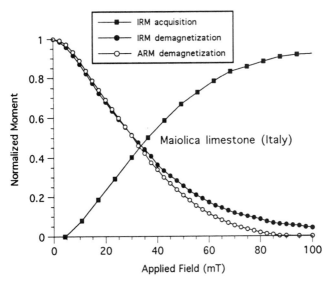

Figure 4.13 The point of intersection of IRM acquisition and IRM demagnetization for a sample of white pelagic limestone is close to 0.5, consistent with noninteracting SD magnetite. According to Cisowski (1981), noninteracting or weakly interacting SD grains do not give the predicted response to the Johnson *et al.* (1975) test based on the relative AF stability of ARM and IRM.

ever, it is now clear that the transition occurs in near-stoichiometric magnetite crystals for all sizes above 37 nm, below which the transition is masked by superparamagnetic unblocking (Özdemir *et al.*, 1993). Small deviations from stoichiometry in magnetite, including oxidation to maghemite, have the effect of suppressing the Verwey transition (Özdemir *et al.*, 1993).

The Morin transition in hematite at $-10°C$, due to a change in crystallographic orientation of the spin axis, is more elusive. It has been observed in iron ore samples (Fuller and Kobayashi, 1967) but not, as far as we know, in sediments. Apparent observation in lake sediments (Creer *et al.*, 1972) has been reinterpreted as due to disruption of grain orientation as a result of ice formation in unconsolidated sediment (Stober and Thompson, 1979). The pyrrhotite transition at 34 K has been recognized in igneous and metamorphic rocks (e.g., Dekkers *et al.*, 1989).

g. Continuous Thermal Demagnetization of IRM

Continuous measurement of remanence during heating has several advantages over stepwise thermal demagnetization. There is less likelihood of mineralogical changes, and no PTRM or VRM will be acquired by the

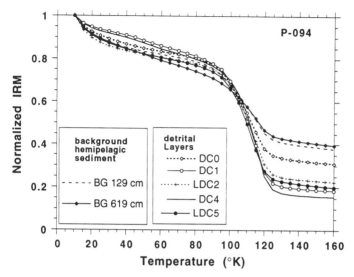

Figure 4.14 Verwey transitions in magnetite-bearing Upper Pleistocene sediments from the Labrador Sea. Coarser grained detrital layers with higher proportion of MD magnetite show larger remanence drop than PSD-dominated background sediments. IRM acquired in 500 mT field at 10 K (after Stoner *et al.*, 1995a).

samples. Cryogenic magnetometers have not been used on a routine basis for continuous thermal demagnetization due to problems of construction of a heating system that does not interfere with the SQUID sensors. The most commonly used magnetometer for continuous thermal demagnetization is a modification of the Digico magnetometer (e.g., Heiniger and Heller, 1976). The loss of sensitivity of the larger fluxgate ring, which provides room for the furnace, relative to the room temperature ring (~factor of 10) is such that, for most sediments, the demagnetization experiments must be carried out on the stronger IRM rather than on the NRM. Continuous demagnetization of superimposed IRM components can be useful for determining the maximum blocking temperatures of two IRM components.

h. Susceptibility/Temperature

Initial susceptibility is often routinely measured after each measurement step during thermal demagnetization in order to monitor changes in magnetic mineralogy during heating. The maghemite-to-hematite transition at about 300°C produces a decrease in susceptibility which is a useful means of identifying the presence of this mineral. The susceptibility of maghemite is 2 to 3 orders of magnitude greater than that of hematite. This susceptibility decrease will be masked if iron sulfides or titanohematite are present, because heating in air oxidizes the iron sulfides and titanohematites to magnetite, which has high susceptibility, comparable to that of maghemite. The formation of magnetite during the thermal demagnetization procedure can result in the development of spurious magnetizations. Note that the most common iron sulfide, pyrite, is paramagnetic and cannot therefore carry remanence.

i. Coercivity/Temperature

An additional method for monitoring the change in mineralogy during heating involves the change of remanence coercivity (H_{cr}). The remanence coercivity of maghemite (and magnetite) is 1 to 2 orders of magnitude less than that of hematite. The maghemite-to-hematite transition at about 300°C will be accompanied by a change in remanence coercivity. The measurements of H_{cr} can be carried out after each heating step with an alternating force gradient magnetometer (μMAG) but the measurement procedure is considerably slower with more standard paleomagnetic equipment. For example, after each heating step, the sample is given 1 T IRM, and a back field IRM (antiparallel to the 1 T IRM) is then imposed in steps of 0.1 T. The back field corresponding to zero IRM corresponds to the remanence coercivity (H_{cr}). The increase in remanence coercivity in the 300–400°C temperature range has been cited as evidence for the presence of maghemite in red pelagic limestones (Channell et al., 1982a).

j. Separation Techniques

Magnetic mineral separation techniques are invariably selective, and not fully representative of the grain size and composition of magnetic minerals present in the sample; however, magnetic separation may be necessary for SEM/TEM, X-ray, Mossbauer, or chemical analyses. The importance of fine SD grains as remanence carriers emphasizes the necessity of making the separation technique as sensitive as possible to the fine grains. As grain shape has become an important criterion for distinguishing detrital and biogenic magnetite, separation procedures to prepare representative extracts for SEM and TEM observation have become more important. A recommended procedure which has been successful for extracting magnetite (including the SD fraction) is as follows: (i) Crush the sample (if necessary) in a jaw crusher with ceramic jaws. (ii) Use a mortar and pestle to produce a powder. (iii) Dissolve carbonate with 1N acetic acid buffered with sodium acetate to a pH of 5, changing the reagent every day until reaction ceases (several weeks). (iv) Rinse the residue with distilled water. (v) Agitate the residue ultrasonically in a 4% solution of sodium hexametaphosphate to disperse the clays. (vi) Extract the magnetic fraction using a high-gradient magnetic separation technique (Schulze and Dixon, 1979), or alternatively, pass the solution (several times, if necessary) slowly past a small rare earth magnet.

5

Fundamentals of Magnetic Stratigraphy

5.1 Principles and Definitions

The purpose of magnetic stratigraphy is to organize rock strata into identifiable units based on stratigraphic intervals with similar magnetic characteristics. Short-period behavior of the magnetic field (secular variation) can be useful for correlation, particularly in Holocene lacustrine sediments and archaeomagnetic studies (Ch. 13). At the other end of the geologic time scale, paleomagnetic pole positions have been used as a means of correlating Precambrian rocks (Irving and Park, 1972). Relative geomagnetic paleointensities as well as a wide range of nondirectional magnetic properties can be used for correlation (Ch. 14). The magnetic characteristic most often used, however, is the polarity of magnetic remanence. The polarity is said to be "normal" (north-seeking magnetization gives a northern hemisphere pole, as today) or "reverse" (north-seeking magnetization gives a southern hemisphere pole). When correlated to the geomagnetic polarity time scale (GPTS), the polarity record observed in stratigraphic sequences allows them to be placed in a chronostratigraphic framework.

The most important geomagnetic property for stratigraphic purposes is aperiodic polarity reversal of the geomagnetic dipole field. The direction of magnetization of a rock is by definition its north-seeking magnetization, and as explained above, the magnetization can usually be designated either normal or reverse polarity. A problem may arise in Early Paleozoic or older rocks, or in displaced terrains, when it is unclear in which hemisphere the rocks acquired their magnetization. An ambiguity may be introduced as to the true polarity of the rock. In this case, an apparent polar wander path (APWP) must be established for the individual tectonic unit being investigated. For instance, the North American APWP crosses the equator in Lower Paleozoic time, introducing some ambiguities as to which is the north pole and which is the south pole, particularly in regions where there has been large-scale local tectonic rotation. In practice, the polarity can be determined by following the motion of the pole sequentially back in time from the present.

The terminology of magnetic polarity stratigraphy has been codified by the IUGS subcommission on the Magnetic Polarity Timescale (Anonymous, 1979). This body introduced the following useful definitions.

1. Magnetic stratigraphy or magnetostratigraphy—the element of stratigraphy that deals with the magnetic characteristics of rock units.

2. Magnetostratigraphic classification—the organization of rock strata into units based on variations in magnetic character.

3. Magnetostratigraphic units (magnetozones)—bodies of rock strata unified by similar magnetic characteristics which allow them to be differentiated from adjacent strata.

4. Magnetostratigraphic polarity classification—the organization of rock strata into units based on changes in the orientation of remanent magnetism in rock strata, related to changes in the polarity of the Earth's magnetic field.

5. Magnetostratigraphic polarity reversal horizons are surfaces or very thin transition intervals in the succession of rock strata, marked by changes in magnetic polarity. In practice, the term magnetostratigraphic polarity reversal horizon may be allowed to stand for a finite transition interval. Less commonly, the polarity change takes place through a substantial interval of strata, and the term magnetostratigraphic polarity transition zone should be used. Magnetostratigraphic polarity reversal horizons and magnetostratigraphic polarity transition zones may be referred to simply as polarity reversal horizons and polarity transition zones, respectively, if in the context it is clear that the reference is to changes in magnetic polarity. Polarity reversal horizons or polarity transition zones provide the boundaries for polarity stratigraphic units.

6. Magnetostratigraphic polarity units (or zones) are bodies of rock strata, in original sequence, characterized by their magnetic polarity, which allows them to be differentiated from adjacent strata.

5.2 Polarity Zone and Polarity Chron Nomenclature

The basic magnetostratigraphic unit is the magnetostratigraphic polarity zone (in accordance with the nomenclature recommended in the International Stratigraphic Guide, pp. 8–14, see Anonymous, 1979). A magneto-stratigraphic polarity zone may be referred to simply as a magnetozone, if in the context it is clear that the reference is to magnetic polarity. Polarity zones are bounded above and below by polarity reversal horizons or polarity transition zones (see definition 5, above). Polarity zones may be of different rank.

Magnetostratigraphic polarity zones may (1) consist of strata with a single polarity throughout or (2) be composed of an intricate alternation of normal and reverse units or (3) be dominantly either normal or reverse magnetozones, but with minor subdivisions of the opposite polarity. Thus a zone of dominantly normal polarity may include lower rank units of reverse polarity.

The words normal and reverse, where it is possible to indicate the polarity unambiguously, may be included as a part of the term name as an aid to clarifying its significance. However, as mentioned above, in pre-middle Paleozoic rock sequences, it is sometimes difficult to determine what is normal and what is reverse. Also, in rare circumstances, excursions of the paleofield may provide polarity units that cannot be classified as normal or reverse, but may be designated as intermediate in direction.

The end of a successful magnetostratigraphic survey is the identification of a series of normally and reversely magnetized polarity zones of varying length arranged in stratigraphic order. At this stage, it is useful to give each magnetozone (polarity zone) a number/letter and geographic designation, such as, from base of the section: San Pedro N1, R1, N2, R2, etc., or Gubbio A−, B+, C−, D+, etc. At Gubbio (Alvarez *et al.*, 1977; Napoleone *et al.*, 1983), polarity zones were designated in ascending alphabetical order, normal zones distinguished by "+" and reverse zones by "−". The thinner polarity zones are given lower hierarchy and are designated as subzones. At Gubbio, subzones within D+ are labeled D1+, D2−, D3+. The labeling of polarity zones (magnetozones) in individual sections (Fig. 5.1) is useful as it is agnostic and separates the observations from the correlation to the geomagnetic polarity time scale. It is a mistake to confuse polarity zone labels with oceanic magnetic anomaly numbers; for example, a polarity zone should not be referred to as "anomaly 34." Oceanic magnetic anomalies and polarity zones are distinct phenomena, and they should not be confused. They may, however, have a common cause and be time correlative.

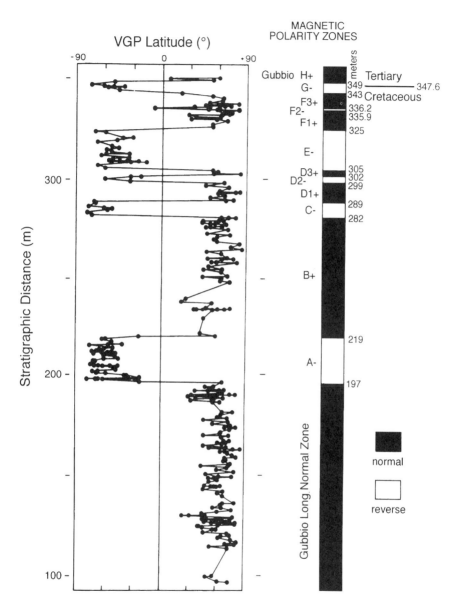

Figure 5.1 Magnetic stratigraphy of the Upper Cretaceous Scaglia Rossa Fm. at Gubbio (Italy). The VGP latitude is presented for each sample. The polarity zones are labeled alphabetically (after Alvarez *et al.*, 1977).

The recommended hierarchy in magnetostratigraphic units and polarity chron (time) units is as follows:

Magnetostratigraphic polarity units	Geochronologic (time) equivalent	Chronostratigraphic equivalent	Approximate duration (yr)
Polarity megazone	Megachron	Megachronozone	$10^8 - 10^9$
Polarity superzone	Superchron	Superchronozone	$10^7 - 10^8$
Polarity zone	Chron	Chronozone	$10^6 - 10^7$
Polarity subzone	Subchron	Subchronozone	$10^5 - 10^6$
Polarity microzone	Microchron	Microchronozone	$<10^5$
Polarity cryptochron	Cryptochron	Cryptochronozone	uncertain existance

Unlike most branches of stratigraphy, the chronologic system in use in magnetic stratigraphy was determined simultaneously with its development. Initially, Cox *et al.* (1963) boldly divided the reversal sequence from 0 to 4 Ma into four magnetic epochs (chrons): Brunhes, Matuyama, Gauss, and Gilbert, named after great geomagnetists. It is noteworthy that, due to techniques used in determining polarity and dating individual lava flows, type sections for the Brunhes, Matuyama, Gauss, and Gilbert were never established. Type localities are known, however, for subchrons such as the Jaramillo, at Jaramillo Creek in New Mexico. In present usage, polarity chron labels, Brunhes, Matuyama, Gauss, and Gilbert, are used for the Pliocene and Pleistocene, with the remainder of the Cenozoic being subdivided into polarity chrons designated by numbers correlated to oceanic magnetic anomalies. For example, polarity chron C26 encompassed the time represented by magnetic anomaly 26 and the reversed interval preceeding it (Fig. 5.2). The numbered magnetic anomalies in the Late Cretaceous and Cenozoic correspond to intervals of normal polarity of the geomagnetic field. The polarity chron nomenclature has evolved progressively to accommodate additional polarity chrons (LaBrecque *et al.*, 1977; Harland *et al.*, 1982; Cande and Kent, 1992a). For example, the three polarity intervals within C23n, from old to young, are labeled C23n.2n, C23n.1r, and C23n.1n (Fig. 5.2). The Cande and Kent (1992a) chron nomenclature (Fig. 5.2) is slightly modified and expanded from that of LaBrecque *et al.* (1977). As an example, let us examine chron 3A (Fig. 5.3) which is divided into a dominantly normal polarity upper segment

Figure 5.2 Magnetic polarity time scale and polarity chron nomenclature for Campanian to Eocene time (after Cande and Kent, 1992a).

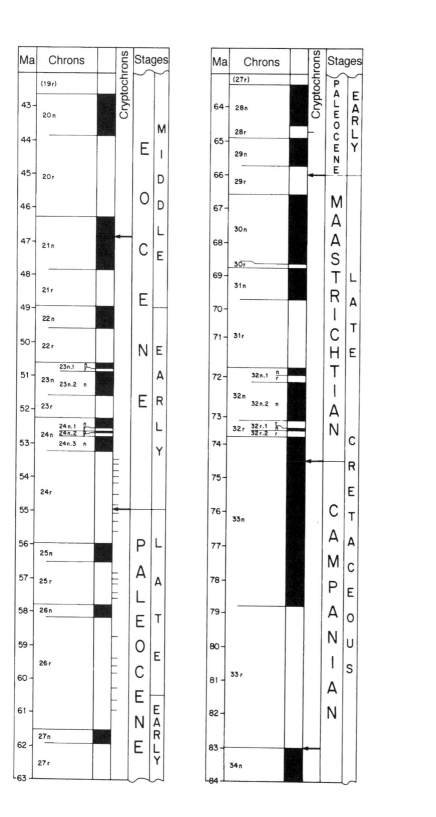

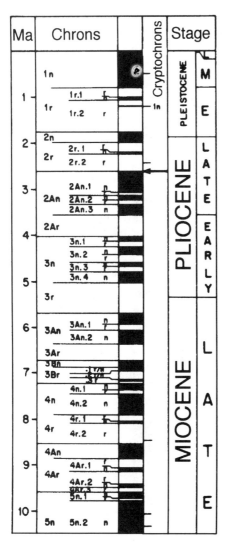

Figure 5.3 Magnetic polarity time scale and polarity chron nomenclature for Late Miocene to recent (after Cande and Kent, 1992a).

(3An) and a preceeding reverse interval (3Ar). The upper normal/reverse interval is designated as 3An.1. The reverse interval would then be 3An.1r. In this way a complete designation of the time scale can be achieved. The location of a geological occurrence in time within such a system may be precisely placed using the chron label with suffix "*x*", where

$$x = (X_e - X_{co})/X_{cy} - X_{co})$$

and X_e = stratigraphic position of event
X_{cy} = location of younger chron boundary
X_{co} = location of older chron boundary,

such that 3An.1r(0.25), for example, is a location within the subchron, 25% from the base of the subchron.

For the Late Jurassic and Early Cretaceous, Harland *et al.* (1982) labeled polarity chrons according to the correlative M-sequence oceanic magnetic anomaly. In contrast to Cenozoic marine magnetic anomalies, most but not all of the Mesozoic (M-sequence) oceanic magnetic anomalies correlate to reverse polarity chrons. Intervening normal polarity chrons are given the number of the next (older) reverse chron with "n" appended. As for the Cenozoic, we recommend use of the prefix "C" when the label refers to a polarity chron (time interval) rather than an oceanic magnetic anomaly.

The most complete record of the reversal pattern of the geomagnetic field since 160 Ma is preserved in the crust of the ocean and detected by seaborne/airborne magnetometer surveys. These magnetic anomalies are interpreted to be due to reversals of the geomagnetic field recorded in the rocks of the ocean floor. The sea floor in a major ocean is quasi-stratigraphic in that it gets younger toward the spreading axis; however, it is not in superposition in the usual sense. The reversal record as preserved in the sea floor has become the template (a sort of type section) for geomagnetic reversal history of the last 160 My and is the basis for the geomagnetic polarity time scale (GPTS). The designation of type sections, in the classical stratigraphic sense, for the geomagnetic reversal pattern since 160 Ma is unnecessary. On the other hand, type sections for magnetobiochronology, the correlation of the biologic record to the magnetic polarity sequence, might be desirable. For example, a potential type section for the Paleogene might be the Contessa section in Italy (Lowrie *et al.,* 1982).

The type section concept for polarity history is the only way in which the polarity record can be built up for time intervals not recorded by present-day sea floor. For periods prior to 160 Ma, for which no oceanic magnetic anomalies are available, the establishment of polarity sequences must be carried out using classical stratigraphic principles involving type sections. An example of a viable magnetostratigraphic type section is that recording the Kiaman Superchron (Interval) in Australia, where the super-chron is thought to be tied directly to the rock record (Irving and Parry, 1963).

5.3 Field Tests for Timing of Remanence Acquisition

Undoubtedly, one of the most important technical challenges in any paleo-magnetic research is to determine the age of magnetization components. It is of utmost importance to determine which, if any, of the magnetization components was acquired near the time of deposition of the sediments or cooling of the igneous rock.

In marine sediments, experiments have shown that the detrital rema-nence (DRM) is acquired at some position below the sediment–water interface, where the water content falls to a level at which the magnetic grains can no longer rotate (Irving and Major, 1964; Kent, 1973; Verosub, 1977). For pelagic sediments, a reasonable estimate of the depth at which the DRM becomes locked is 10–20 cm below the sediment–water interface (3.3c). The age of the sediment magnetization will not, therefore, coincide with the age of sediment deposition but will lag behind by intervals dictated by the lock-in depth.

In marine sediments retrieved from the bottom of the ocean, the stabil-ity of the record can be established by laboratory demagnetization experi-ments, by the internal coherence of the reversal pattern, and by a reversal test. A reversal test is based on the fact that normal and reverse intervals of the core should yield mean directions 180° apart (Fig. 5.4). This is an important test since any magnetization component acquired later will tend to affect the reverse and normal directions differently. It is usual to test whether the circles of confidence at the 95% level overlap for the reverse and normal mean directions. The results of this test depend on sample size and dispersion of the data, and if one of the sets of directions has a large

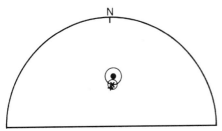

● Mean of reverse sites flipped through 180°
x Mean of normal sites
+ Dipole field at sampling site

Figure 5.4 Mean directions of normally and reversely magnetized sites from Curtis Ranch, San Pedro Valley (Arizona) with overlapping circles of 95% confidence. Reverse mean direction is flipped through 180° (after Johnson *et al.*, 1975).

α_{95} then no significant difference will be found. McFadden and McElhinny (1990) have proposed a simple version of the reversal test in which the mean directions of the reversed and normal data sets and α_{95} are utilized. It is assumed that both sets have a common precision. In the case where both sets of data have N greater than 5, the following equation can be used to calculate the critical angle γ_c with $p = 0.05$, where $N = N_1 + N_2$ and R_1 and R_2 are the vector sums for the two polarities.

$$\cos \gamma_c = 1 - \frac{(N - R_1 - R_2)(R_1 + R_2)}{R_1 R_2} \left[\left(\frac{1}{p}\right)^{\frac{1}{N-2}} - 1 \right] \qquad (5.1)$$

If the angle γ_o between the two mean directions is greater than the angle γ_c then the hypothesis of a common mean direction may be rejected at the 95 percent confidence limit and the reversal test fails. The "positive" reversal tests are classified on the basis of γ_c, as A if $\gamma_c \leq 5°$, as B if $5° < \gamma_c \leq 10°$, as C if $10° \leq \gamma_c \leq 20°$.

Very few data are classified as A since the requirements are so stringent. The data presented in Fig. 5.4 are classified as B, since the angle γ_c is greater than 10° and less than 20°. If γ_c exceeds 20° the reversal test is considered to be negative.

Sedimentary rocks often become involved in tectonic processes and are then exposed through erosion at the Earth's surface. This makes possible other tests for stability such as the bedding tilt test suggested by Graham (1949). Graham pointed out that if the direction of remanent magnetization was stable from the time of rock formation, the direction of remanence determined from tilted rock strata would be dispersed but would coincide after correction for bedding tilt (Fig. 5.5). It is now well known that directions of magnetization can be acquired before, during, and after the rocks have been deformed. It is therefore necessary to determine the stage in the tilting process at which a particular magnetization component reaches maximum concentration. This is usually done by incrementally untilting the beds and calculating the Fisher (1953) concentration parameter (k) at each step. The stage in this process at which the Fisher "precision" or concentration parameter (k) is maximized is usually taken to represent the attitude of the beds at the time that the magnetization component was acquired. In Fig. 5.5, it can be seen that the highest values of k for component A occur when the beds are in their present position (0% untilting). Therefore, component A was acquired after folding of the rocks. The B component of magnetization reaches its greatest concentration when the beds are unfolded by ~50%; this component was therefore acquired during the deformation process and is a synfolding magnetization. Component C reaches its highest concentration after 100% unfolding. The C component was therefore acquired before folding of the beds and is the only component which can be used for magnetostratigraphic

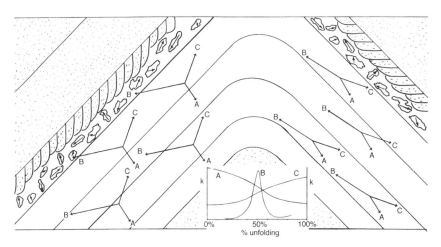

Figure 5.5 Schematic presentation of fold and conglomerate test. Three magnetization components are present in these sediments. The A component is postfolding and was acquired with the sediments in their present position. Component B is synfolding and was acquired when the rocks were tilted by 50% of their present attitude. Component C is prefolding and the directions group best after restoring the beds to horizontal. The magnetization directions in syndepositional rip-up clasts show randomly directed C components of magnetization, indicating that the C component was acquired very early in the history of the sediment. k is the Fisher (1953) concentration parameter.

purposes. The improvement in grouping may be tested statistically (Mc-Elhinny, 1964). In this test, the precision of the mean direction before tilt correction (k_1) is compared to the precision after tilt correction (k_2). The ratio k_2/k_1 can be compared with difference ratio tables with equal degrees of freedom ($2N - 1$). This test has been widely applied by paleomagnetists for many years and is still in use.

McFadden and Jones (1981) pointed out that the McElhinny (1964) tilt test was invalid because the two estimates of k are related through the bedding attitude. The McElhinny (1964) fold test is often said to be conservative; however, it is clear that in some cases it can be misleading, particularly when the magnetization being tested was acquired during folding (e.g., Kent and Opdyke, 1985). McFadden (1990) proposed a fold test based on correlations between the distribution of magnetization directions and bedding attitudes.

Recently two new tilt tests have been proposed that have considerable merit. Watson and Enkin (1993) argued that the problem could be solved by parameter estimation. This method legitimizes the use of Fisher's precision parameter (k). Their method involves parametric resampling of the original data and plotting k as a function of the degree of untilting (Fig. 5.6). The number of samples, observed mean direction, and k are used to create synthetic data sets which are sequentially untilted around a horizontal axis generating, say, 1000 synthetic curves. The technique permits the calculation of the best estimate of the amount of unfolding at the time of magnetization and a calculation of 95% confidence limits for this value. Another tilt test has been proposed by Tauxe and Watson (1994) which is based on eigenvector analysis and parameter estimation techniques. This fold test has the advantage that the polarity of the resultant vector is not critical to the analysis, so no assumptions are required as to polarity or whether the data have a Fisherian distribution. A bootstrap method is employed to determine 95% confidence limits on parameters.

The conglomerate test utilizes the fact that an early magnetization in conglomerate clasts will be randomly distributed (Fig. 5.5). The most powerful conglomerate test is that applied to syndepositional conglomerates, such as rip-up clasts. These synsedimentary conglomerates can be formed by river channels meandering into floodplains containing previously deposited muds and silts. In some cases coherent blocks of sediment are incorporated into the sandstone of an ancient river channel. If magnetization directions in these clasts have magnetic properties similar to those of the parent sediment and possess a stable component which is randomly directed, then the investigator has demonstrated that the component in question was acquired early in the history of the rock (Fig. 5.7). The randomness of the clast magnetization directions can be tested using the Watson (1956b) test for randomness.

Another important test of early acquisition of magnetization (Graham, 1949) is the "slump" test. Syndeposition slumps are often present in rock sequences; however, application of this test is uncommon. If magnetization was acquired during deposition, remanence directions from slumps should be different from the unslumped parent rock. Alvarez and Lowrie (1984) have demonstrated that Upper Cretaceous pelagic limestones in Italy contain large slumps that yield directions which deviate widely from those in the unslumped parent rock, indicating that these sediments were magnetized early in their history.

5.4 Field Sampling for Magnetic Polarity Stratigraphy

The geomagnetic field reverses in an aperiodic fashion, with intervals of constant polarity having durations ranging from 20 ky to 50 My. An ideal sampling scheme would be one that is designed to monitor the geomagnetic field every ~5000 years. This is often impractical or impossible due to the

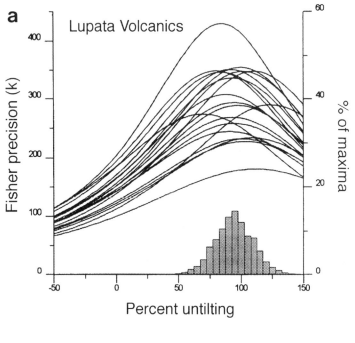

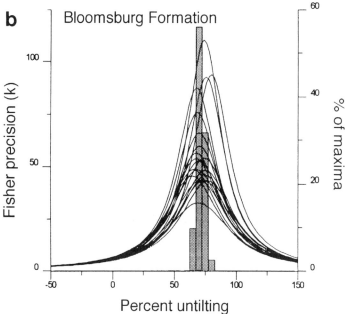

Figure 5.6 The value of the Fisher (1953) concentration parameter (k) as a function of the degree of untilting for 20 examples of parametric resampling of the original data. The histogram shows the distribution of the maxima from 1000 trials in groups of 5% ranges of unfolding. (a) Data from Gough and Opdyke (1963). (b) Data from Stamatakos and Kodama (1991).

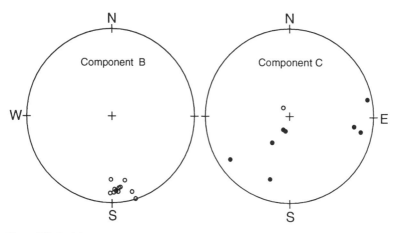

Figure 5.7 Positive conglomerate test on Carboniferous Mauch Chunk sediments from Pennsylvania. Magnetization directions from individual clasts indicate a C component predating deposition of the conglomerate and a B component postdating the deposition of the conglomerate.

type of rock being sampled. In sampling marine sedimentary cores, it is common practice to remove samples at 10-cm intervals by sawing cubes if the sediments are dry, or by pressing plastic containers into the sediment if the samples are wet. Sampling at this density for constant pelagic sedimentation rates of 10 m/My would resolve polarity chrons with duration greater than ~10 ky. Usually, sample volume is about 7–10 cm^3. The sensitivity of modern magnetometers is such that much smaller samples could be measured; however, small samples are more difficult to orient and handle. In the study of polarity transitions, thin 3-mm slices of sediment have been used (e.g., Clement *et al.,* 1982).

The time between reversals of the geomagnetic field is variable. Polarity subchrons as short as 20 ky in duration are present in the polarity record (e.g., Cobb Mountain and Réunion subchrons). Ideally, any sampling scheme should attempt to sample intervals of this duration. At pelagic sedimentation rates, such resolution would require continuous back-to-back sampling with 7-cm^3 plastic cubes. Pass-through cryogenic magnetometers can be used to measure split half-cores or whole cores, without subsampling. One drawback of continuous measurement is that it is often desirable to do progressive demagnetization using higher alternating fields than are available in these instruments, or to do thermal demagnetization. In these cases, discrete and separate samples are required. A pass-through magnetometer, which can produce independent measurements at ~20 cm spacing, has been used for routine core analysis aboard the ODP vessel R/V *Joides Resolution* since

1983. Weeks *et al.* (1993) describe a modified 2G-755R magnetometer in which the SQUID sensing coils are arranged for high resolution measurement of magnetization from continuous "u-channels" tracked through the sensing region. The width of the response function of the pick-up coils is about 3.5 cm. Deconvolution of the signal in the time domain (Constable and Parker, 1991) can give comparable resolution to that achieved by back-to-back discrete sampling. The u-channel (see Tauxe *et al.*, 1983b; Nagy and Valet, 1993), typically 1.5 m in length with a cross sectional area of about 4 cm^2, allows continuous sampling of the core with less sediment distortion than for discrete back-to-back sampling.

The rate of sediment accumulation is generally variable through time. Sadler (1981) emphasized this variability and termed the process inherently "unsteady," characterized by large changes in rates of accumulation over short time periods. In some environments, measured rates of sediment accumulation during floods may be $\sim 10^2$ cm/hour, whereas accumulation rates integrated over 10^6 years would give much lower values, implying large changes in rates of sedimentation, abundant hiatuses, and unconformities. The ideal condition for a continuous record of geomagnetic polarity would be a continuous rain of sedimentary particles accumulating uninterrupted through time. Such ideal conditions are rare, but in certain cases apparently continuous sediment accumulation has been recorded for marine sediments over significant periods of time (5–10 My) and for lacustrine sediments over intervals of $\sim 10^4$ years. Variable rates of sedimentation caused by tectonics, or changes in climate or paleoceanographic circulation, are well known. When conditions are favorable, complete and continuous reversal sequences have been recorded in marine cores (Opdyke *et al.*, 1966; Foster and Opdyke, 1970; Tauxe *et al.*, 1983c) and in outcrop (Lowrie and Alvarez, 1977b; Channell and Erba, 1992). Some marine sediments preserve such excellent records of the geomagnetic field that even short-period behavior of the field, such as the process of field reversal, can be studied (Opdyke *et al.*, 1973; Clement and Kent, 1984). Lacustrine sediments can also be excellent recorders of the geomagnetic field and they have often been used to study secular variation and magnetic excursions (Ch. 13).

Unlike deep-sea sediments, terrestrial sediments are rarely, if ever, homogeneous with respect to lithology or sedimentation rates. Terrestrial sections are usually selected for sampling for particular reasons, such as the presence of important vertebrate faunas, radiometrically dated horizons, or climatically indicative levels in loess, tills, or pollen-rich deposits. In the Siwaliks of Pakistan and India, the sedimentary rocks comprise mudstones, sandstones, and gravels deposited in the subsiding foreland basin in the front of the rising Himalayan mountain chain (Opdyke *et al.*, 1979; Johnson *et al.*, 1982, 1985; Tauxe, 1980). The sediments were deposited by meandering river systems across the floodplain and consist of fining upward sequences of river channel sands and overbank silts and muds on which soils developed. Such a sequence must contain many disconformities. An analysis of the variability of sedimentation in this facies indicates that time gaps of thousands of years are present (Friend *et al.,* 1989; McRae, 1990a,b). Nevertheless, Johnson *et al.* (1985) have shown that over periods of the order of 10^5 to 10^6 years the magnetic polarity record is amazingly complete. Sampling in the Siwaliks foredeep sediments was confined to the fine-grained mudstones and siltstones, with sandstone intervals not sampled. The sites were, therefore, distributed at random intervals determined by lithology and outcrop availability, with stratigraphic spacing of sites of 5 to 10 m. Studies on sediments in Argentina deposited by unchannelized flows on sandflats and distal alluvial fans in an intermountain basin indicate that the record is fairly complete on time scales greater than 10^4 years, although in this case the stratigraphic spacing of sites was ~25 m (Beer, 1990). Most investigations of intermontane fluvial systems have shown them to be faithful recorders of the geomagnetic field for features with duration greater than ~10^5 years, although obviously with lower rates of sedimentation shorter sampling intervals are required. Several studies in Pakistan and Argentina have dealt with the variability and reproducibility of polarity zones along strike (Behrensmeyer and Tauxe, 1982; Johnson *et al.,* 1988; Beer, 1990). Johnson *et al.* (1988) have shown that individual polarity zones (Fig. 5.8) may vary along strike by up to a factor of two in this type of sedimentary environment.

It is a great advantage in magnetostratigraphic studies if sections of appreciable thickness (10^2–10^3 m) are available for sampling. Thick sections increase the probability that more than one polarity zone is present in the section and increases the chances that the reversal pattern can be correlated to the GPTS. The following technique for sampling unindurated sediments was developed by Johnson *et al.* (1975) for study of Plio–Pleistocene non-marine sediments of the San Pedro Valley of Arizona. During the initial field work a suitable sampling interval is chosen. Sites are preferentially selected in finer grained sediments (silt size or finer) at stratigraphic spacing of 2–20 m. Sections are often sited in arroyos where erosion has exposed fresh unweathered material well away from local peaks or promontories which might be the sites of lightning strikes. At individual sites, it is often necessary to excavate a pit to obtain unweathered material, using a pick-mattock, spade, or some other suitable implement. After unweathered material is exposed, a horizontal or vertical face is fashioned on a fist-sized sample, using a hand rasp. The strike and dip of the surface are measured and marked with fine pen or pencil. Three or more independently oriented samples are generally chosen at each site, so that the polarity can be determined unambiguously and rudimentary site statistics can be deter-

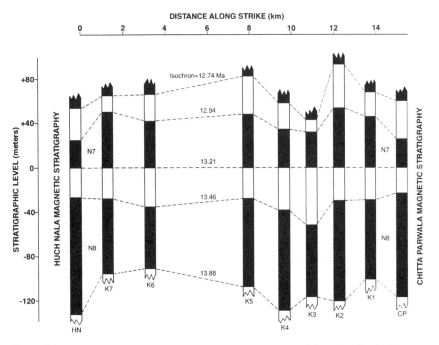

Figure 5.8 Variation of the thickness of polarity zones along 15 km of strike in Miocene molasse sediments in Pakistan. An individual polarity zone may vary in thickness by a factor of 2. Time scale is from Mankinen and Dalrymple (1979) (after Johnson *et al.*, 1988).

mined. After the sample is returned to the laboratory it is fashioned into a cube (approximately 2.5 cm on a side) using a band saw or grinding wheel. Although the technique described above is laborious, once the cubes have been fashioned they are easy to handle and measure. Unindurated sediments are often sampled by pressing the samples into plastic boxes or cylinders in the field. The technique is rapid and is convenient for sample transportation and handling. If thermal demagnetization is necessary, the samples must be dried and/or indurated, removed from the plastic contain-ers, and wrapped in aluminum foil for heating and measurement.

Lava flows or indurated sediments in outcrop can be drilled using a portable gasoline-powered drill. Pelagic limestones in land section, such as those at Gubbio (Italy), are usually sampled by drilling individual oriented cores (2.5 cm diameter, 5–10 cm in length) spaced at 20–50 cm stratigraphic intervals. In lava sequences, three or more cores are usually taken from each lava flow. The most intensive study of this type has been performed by Watkins and Walker (1977) on the Icelandic lavas.

5.5 Presentation of Magnetostratigraphic Data

Magnetostratigraphic data can be presented as plots of declination and inclination of the primary magnetization components as a function of stratigraphic position in the section. Magnetization directions from individual stratigraphic horizons are often represented as virtual geomagnetic poles (VGPs) computed using the dipole formula (Ch. 2). VGPs from individual samples/horizons are compared with the overall paleomagnetic pole computed from the mean magnetization direction for the entire section. Variation in VGP in the section is then expressed as a VGP latitude which represents the latitude of individual VGP in the coordinates of the overall paleomagnetic pole. VGP latitudes close to $+90°$ represent normal polarity, and VGP latitudes close to $-90°$ represent reverse polarity (Fig. 5.1). VGP latitudes are plotted as a function of stratigraphic position of the samples/horizons. The polarity interpretation is usually presented as black (normal) and white (reverse) bar diagrams.

5.6 Correlation of Polarity Zones to the GPTS

The investigator is often confronted with a series of normal and reverse polarity zones, the pattern of which may correlate with more than one segment of the GPTS. Usually, it is initially assumed that the sedimentation rate, or rate of extrusion of the lavas, is more or less constant, and trial correlations are made on the basis of the fit of the observed polarity pattern to the GPTS. In practice, correlations to the GPTS are aided by radiometric ages or biostratigraphic events in the section, which have been correlated to the GPTS elsewhere. For example, in the Haritalyangar section in India (Fig. 5.9), the correlation of polarity zones to the GPTS is aided by magnetostratigraphic data from Pakistan where some of the same Miocene vertebrate fossils are correlated to radiometric ages (on volcanic ashes) and hence to the GPTS. The fossil assemblage at Haritalyangar is equivalent to that in Pakistan which is correlated to the chron 3A to chron 4A interval, and on this basis the reversal sequence can be unequivocally correlated to the GPTS on the basis of pattern fit (Fig. 5.9). The resulting regression is significant at the 99% confidence level, allowing sediments in India to be securely correlated to the GPTS and to similar sediments in Pakistan. It also indicates that the rate of sedimentation is reasonably constant.

The Haritalyangar section is ideal in that it is long and continuously exposed. Short sections which contain few polarity zones are much more

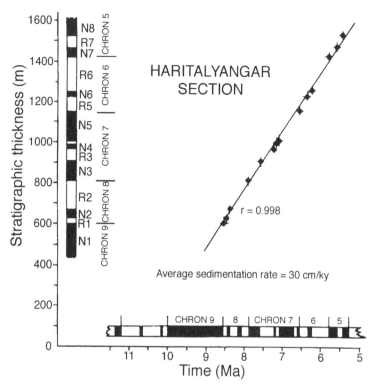

Figure 5.9 Magnetostratigraphy of the Haritalyangar section in India and correlation to the GPTS of Mankinen and Dalrymple (1979). The high linear regression value ($r = 0.998$) indicates that the correlation to the GPTS is feasible and that the sedimentation rate was more or less constant (after G. D. Johnson *et al.*, 1983).

difficult to correlate. In some cases, prominent subchrons may be missing because of hiatuses. Spurious apparent polarity subchrons can be an artifact of a hard secondary magnetization not (completely) removed by the demagnetization techniques.

Cross-correlation techniques can be used to test the correlation of polarity zones to the GPTS (e.g., Lowrie and Alvarez, 1984; Langereis *et al.*, 1984). The polarity zone pattern in stratigraphic section can be compared to the GPTS and the correlation coefficient computed for each successive match. The minimum value of the correlation coefficient which is significant at the 95% confidence level can be used to screen possible correlations. The optimal correlation can then be chosen based on stratigraphic age constraints. It is important to note that this technique is most useful in sections which record multiple polarity chrons, and where sedimentation rates are fairly constant such that the polarity zone pattern reflects the polarity chron pattern in the GPTS.

5.7 Quality Criteria for Magnetostratigraphic Data

Van der Voo (1990) has recently suggested a reliability index for paleomagnetic studies. The index is based on seven reliability criteria, and is designed for paleomagnetic studies that contribute to construction of apparent polar wander paths. We propose a reliability index for magnetostratigraphic studies based on ten criteria. Some of the criteria are the same as those proposed by Van der Voo (1990); however, some are unique to magnetostratigraphy. The criteria and their justification are outlined below:

1. Stratigraphic age known to the level of the stage, and associated paleontology presented adequately. The placement of fossil occurrences and ranges with respect to the magnetic polarity determinations is often critical to interpretation.

2. Sampling localities placed in a measured stratigraphic section. Unless the section is accurately measured, correlation of the polarity zone record to the GPTS will be compromised.

3. Complete thermal or alternating field demagnetization performed and analysis of magnetization components carried out using orthogonal projections (Wilson, 1961; Zijderveld, 1967). All modern studies employ demagnetization, and vector analysis allows the identification of magnetization components present in the samples. Blanket demagnetization at an individual temperature or alternating field value is no longer considered adequate.

4. Directions determined from line-fitting least squares analysis (Kirschvink, 1980). This type of analysis has been widely applied in recent years.

5. Numerical data published completely. Black-and-white bar diagrams are not adequate. Data are best presented as VGP latitude/stratigraphic distance plots and/or as declination/stratigraphic distance and inclination/stratigraphic distance plots. The statistical parameters (e. g., Fisher, 1953) should be fully documented.

6. Magnetic mineralogy determined. It is often useful to know the magnetic mineralogy, which may constrain the timing of remanence acquisition.

7. Field tests for the age of the magnetization are presented (tilt tests, conglomerate tests). If possible, these should always be attempted since they constrain the timing of remanence acquisition.

8. Reversals are antipodal. The field should reverse through 180°. If this reversal test fails, the investigator probably has not fully removed the secondary magnetization components.

9. Radiometric ages are available in the stratigraphic sections. Radiometric dates, especially $^{40}Ar/^{39}Ar$ or U-Pb ages from volcanic ashes or bentonites, are very useful for correlating polarity zones to the GPTS.

10. Multiple sections. Magnetic stratigraphy is more convincing if multiple sections yield similar or overlapping polarity zone patterns.

Quite clearly, few studies will be able to meet all these criteria since, for example, radiometric dating is often not possible, and field tests are impossible in flat-lying beds or deep-sea cores. However, ratings of at least 5 (out of 10) should be achieved by modern magnetostratigraphic studies.

6

The Pliocene–Pleistocene Polarity Record

6.1 Early Development of the Plio–Pleistocene GPTS

The Plio–Pleistocene geomagnetic polarity time scale (GPTS) developed from the early studies of Cox *et al.* (1963a) and McDougall and Tarling (1963a) which coupled K-Ar age determinations and magnetic remanence data in volcanic rocks from California and the Hawaiian Islands. By the end of the 1960s, most of the Plio–Pleistocene polarity chrons had been identified in radiometrically dated volcanic rocks (Fig. 1.2). As more K-Ar age determinations became available the chronogram technique for optimizing the age of polarity chron boundaries was introduced (Cox and Dalrymple, 1967). The chronogram is a plot of standard deviation or error function of available ages close to a polarity chron boundary against trial ages for the boundary (see Harland *et al.*, 1990, pp. 106–109). The minimum error function corresponds to the best estimate of the age of the polarity chron boundary.

Subchrons within the Plio–Pleistocene were initially identified by their associated radiometric ages. It is important to note that the early studies of Plio–Pleistocene geomagnetic polarity coupled K-Ar and magnetic data from widely spaced locations and, apart from the work in Iceland (e.g., Watkins *et al.*, 1977), were not carried out in continuous stratigraphic sections. Plio–Pleistocene polarity subchrons were initially documented in the following publications: Jaramillo (Doell and Dalrymple, 1966), Cobb Mountain (Mankinen *et al.*, 1978), Olduvai (Grommé and Hay, 1963), Kaena and Réunion (McDougall and Chamalaun, 1966), Mammoth (Cox *et al.*, 1963), Cochiti (Dalrymple *et al.*, 1967), and Nunivak (Hoare *et al.*, 1968). The two oldest subchrons in the Gilbert were initially identified by Hays and Opdyke (1967) from marine cores and by Pitman and Heirtzler (1966) from marine magnetic anomalies. Eventually they were identified from lava sequences in Iceland and named the Sidufjall and Thvera subchrons (Watkins *et al.*, 1977).

6.2 Subchrons within the Matuyama Chron

Subsequent to the recognition of the existence of a normally magnetized lava at 1.8 Ma by Grommé and Hay (1963) from Olduvai Gorge, McDougall and Wensink (1966) reported a series of polarity zones (from base: R-N-R-N) in a sequence of lavas from Iceland with ages of $1.60 \pm .05$ Ma. Due to the significant age difference between these normal polarity zones and that observed at Olduvai, the normal intervals from Iceland were considered to be distinct from the Olduvai subchron and were called the Gilsa subchrons. Magnetic stratigraphy in deep-sea cores and oceanic magnetic anomaly profiles, however, appeared to indicate a single normal polarity subchron in the lower middle part of the Matuyama Chron, closely coincident with the Pliocene–Pleistocene boundary. The question arose, therefore, as to whether this interval of normal polarity should be called the Gilsa or Olduvai subchron. This led to a great deal of confusion and controversy (Watkins *et al.*, 1975) which was finally resolved in favor of the name Olduvai. Recently, however, a short-lived normal subchron above the Olduvai has been documented by Clement and Kent (1987) and was called by them the Gilsa subchron (Fig. 6.1).

Two short intervals of normal polarity below the Olduvai were identified in sea floor spreading anomalies and called anomalies "x" and "y" by Heirtzler *et al.* (1968). The youngest of these normal intervals probably coincides with a normally magnetized lava dated by McDougall and Chamalaun (1966) at 2 Ma and named the Réunion Event by Grommé and Hay (1971). This subchron has a duration of approximately 20 ky (Clement and Kent, 1987; Khan *et al.*, 1988).

Mankinen *et al.* (1978) detected a short interval of normal polarity preceding the Jaramillo which they called the Cobb Mountain subchron. This subchron has been observed in high sedimentation rate cores from the Caribbean Sea (Kent and Spariosu, 1983) and the North Atlantic (Fig. 6.1; Clement and Kent, 1987). Another geomagnetic excursion has been reported between the Jaramillo subchron and the Brunhes/Matuyama boundary by Maenaka (1983) and Mankinen and Grommé (1982) and it

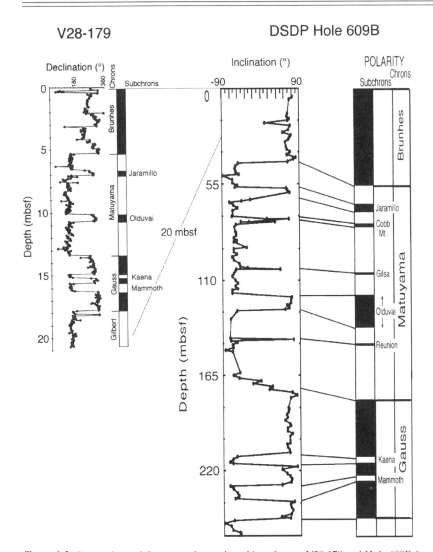

Figure 6.1 Comparison of the magnetic stratigraphies of cores V28-179 and Hole 609B from the North Atlantic. The cores cover the same time intervals but they have sedimentation rates differing by an order of magnitude (after Shackleton and Opdyke, 1977; Clement and Kent, 1987).

Table 6.1
Plio–Pleistocene

	Rock unit	Lo Age Hi	Region	λ	Φ	NSE	NSI	NSA	M	D	RM	DM	AD	A	NMZ	NCh	%R	RT	F.C.T.	Q	References
1	Calabria	Mat.-Ga	Italy	+38.2	+16.2	4	416	2	235	As	–	T-A	Z-P	D-I	12	2	86	–	–	7	Zijderveld et al. (1991)
2	Rossello	Mat.-Gil	Italy	+37.31	+13.46	4	.5 m	2	135	As	–	T-A	Z-P	D-I	16	3	47	–	–	7	Langereis and Hilgen (1991)
3	Site 609	Br-C3A	N. Atlantic	+49.88	−24.24	6	1.5 m	1	360	–	–	A	Z-B	I	27	4	46	–	–	5	Clement and Robinson (1986)
4	Site 644	Br-Ga	N. Atlantic	+66.7	+4.56	8	.4 m	1	255	–	–	A	Z-V	I	9	3	46	–	–	6	Bleil (1989)
5	Hole 704A	Br-Gil	S. Atlantic	−46.88	+7.42	2	1.5 m	1	233	O	I	A	Z-V	I	17	4	45	–	–	7	Hailwood and Clement (1991)
6	San Salvador	Br-Gil	Bahamas	+24.1	−74.5	1	174	1	91	O	T	A	Z-P	I	15	4	37	–	–	6	McNeill et al. (1988)
7	Site 758	Br-C4	Indian O.	+5.38	+90.36	2	P	–	96	O	K	A	B	D	22	5	60	–	–	6	Shipboard Party (1988)
8	Site 711	Br-C3A	Indian O.	−1.25	+60.00	2	P	–	100	–	–	A	Z	D	24	5	52	–	F+	6	Schneider and Kent (1990b)
9	Site 577	Br-C3A	N. Pacific	+32.44	+157.72	2	.5 m	1	62	–	–	A	Z	I	21	5	52	–	–	5	Bleil (1985)
10	RC12-66	Br-C3A	E. Pacific	+4.65	−144.97	1	.1 m	1	24	–	–	A	B	D	27	5	48	–	–	3	Foster and Opdyke (1970)
11	Site 594	Br-C3A	S. Pacific	−45.52	+174.95	1	1.5 m	1	22	–	V	A	Z-B	I	19	5	47	–	–	5	Barton and Bloemendal (1986)
12	E113-17	Br-C3A	S. Pacific	−65.68	−124.1	1	.1 m	1	26	–	–	A	B	I	19	5	44	–	–	3	Hays and Opdyke (1967)
13	Site 573	Br-Ga	E. Pacific	+.5	−133.31	1	.5 m	1	45	–	–	A	F	D	11	3	47	–	–	4	Weinreich and Theyer (1985)
14	Site 574	Ga-C3A	E. Pacific	+4.21	−133.33	1	.5 m	1	85	–	–	A	F	D	26	5	51	–	–	4	Weinreich and Theyer (1985)
15	Site 844	B-C3A	E. Pacific	+7.92	−90.48	4	P	–	75	–	–	A	Z-B	D	41	8	48	–	–	5	Shipboard Party (1992)
16	Site 845	Ga-C3A	E. Pacific	+9.58	−94.59	2	P	–	160	–	I-T	A	Z-B	D-I	49	9	49	–	–	5	Schneider (1995)
17	Mururoa	Br-Gil	F. Polynesia	−21.83	−138.83	1	.5 m	1	100	–	–	A-T	Z-P	I	11	4	24	–	–	6	Aissaoui et al. (1990)
18	Suva Marls	Ga-Gil	Fiji	−18.13	+178.44	1	95	5	100	A	–	A	P	D-I	7	1	68	–	–	5	Rhodda et al. (1985)

is thought to have occurred between 0.84 and 0.85 Ma (Champion *et al.*, 1988). A subchron of comparable age was observed in a loess sequence from China by Heller and Liu (1984). In the time scale presented here (Fig. 6.2), only those subchrons recorded in oceanic magnetic anomaly records and in magnetostratigraphic section are included.

6.3 Magnetic Stratigraphy in Plio–Pleistocene Sediments

The intensive magnetostratigraphic study of marine sediments of Plio–Pleistocene age began in the early 1960s utilizing the piston core collections at Lamont Doherty Geological Observatory (LDGO) and cores from the Eltanin collection stored at Florida State University (Opdyke *et al.*, 1966; Ninkovitch *et al.*, 1966; Opdyke and Foster, 1970; Hays *et al.*, 1969; Opdyke and Glass, 1969; Hunkins *et al.*, 1971; Goodell and Watkins, 1968). The maximum age of the sediments in these studies was limited by the penetration of piston coring devices. The longest conventional piston core in the LDGO collection is about 26 m and, therefore, a stratigraphic record to the basal Pliocene was extremely rare. Cores recording most of the Plio–Pleistocene magnetic polarity record included E13-17 (Hays and Opdyke, 1967), RC 12-66 (Foster and Opdyke, 1970) (Fig. 6.2), and V28-179 (Fig. 6.1) (Shackleton and Opdyke, 1977). In early studies, the durations of the Plio–Pleistocene subchrons were estimated through the magnetic stratigraphy of deep-sea cores by extrapolation of sedimentation rates from reversal transitions or by interpolation assuming constant sedimentation rates and adopting K-Ar ages for polarity chron boundaries (Fig. 6.3). Estimates were compared with those obtained from high-resolution oceanic magnetic anomaly profiles assuming constant sea floor spreading rates.

The last decade has witnessed a dramatic improvement in the quantity and quality of the Plio–Pleistocene magnetostratigraphic records due, in large part, to the advent in the early 1980s of the hydraulic piston corer (HPC) and, subsequently, the advanced piston corer (APC). The HPC, first used during Leg 68 of the Deep Sea Drilling Project (DSDP), allowed recovery of several hundred meters of sediment by piston coring, increasing the penetration of piston coring by more than an order of magnitude. The HPC was a considerable improvement over the standard rotary drilling techniques, which often resulted in poor core recovery and drilling-related deformation. Notable among the Plio–Pleistocene magnetostratigraphies recovered by HPC/APC are those from DSDP Leg 68 in the Carribbean (Kent and Spariosu, 1983), DSDP Leg 73 in the South Atlantic (Tauxe *et al.*, 1983c, 1984; Poore *et al.*, 1984), and DSDP Leg 94 in the North Atlantic (Weaver and Clement, 1986). Apart from their value for biomagnetostrati-

Figure 6.2 Selected Plio–Pleistocene magnetic stratigraphies. For key, see Table 6.1.

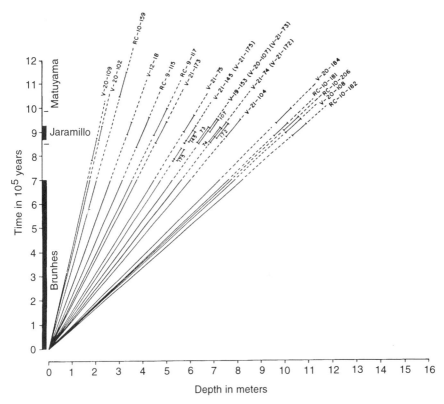

Figure 6.3 Time vs. depth plot showing the position and duration of the Jaramillo subchron in deep-sea piston cores (after Opdyke, 1972).

graphic correlations, the Plio–Pleistocene magnetic stratigraphies from DSDP Leg 94 were important for two additional reasons. First, as mentioned above, they resulted in the documentation of two short-duration normal polarity subchrons within the Matuyama chron (Gilsa and Cobb Mountain subchrons) which had not previously been adequately recorded in sediments (Fig. 6.1) (Clement and Kent, 1987). Second, the correlation of oxygen isotope records to Leg 94 Plio–Pleistocene magnetic stratigraphies resulted in a first attempt to astronomically tune this part of the GPTS (Ruddiman *et al.*, 1986, 1989; Raymo *et al.*, 1989).

The vast majority of Plio–Pleistocene marine magnetostratigraphic records are from pelagic deep-sea sediments. Plio–Pleistocene shallow water platform carbonates recovered from boreholes in the Bahamas (McNeill *et al.*, 1988) and Mururoa Atoll (Aissaoui *et al.*, 1990) have yielded excellent

magnetic stratigraphies despite pervasive dolomitization. These magnetostratigraphies provide a powerful chronostratigraphic tool in an environment where biostratigraphic control is inherently poor. It is important to
balance these important magnetostratigraphic studies in Plio–Pleistocene
shallow water limestones with the many unsuccessful (and usually unpub-
lished) magnetostratigraphic studies in ancient shallow water limestones
which are generally thwarted by remagnetization or very weak remanence intensities.

The Plio–Pleistocene stages in most geological time scales are derived
from stratotype sections located in Italy (see Rio *et al.,* 1991). Correlation
of these stratotype sections to marine sequences outside the Mediterranean
is complicated by provinciality of Mediterranean flora and fauna and by
changes in lithofacies in the stratotype sections. Magnetic stratigraphies in
the stratotype region provide the potential means of global correlation.
The focus of Early Pliocene Mediterranean magnetostratigraphies has been
the Trubi Limestones, a fine-grained pelagic limestone recording the post-
Messinian flooding of the Mediterranean and cropping out in Sicily and
Calabria (southern Italy). Magnetostratigraphic records of the Miocene–
Pliocene boundary and the Gilbert Chron in Calabria (Zijderveld *et al.,*
1986; Channell *et al.,* 1988) were expanded by the development of a composite reference section in Sicily (Rossello composite) from the Miocene–
Pliocene boundary to the base of the Matuyama Chron (Fig. 6.4) (Langereis
and Hilgen, 1991). The Trubi Limestones are overlain by a more marly
facies (Monte Narbone Formation), the transition occurring in the middle
part of the Gauss Chron. Upward extension of magnetic stratigraphies into
the Late Pliocene and Early Pleistocene has been documented in Calabria
(Zijderveld *et al.,* 1991) and Sicily (Zachariasse *et al.,* 1989, 1990). The
Plio–Pleistocene boundary stratotype at Vrica (Calabria) originally studied
magnetostratigraphically by Tauxe *et al.* (1983a) has been restudied by
Zijderveld *et al.* (1991), and the stage boundary placed close to, but just
above, the top of the Olduvai subchron. The land section magnetostratigraphic data in southern Italy are far better quality than those obtained
from rotary cores during ODP Leg 107 in the Tyrrhenian Sea (Channell
et al., 1990b), although biostratigraphic control is better in the Tyrrhenian
cores due to enhanced nannofossil preservation (Rio *et al.,* 1990). The
biomagnetostratigraphic correlations which have resulted from land section
studies in Italy and ODP Leg 107 cores (Fig. 6.5) facilitate export of Plio–
Pleistocene stages from the Mediterranean and are an important basis for
global Plio–Pleistocene correlations.

Although the Early Pliocene (Zanclean) type section is within the
Sicilian Trubi Limestones, other Pliocene stratotype sections (Tabianian
and Piacenzian) are in the terrigeneous "argille azzurre" (blue clay) facies

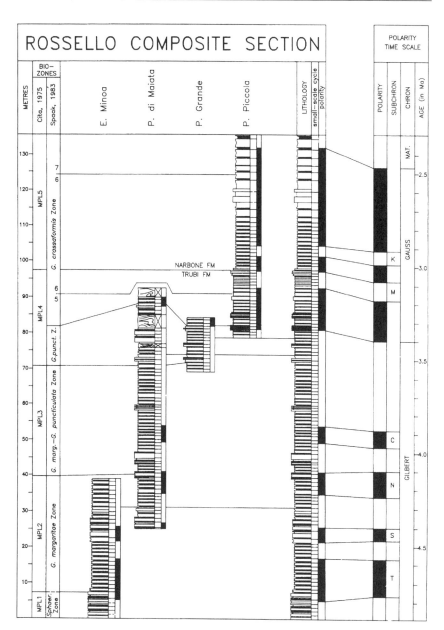

Figure 6.4 Composite magnetic stratigraphy from Capo Rossello region, in Sicily (after Langereis and Hilgen, 1991).

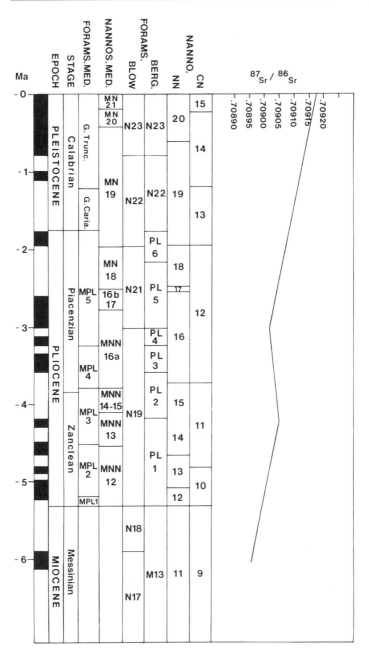

Figure 6.5 Correlation of the Mediterranean and extra-Mediterranean tropical biozonations to the GPTS. The variation of $^{87}Sr/^{86}Sr$ after Hodell *et al.* (1991).

of the Northern Apennines (see Rio *et al.*, 1991). This is a more problematic facies for magnetic stratigraphy due to the ubiquitous presence of iron sulfides, although Pliocene magnetic stratigraphies are available from a few sections in this region (Mary *et al.*, 1993; Channell *et al.*, 1994). The available magnetic stratigraphies indicate that sedimentation in the Northern Apennine foredeep in the vicinity of the stratotype sections was very discontinuous and that the region is therefore not suitable for stratotype designations.

The Miocene–Pliocene boundary stratotype is a lithologic boundary in the stratigraphic sequence at Capo Rosello (Sicily) (Fig. 6.4). Benson and Hodell (1994) argue that it would be better to redefine the Miocene/Pliocene boundary and correlate it to the base of the Gilbert Chron in Morocco, because of the inadequate biostratigraphic definition of the boundary in Sicily. The Pliocene–Pleistocene boundary at the stratotype section (Vrica, Italy) is defined by the first occurrence of migrant taxa such as *Artica islandica* which cannot be easily correlated to the open ocean. These species first appear in the Vrica section just above the Olduvai subchron (Tauxe *et al.*, 1983a; Zijderveld *et al.*, 1991). A definition of the Pliocene–Pleistocene boundary at the top of the Olduvai subchron should perhaps be reconsidered.

Berggren *et al.* (1980) gave 37 biostratigraphic datums for the Plio–Pleistocene of the world ocean and many more have subsequently been added (Fig. 6.6). The apparent correlation of bioevents to the GPTS can vary from site to site due to disconformities, fossil preservation, and the operational method used to define the bioevent and the species itself. Many of these datums have been correlated directly to the $\delta^{18}O$ time scale. Emiliani (1955) produced the first long records of the change in $\delta^{18}O$ in the oceans through time in foraminiferal tests relative to a standard (a belemnite from the Pee Dee Formation in South Carolina). The ocean water was shown to oscillate between times when the ocean had $\delta^{18}O$ ratios similar to those today (interglacial conditions) to times when the oceans were enriched in ^{18}O (glacial conditions). Emiliani numbered these changes beginning at the present interglacial (stage 1) and the last glacial (stage 2). In this scheme, the interglacials are indicated by odd numbers and the glacial stages by even numbers. Shackleton and Opdyke (1973) extended this nomenclature to below the Brunhes/Matuyama boundary and correlated this boundary to isotopic stage 19. The $\delta^{18}O$ record was extended to the entire Pleistocene by Shackleton and Opdyke (1976) and to the base of the Matuyama by Ruddiman *et al.* (1986), who extended the numbering system to isotopic stage 102 at the Gauss/Matuyama boundary. The isotopic variations since ~1 Ma were shown to have a dominant periodicity of ~100 ky (Hays *et al.*, 1976). For the time from the Late Jaramillo to the base of the Matuyama, the variations in $\delta^{18}O$ have been shown to have a dominant

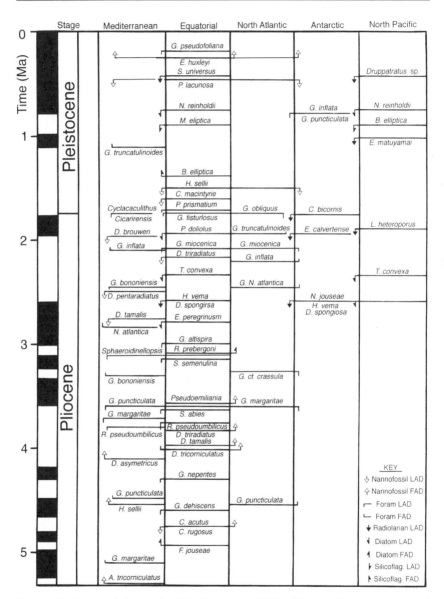

Figure 6.6 The correlation to the Plio–Pleistocene GPTS of first and last appearances of selected fauna and flora. FAD = first appearance datum, LAD = last appearance datum.

periodicity of ~42 ky (Ruddiman *et al.*, 1986). The isotopic sequence has been extended by Shackleton *et al.* (1995b) to the base of the Gilbert Chron at Site 846 from the eastern Pacific. The sequential numbering system was discontinued below the Gauss/Matuyama boundary and a letter prefix was used to label isotopic stages in the Gauss and Gilbert Chrons (Fig. 6.7).

6.4 Astrochronologic Calibration of the Plio–Pleistocene GPTS

Hays *et al.* (1976) were the first to demonstrate the presence of orbital cycles of eccentricity, obliquity, and precession in paleoclimate proxies in deep-sea sediments. These authors adjusted their initial time scale to bring the peak variance of the obliquity cycles to the same frequency calculated for obliquity by astronomers, and were thus the first to use cyclostratigraphy as a means of tuning time scales. During the 1980s, astrochronology was largely restricted to the Brunhes Chron and was utilized successfully to constrain the age of the oxygen isotope record with a precision of a few thousand years (e.g., Martinson *et al.*, 1987). The turning point for astrochronology/cyclostratigraphy came with DSDP/ODP hydraulic piston core recovery at multiple holes for single sites with adequate sedimentation rates, oxygen isotopic records, and magnetic stratigraphies. The hydraulic piston corer allowed good undisturbed sediment core recovery, and the drilling of multiple holes allowed complete recovery of composite sections. Ruddiman *et al.* (1989) and Raymo *et al.* (1989) compiled oxygen isotope data for the Pleistocene and Upper Pliocene at DSDP Leg 94 Site 607 (North Atlantic). Orbital tuning using the strong obliquity signal for the Matuyama Chron did not reveal any significant discrepancy with the standard (Mankinen and Dalrymple, 1979) K-Ar polarity chron ages, apart from a significantly shorter duration for the Olduvai subchron.

For over ten years, the Mankinen and Dalrymple (1979) compilation of K-Ar ages of Plio–Pleistocene polarity chrons was generally accepted in spite of astrochronology indicating a significantly older age for the Brunhes/ Matuyama boundary (Johnson, 1982). The realization that the Mankinen and Dalrymple (1979) polarity chron ages may be in error came with the study of ODP Site 677 in the eastern equatorial Pacific (Shackleton *et al.*, 1990). At this site, the oxygen isotopic record from benthic foraminifera is dominated by the obliquity signal and the planktic record by precession. Because the effect of precession is modulated by the eccentricity cycle, the precession-dominated planktic record provided a better means of correlation to the astronomical data than the obliquity record. The Site 677 oxygen isotope data could therefore be matched to the astronomical data with more confidence than the poorly modulated obliquity signal at Site

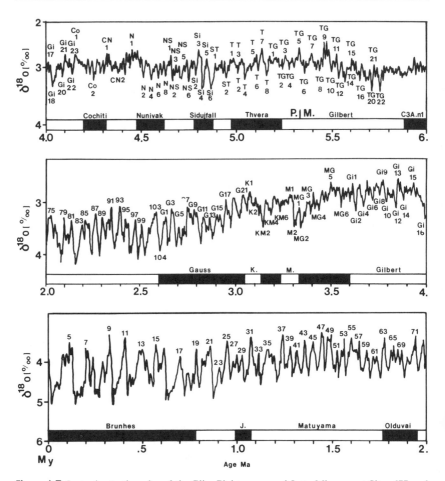

Figure 6.7 Isotopic stratigraphy of the Plio–Pleistocene and Late Miocene at Sites 677 and 841 (after Shackleton *et al.*, 1990, 1995b).

607. Indeed at Site 607, Ruddiman *et al.* (1989) apparently miscounted the number of obliquity cycles in the lower Brunhes and early Pleistocene partly because they accepted the K-Ar age for the Brunhes/Matuyama boundary. The Site 677 astrochronology (Shackleton *et al.*, 1990) yielded an age for the Brunhes/Matuyama boundary (0.78 Ma) at odds with the K-Ar age (0.73 Ma).

The older astrochronological age for the Brunhes/Matuyama boundary from Site 677 coincided with the observation of discrepancies between conventional K-Ar ages and Pliocene astrochronology derived from carbon-ate cyclostratigraphy in the southern Italian Trubi Marls (Hilgen and Lang-ereis, 1989). Spectral analyis of $CaCO_3$ content in these Pliocene carbonates yielded periodicities of 15.5, 18.5, 35.0, and 335 ky, indicating a consistent discrepancy with the astronomical solutions of 19, 23, 41, and 413 ky. Hilgen and Langereis (1989) computed the age of the Gilbert and Gauss reversals using an age of 3.40 Ma for the Gauss/Matuyama boundary and tuning the $CaCO_3$ cycles to the astronomical solution.

The Late Pliocene and Early Pleistocene of southern Italy is character-ized by sapropels, dark organic-rich laminated layers. Hilgen (1991a) showed that for Late Pleistocene sapropels from the eastern Mediterranean (core RC 9-181), individual sapropels correlate to minimum peak values of the precession index and small- and large-scale sapropel clusters correlate to eccentricity maxima related to the 100 ky and 400 ky eccentricity cycles. Hilgen (1991a) applied this observation to the sapropel-bearing Late Plio-cene–Early Pleistocene sections in southern Italy which had magneto-stratigraphic age control. The sapropel occurrences and sapropel clusters could be matched to the astronomical solution leading to a new time scale for the 1.8–3.2 Ma interval. Hilgen (1991a) illustrated how the time scales of Berggren *et al.* (1985b) and Raymo *et al.* (1989) lead to a mismatch of the sapropel occurrences with the astronomical solution.

Below the main interval of sapropel occurrence, Hilgen (1991b) used the $CaCO_3$ cycles in the same southern Italian sections to generate an astrochronology for the Pliocene. Hilgen (1991b) used the relationship between the sapropel occurrences and $CaCO_3$ content in the Upper Plio-cene, to infer that the gray marl beds denoting the small-scale $CaCO_3$ minima correspond to minima in the precession index and that the larger scale $CaCO_3$ minima correspond to maximum amplitude variations of the precession index related to the eccentricity modulation. The recognition of both a precession and an eccentricity signal in the $CaCO_3$ record allowed the observations to be tuned to the astronomical solution with considerably more confidence than would have been the case in the absence of the eccentricity modulation. This important paper resulted in astrochronologi-cal ages for Gauss and Gilbert reversals (Table 6.2).

Neogene astrochronology has been extended into the Late Miocene using cores collected during ODP Leg 138 (Shackleton *et al.*, 1992, 1995a). Multiple cores at each site allowed composite sections to be compiled at each site using magnetic susceptibility and GRAPE (Gamma Ray Attenua-tion Porosity Evaluator) data (Hagelberg *et al.*, 1992). The GRAPE records reflect the ratio of calcite to biogenic opal, and the GRAPE records from the composite sections were matched with the orbital insolation solution of Berger and Loutre (1991). This matching has led to astrochronological

Table 6.2
Pliocene–Pleistocene Time Scales

Polarity chron boundary	Mankinen and Dalrymple [1979] (Ma)	Berggren et al. [1985] (Ma)	Shackleton et al. [1990] (Ma)	Hilgen [1991a,b] (Ma)	Cande and Kent [1992a] (Ma)	Schackleton et al. [1995a] (Ma)	Baksi [1994] (Ma)	Cande and Kent [1995]
Brunhes/Matuyama	0.73	0.73	0.78		0.780		0.78	0.78
Jaramillo (top)	0.90	0.91	0.99		0.984		0.99	0.99
Jaramillo (base)	0.97	0.98	1.07		1.049		1.05	1.07
Olduvai (t)	1.67	1.66	1.77		1.757		1.78	1.77
Olduvai (b)	1.87	1.88	1.95		1.983		2.02	1.95
Matuyama/Gauss	2.48	2.47	2.60	2.57/2.62	2.600		2.64	2.58
Kaena (t)	2.92	2.92		3.04	3.054	3.046	3.10	3.04
Kaena (b)	3.01	2.99		3.11	3.127	3.131	3.17	3.11
Mammoth (t)	3.05	3.08		3.22	3.221	3.233	3.27	3.22
Mammoth (b)	3.15	3.18		3.33	3.325	3.331	3.38	3.33
Gauss/Gilbert	3.40	3.40		3.58	3.553	3.594	3.61	3.58
Cochiti (t)	3.80	3.88		4.18	4.033	4.199	4.12	4.18
Cochiti (b)	3.90	3.97		4.29	4.134	4.316	4.23	4.29
Nunivak (t)	4.05	4.10		4.48	4.265	4.479	4.37	4.48
Nunivak (b)	4.20	4.24		4.62	4.432	4.623	4.55	4.62
Sidufjall (t)	4.32	4.40		4.80	4.611	4.781	4.74	4.80
Sidufjall (b)	4.47	4.47		4.89	4.694	4.878	4.82	4.89
Thvera (t)	4.85	4.57		4.98	4.812	4.977	4.94	4.98
Thvera (b)	5.00	4.77		5.23	5.046	5.232		5.23
Gilbert (b)		5.35			5.705	5.882		5.89

age estimates for geomagnetic reversals in the 3–6.2 Ma (top Kaena to C3A.n2) interval. Consistent estimates were derived from several holes/ sites. The estimates are within a few tens of thousands of years of the estimates given by Hilgen (1991a,b) from southern Italian sections and differ from those given by Cande and Kent (1992a) by up to several hundred thousand years. Finally, these authors (Shackleton *et al.*, 1995a) generated a sea floor anomaly time scale beyond C3A.2n by utilizing their age for the top of C3A.1n (5.875 Ma), the radiometric age of Baksi (1993) for the top of C5n.1n (9.639 Ma), and the sea floor distances given by Cande and Kent (1992a). How far back in time the astrochronological technique can be carried will rest on the availability of a suitable proxy climatic records in the oceans or cropping out on land (see Hilgen *et al.*, 1995).

6.5 ^{40}Ar/^{39}Ar Age Calibration of the Plio–Pleistocene GPTS

The advent of high-precision ^{40}Ar/^{39}Ar age dating techniques and the implication from cyclostratigraphy that the conventional K-Ar ages for the Plio– Pleistocene (e.g., Mankinen and Dalrymple, 1979) were too young have resulted in a large number of ^{40}Ar/^{39}Ar studies aimed at testing the astro- chronological ages of Plio–Pleistocene polarity chrons. The conventional K-Ar age for the Brunhes/Matuyama boundary (0.73 Ma) has been modified to 0.78 Ma by a number of independent ^{40}Ar/^{39}Ar studies (Izett and Obradovich, 1991; Baksi *et al.*, 1992; Spell and McDougall, 1992; Tauxe *et al.*, 1992; Hall and Farrell, 1993). This age is consistent with the cyclostrati- graphic estimates giving strong support to the new chronology (recent ^{40}Ar/ ^{39}Ar ages summarized by Baksi, 1994, see Table 6.2). The astrochronological ages derived by Shackleton *et al.* (1990) and Hilgen (1991a,b) have now been supported by ^{40}Ar/^{39}Ar ages for the Jaramillo (Glass *et al.*, 1991; Spell and McDougall, 1992; Tauxe *et al.*, 1992), Cobb Mountain (Turrin *et al.*, 1994), Réunion (Baksi *et al.*, 1993b), Olduvai (Walter *et al.*, 1991; Baksi, 1994). Kaena and Mammoth (Renné *et al.*, 1993; Walter, 1994; Walter and Aronson, 1993), and Gilbert (McDougall *et al.*, 1992; Baksi *et al.*, 1993a) (Fig. 6.8). The ^{40}Ar/^{39}Ar ages of Plio–Pleistocene polarity chrons (reviewed by Baksi, 1994) have in general confirmed the astrochronological determina- tions, thereby ratifying the astrochronology (Table 6.2). The ages for Plio– Pleistocene reversals based on oceanic magnetic anomalies given by Cande and Kent (1992a) are not consistent with the ^{40}Ar/^{39}Ar or astrochronological estimates, particularly for the Gilbert Chron (Table 6.2). D. S. Wilson (1993) analyzed the magnetic anomalies at various spreading centers and concluded that the anomaly spacing is, in fact, consistent with the astrochro- nological estimates. Cande and Kent (1992a) adopted the spacing of Plio–

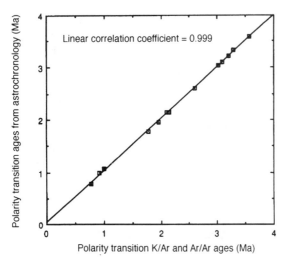

Figure 6.8 Comparison of geomagnetic polarity reversal ages based on orbital tuning (Shackleton *et al.*, 1990; Hilgen 1991a,b) with reversal ages based on K/Ar and ^{40}Ar/^{39}Ar feldspar ages from the Turkana Basin (Spell and McDougall, 1992; McDougall *et al.*, 1992).

Pleistocene anomalies given by Klitgord *et al.* (1975), which appears to be the source of the error. In Cande and Kent (1995), the astrochronological estimates for the age of the Plio–Pleistocene reversals were adopted (Table 6.2). The consistency of the astronchronological, ^{40}Ar/^{39}Ar, and sea floor age estimates for Plio–Pleistocene anomalies has ratified the astrochronological techniques. The resolution of astrochronological estimates for the age of reversal boundaries in the Plio–Pleistocene has reached one precession cycle (~20 ky), only a factor of two greater than the time taken for field reversal to occur. Renné *et al.* (1994) have taken the next step and used the astrochronological estimates to calibrate the ^{40}Ar/^{39}Ar dating standard (Fish Canyon sanidine, Mmhb-1), thereby reducing the uncertainty to 0.6% for ^{40}Ar/^{39}Ar ages calibrated against this standard.

7

Late Cretaceous–Cenozoic GPTS

7.1 Oceanic Magnetic Anomaly Record

Vine and Matthews (1963) proposed that lineated magnetic anomalies observed at sea using towed magnetometers resulted from the permanent magnetization of the oceanic crust. They suggested that this magnetic signature was a result of normal and reverse polarity geomagnetic fields and the sea floor spreading process outlined by Hess (1962). This hypothesis was essentially confirmed by Vine and Wilson (1965), Pitman and Heirtzler (1966), and Vine (1966). These papers demonstrated that: (1) the magnetic anomaly pattern was generally symmetric on both sides of a spreading ridge, and (2) the anomaly pattern emanating from the ridge reproduces in detail the reversal sequence obtained by the radiometric dating of lava flows on land.

As a result, Vine (1966) and Pitman and Heirtzler (1966) extended the magnetic reversal pattern to 10 Ma and dated it by assuming constant spreading rate and an age for the base of the Gauss Chron. It rapidly became apparent that correlatable magnetic anomalies generated by sea floor spreading extend for thousands of km away from the ridge crests of the North Pacific, South Pacific, and South Atlantic oceans (Fig. 7.1). This led to the realization that it might be possible to extend the geomagnetic polarity time scale (GPTS), as derived from these anomalies, back in time to the late Cretaceous with reasonable accuracy if the spreading rates remained constant. Heirtzler *et al.* (1968) compared anomaly patterns from all oceans. Taking into consideration what was known about the age of the sea floor at that time, they took a courageous step by deciding that the South Atlantic Ocean had spread at an almost constant rate since the Late

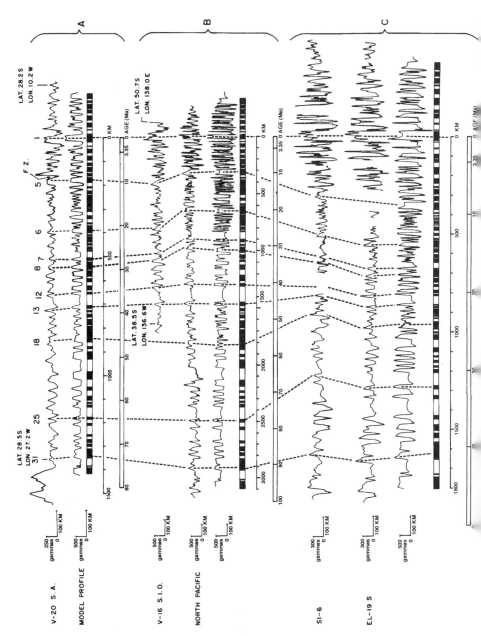

Figure 7.1 Oceanic magnetic anomaly profiles from South Atlantic (V-20), North Pacific (V-16), and South Pacific (SI-6, EL-19S) with synthetic model profiles. Time scale constructed assuming constant seafloor spreading rates and an age of 3.35 Ma for the base of the Gauss Chron (after Heirtzler et al., 1968).

Cretaceous. They then simulated the polarity pattern necessary to give the observed pattern of magnetic anomalies, assumed an age of 3.35 Ma for the base of the Gauss, and assigned ages by extrapolation assuming a constant spreading rate. With this bold and imaginative step, Heirtzler *et al.* (1968) produced a dated polarity pattern for the entire Cenozoic and part of the Cretaceous and deduced an age of 60 Ma for the K/T boundary. This geomagnetic polarity time scale (GPTS) became a vital tool for deciphering the age and tectonic history of the ocean basins.

In their original treatment of magnetic anomalies, Heirtzler *et al.* (1968) assigned numbers to certain prominent magnetic anomalies, from anomaly 1 (Brunhes Chron) at the ridge crest to anomaly 32 (Late Cretaceous) (Fig. 7.1). The labeling of magnetic anomalies was found to be very useful as an aid to correlation. As more detailed surveys became available, however, it became clear that a finer division of the anomaly pattern was desirable. A study by Blakely (1974) of closely spaced magnetic anomalies in the Neogene Pacific led to the adoption of a number/letter scheme for magnetic anomalies in the Miocene, a change made necessary by the high reversal rate during that time.

Since the publication of Heirtzler *et al.* (1968), most published geomagnetic polarity time scales (e.g., LaBrecque *et al.*, 1977; Berggren *et al.*, 1985a,b; Harland *et al.*, 1990) utilized the basic Heirtzler *et al.* (1968) anomaly sequence interpretation (Fig. 7.2), albeit with some important modifications for anomalies 4A to 6 (Blakely, 1974), anomalies 1 to 3A (Klitgord *et al.*, 1975), and anomalies 30 to 34 (Cande and Kristoffersen, 1977). The most recent and extensive revision of the geomagnetic polarity time scale for the Late Cretaceous and Cenozoic has been made by Cande and Kent (1992a, 1995).

In most recently derived Late Cretaceous–Cenozoic geologic time scales, absolute ages are correlated to the GPTS, which is then utilized to interpolate between absolute age estimates. The template for the GPTS is derived from oceanic magnetic anomalies using the constant spreading rate assumption. The absolute ages of geologic stage boundaries can then be estimated if the stage boundaries are correlated to the GPTS. In this way, the GPTS is central to the construction of geologic time scales and provides the means to correlate among the diverse measures of geologic time, such as biostratigraphy, isotope stratigraphy, and absolute ages. In some time scales, the link between geologic stages and absolute ages has been attempted without use of the GPTS as a means of interpolation (e.g., Odin *et al.*, 1982); however, it is now generally accepted that the GPTS should be the central thread of Late Cretaceous–Cenozoic time scales.

Cande and Kent (1992a), after surveying oceanic magnetic anomalies from the world's oceans, concluded, as had Heirtzler *et al.* (1968), that the

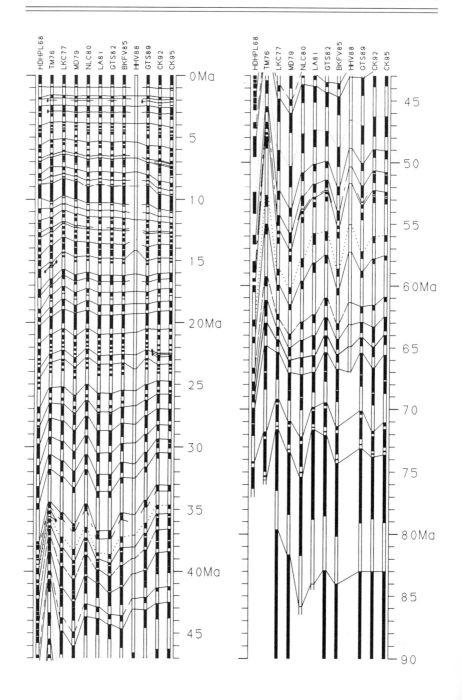

South Atlantic magnetic anomalies were the most appropriate as the basis for a new GPTS. The South Atlantic has a well studied set of magnetic anomalies (Cande *et al.*, 1988) to anomaly 34. The South Atlantic also has both flanks of the ridge crest preserved and has not suffered from major plate reorganization. It does, however, have a rather slow rate of spreading. The procedure of Cande and Kent (1992a) was to use a set of nine finite rotation poles and determine the distance to nine conjugate magnetic anomalies. These were positioned on a flow line at approximately 30°S latitude. The chosen anomalies (4A, 5C, 7, 13, 20, 24, 30, 33, and 34) were positioned at intervals of 150 to 300 km along the flow line. These segments were designated category 1 intervals. The details of the anomaly pattern between these category 1 intervals were filled in by choosing undisturbed segments of anomaly sequences from individual profiles elsewhere in the South Atlantic. These sequences were then continued downward 1.5 to 2 km, deskewed, and the crossover points on the profiles were then used to indicate the position of the reversal boundaries on the sea floor record (Fig. 7.3). Five to nine individual records were stacked after adjustment to the previously determined tie points (category 1 intervals) (e.g., Fig. 7.4). The positions and widths of the anomalies were then determined by averaging. This process yielded an average error of about 7%. In order to obtain the finer detail of the anomaly sequence, known from fast spreading ridges in the Pacific and Indian oceans, individual studies of anomaly sequences from these oceans were inserted between the tie points. For instance, the studies of Klitgord *et al.* (1975) and Blakely (1974) (Fig. 7.5) were used to determine the fine-scale reversal sequence between the ridge crest and anomaly 6.

Very low amplitude, but correlatable, magnetic anomalies have been called "tiny wiggles" (Cande and LaBrecque, 1974; Blakely, 1974) which can be modeled as short polarity chrons or as intensity fluctuations of the geomagnetic field. Some have already been identified as short polarity chrons, such as the Réunion subchron (Grommé and Hay, 1971) and the Cobb Mountain subchron (Mankinen *et al.*, 1978, Mankinen and Grommé, 1982; Clement and Kent, 1987), which have durations of about 20 ky. Cande and Kent (1992b) have suggested that many of the correlatable tiny wiggles

---

Figure 7.2 Evolution of Late Cretaceous–Cenozoic geomagnetic polarity time scales from Hiertzler *et al.* (1968) (HDHPL68) to Cande and Kent (1995) (CK95) (after Mead, 1996). Key: HDHPL68: Heirtzler *et al.* (1968), TM76: Tarling and Mitchell (1976), LKC77: LaBrecque *et al.* (1977), MD79: Mankinen and Dalrymple (1979), NLC80: Ness *et al.* (1981), LA81: Lowrie and Alvarez (1981), GTS82: Harland *et al.* (1982), BKFV85: Berggren *et al.* (1985a,b), HHV88: Haq *et al.* (1987), GTS89: Harland *et al.* (1990), CK92: Cande and Kent (1992a), CK95: Cande and Kent (1995).

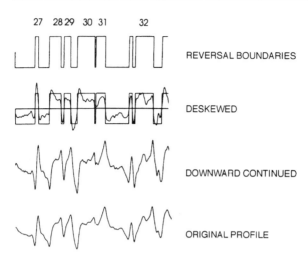

Figure 7.3 Procedure for constructing block models from oceanic magnetic anomaly profiles. Original profile is projected perpendicular to the strike of the lineations, bandpassed, downward continued, deskewed, and reversal boundaries determined from zero crossings (after Cande and Kent, 1992a).

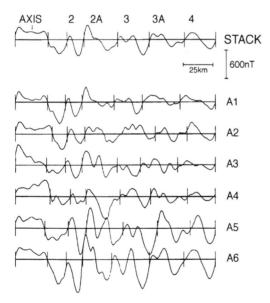

Figure 7.4 South Atlantic stack for anomaly 1 to anomaly 4. Distances in the stack (category 2 distances) are determined by averaging the widths of the subintervals on the individual profiles (after Cande and Kent, 1992a).

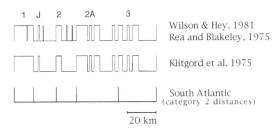

Wilson & Hey, 1981
Rea and Blakeley, 1975

Klitgord et al. 1975

South Atlantic
(category 2 distances)

20 km

Figure 7.5 Block models for anomaly 1 to anomaly 3A from fast-spreading centers integrated within the South Atlantic (category 2) distance intervals (after Cande and Kent, 1992a).

are related to geomagnetic intensity variations and have coined the term "cryptochron" for these short duration intervals.

7.2 Numerical Age Control

The Deep Sea Drilling Project, which began in 1968, provided a wealth of biostratigraphic data for sediments directly overlying magnetic anomalies, providing a minimum paleontological age for many anomalies. Improved correlations of magnetic anomalies to the geologic time scale have subsequently been achieved by correlations of magnetic anomalies to polarity zones in magnetostratigraphic sections with good biostratigraphic control. It is generally not possible to obtain radiometric ages on the sea floor lavas themselves, since they have suffered extensive alteration and argon does not completely outgas at deep ocean floor pressures. Absolute ages are generally correlated to magnetic anomaly records by a two-step process: correlation of magnetic anomalies to magnetostratigraphic sections and correlation of absolute ages to magnetostratigraphic sections.

The Cande and Kent (1992a) time scale (Table 7.1) utilized the following nine absolute age tie points for the last 83 Ma. (1) Gauss/Matuyama boundary at 2.60 Ma from astrochronology (Shackleton *et al.*, 1990). (2) C5Bn at 14.8 Ma through the correlation of foraminiferal zonal boundary N9/N10 to the GPTS (Miller *et al.*, 1985; Berggren *et al.*, 1985b) and the age estimates for this foraminiferal boundary from Japan at 14.5 ± 0.4 Ma (Tsuchi *et al.*, 1981) and from Martinique at 15.0 ± 0.3 Ma (Andreieff *et al.*, 1976). (3) Oligocene–Miocene boundary correlated to the middle part of C6Cn (Berggren *et al.*, 1985b), chronogram ages for this stage boundary (23.8 Ma) from Harland *et al.* (1990). (4) Eocene–Oligocene boundary correlated to chron C13r (.14) in the Apennines (Nocchi *et al.*, 1986), with absolute age estimate (33.7 ± 0.4 Ma) from

Table 7.1
Normal Polarity Intervals (Ma)

Polarity chron	Cande and Kent (1992a)	Cande and Kent (1995)	Wei (1995)
C1n	0.000–0.780	0.000–0.780	0.000–0.780
C1r.1n	0.984–1.049	0.990–1.070	0.990–1.070
C2n	1.757–1.983	1.770–1.950	1.770–1.950
C2r.1n	2.197–2.229	2.140–2.150	2.140–2.150
C2An.1n	2.600–3.054	2.581–3.040	2.580–3.040
C2An.2n	3.127–3.221	3.110–3.220	3.110–3.220
C2An.3n	3.325 3.553	3.330–3.580	3.330–3.580
C3n.1n	4.033–4.134	4.180–4.290	4.180–4.290
C3n.2n	4.265–4.432	4.480–4.620	4.480–4.620
C3n.3n	4.611–4.694	4.800–4.890	4.800–4.890
C3n.4n	4.812–5.046	4.980–5.230	4.980–5.230
C3An.1n	5.705–5.946	5.894–6.137	5.829–6.051
C3An.2n	6.078–6.376	6.269–6.567	6.173–6.450
C3Bn	6.744–6.901	6.935–7.091	6.795–6.943
C3Br.1n	6.946–6.981	7.135–7.170	6.986–7.019
C3Br.2n	7.153–7.187	7.341–7.375	7.183–7.216
C4n.1n	7.245–7.376	7.432–7.562	7.271–7.398
C4n.2n	7.464–7.892	7.650–8.072	7.483–7.902
C4r.1n	8.047–8.079	8.225–8.257	8.055–8.088
C4An	8.529–8.861	8.699–9.025	8.543–8.887
C4Ar.1n	9.069–9.146	9.230–9.308	9.106–9.191
C4Ar.2n	9.428–9.491	9.580–9.642	9.490–9.560
C5n.1n	9.592–9.735	9.740–9.880	9.670–9.827
C5n.2n	9.777–10.834	9.920–10.949	9.874–11.089
C5r.1n	10.940–10.989	11.052–11.099	11.214–11.273
C5r.2n	11.378–11.434	11.476–11.531	11.738–11.806
C5An.1n	11.852–12.000	11.935–12.078	12.314–12.495
C5An.2n	12.108–12.333	12.184–12.401	12.628–12.904
C5Ar.1n	12.618–12.649	12.678–12.708	13.256–13.294
C5Ar.2n	12.718–12.764	12.775–12.819	13.379–13.436
C5AAn	12.941–13.094	12.991–13.139	13.654–13.843
C5ABn	13.263–13.476	13.302–13.510	14.050–14.312
C5ACn	13.674–14.059	13.703–14.076	14.555–15.021
C5ADn	14.164–14.608	14.178–14.612	15.147–15.677
C5Bn.1n	14.800–14.890	14.800–14.888	15.901–16.007
C5Bn.2n	15.038–15.162	15.034–15.155	16.178–16.320
C5Cn.1n	16.035–16.318	16.014–16.293	17.289–17.592
C5Cn.2n	16.352–16.515	16.327–16.488	17.628–17.800
C5Cn.3n	16.583–16.755	16.556–16.726	17.872–18.051
C5Dn	17.310–17.650	17.277–17.615	18.617–18.955
C5En	18.317–18.817	18.281–18.781	19.603–20.074
C6n	19.083–20.162	19.048–20.131	20.321–21.295
C6An.1n	20.546–20.752	20.518–20.725	21.633–21.814
C6An.2n	21.021–21.343	20.996–21.320	22.047–22.324
C6AAn	21.787–21.877	21.768–21.859	22.703–22.780
C6AAr.1n	22.166–22.263	22.151–22.248	23.025–23.107

Table 7.1 *continued*

Polarity chron	Cande and Kent (1992a)	Cande and Kent (1995)	Wei (1995)
C6AAr.2n	22.471–22.505	22.459–22.493	23.284–23.312
C6Bn.1n	22.599–22.760	22.588–22.750	23.392–23.527
C6Bn.2n	22.814–23.076	22.804–23.069	23.572–23.794
C6Cn.1n	23.357–23.537	23.353–23.535	24.031–24.183
C6Cn.2n	23.678–23.800	23.677–23.800	24.302–24.405
C6Cn.3n	23.997–24.115	23.999–24.118	24.573–24.673
C7n.1n	24.722–24.772	24.730–24.781	25.191–25.235
C7n.2n	24.826–25.171	24.835–25.183	25.281–25.579
C7An	25.482–25.633	25.496–25.648	25.850–25.982
C8n.1n	25.807–25.934	25.823–25.951	26.136–26.247
C8n.2n	25.974–26.533	25.992–26.554	26.284–26.784
C9n	27.004–27.946	27.027–27.972	27.214–28.100
C10n.1n	28.255–28.484	28.283–28.512	28.400–28.625
C10n.2n	28.550–28.716	28.578–28.745	28.690–28.854
C11n.1n	29.373–29.633	29.401–29.662	29.514–29.779
C11n.2n	29.737–30.071	29.765–30.098	29.886–30.228
C12n	30.452–30.915	30.479–30.939	29.576–31.103
C13n	33.050–33.543	33.058–33.545	33.313–33.812
C15n	34.669–34.959	34.655–34.940	34.922–35.200
C16.1n	35.368–35.554	35.343–35.526	35.586–35.760
C16n.2n	35.716–36.383	35.685–36.341	35.909–36.518
C17n.1n	36.665–37.534	36.618–37.473	36.771–37.543
C17n.2n	37.667–37.915	37.604–37.848	37.660–37.877
C17n.3n	37.988–38.183	37.920–38.113	37.941–38.112
C18n.1n	38.500–39.639	38.426–39.552	38.389–39.382
C18n.2n	39.718–40.221	39.631–40.130	39.451–39.892
C19n	41.353–41.617	41.257–41.521	40.898–41.135
C20n	42.629–43.868	42.536–43.789	42.064–43.245
C21n	46.284–47.861	46.264–47.906	45.731–47.511
C22n	48.947–49.603	49.037–49.714	48.778–49.540
C23n.1n	50.646–50.812	50.778–50.946	50.734–50.921
C23n.2n	50.913–51.609	51.047–51.743	50.034–51.802
C24n.1n	52.238–52.544	52.364–52.663	52.478–52.800
C24n.2n	52.641–52.685	52.757–52.801	52.897–52.947
C24n.3n	52.791–53.250	52.903–53.347	53.057–53.527
C25n	55.981–56.515	55.904–56.391	56.113–56.584
C26n	57.800–58.197	57.554–57.911	57.691–58.027
C27n	61.555–61.951	60.920–61.276	60.850–61.188
C28n	63.303–64.542	62.499–63.634	62.359–63.479
C29n	64.911–65.732	63.976–64.745	63.821–64.613
C30n	66.601–68.625	65.578–67.610	
C31n	68.745–69.683	67.735–68.737	
C32n.1n	71.722–71.943	71.071–71.338	
C32n.2n	72.147–73.288	71.587–73.004	
C32r.1n	73.517–73.584	73.291–73.374	
C33n	73.781–78.781	73.619–79.075	
C34n	83.000–	83.000–	

Odin *et al.* (1991). (5) An age of 46.8 ± 0.5 Ma by Bryan and Duncan (1983) correlated to C21n by Berggren *et al.* (1983a). (6) An age of 55 Ma for the NP9/NP10 nannofossil boundary (Swisher and Knox, 1991) correlated to the Paleocene/Eocene boundary (Berggren *et al.*, 1985b). (7) Cretaceous–Paleocene boundary at 66 Ma (Harland *et al.*, 1990). (8) Campanian–Maastrichtian boundary at 74.5 Ma (Obradovich and Cobban, 1975; Obradovich *et al.*, 1986) based on the correlation of this stage boundary to the late part of C33n (Alvarez *et al.*, 1977). (9) Campanian–Santonian boundary at 84 Ma (Obradovich *et al.*, 1986; Alvarez *et al.*, 1977).

In the revised version of their time scale, Cande and Kent (1995) adopted the astrochronological estimates (Shackleton *et al.*, 1990; Hilgen, 1991a,b) for all Plio–Pleistocene reversals, and the 65 Ma estimate (as opposed to 66 Ma) for the Cretaceous–Tertiary boundary (Swisher *et al.*, 1992). Otherwise the absolute age tie points are as for Cande and Kent (1992a) (Table 7.1, Fig. 7.6).

Wei (1995) has taken issue with calibration points (2), (3), (4), (5), and (6) used by Cande and Kent (1992a, 1995). In place of calibration point (2) for C5Bn, Wei (1995) proposed a calibration point at 9.67 Ma for C5n (Baksi *et al.*, 1993a) and 16.32 Ma for C5Br (Baksi, 1993). In place of calibration point (3) for the middle part of C6Cn at the Oligocene–Miocene boundary, Wei (1995) proposed a calibration point at 28.1 Ma for C9r (Odin *et al.*, 1991). In place of calibration point (4) for the Eocene–Oligocene boundary, Wei (1995) proposed a calibration point of 35.2 Ma for C15r. This is based on a linear regression of nine age determinations in the C13r to C16n interval. In place of calibration points (5) and (6), Wei (1995) proposed a calibration point of 52.8 Ma for C24n.1r. This is based on an ^{40}Ar/^{39}Ar age within a polarity chron interpreted as C24n.1r in terrestrial sediments in Wyoming (Wing *et al.*, 1991; Tauxe *et al.*, 1994; Clyde *et al.*, 1994). The resulting time scale differs significantly from the Cande and Kent (1992a, 1995) time scales, particularly for the Miocene (Table 7.1).

Obradovich (1993) has reviewed ^{40}Ar/^{39}Ar age control on Late Cretaceous ammonite zones in the U.S. Western Interior. These zones can be correlated to the European ammonite zones to provide the best available numerical age estimates of Albian/Cenomanian and younger Late Cretaceous stage boundaries (Table 7.2). Building on earlier work (De Boer, 1982; Schwarzacher and Fischer, 1982; Weissert *et al.*, 1985; Herbert and

Figure 7.6 Geomagnetic polarity time scale for Late Cretaceous and Cenozoic (after Cande and Kent, 1995). Polarity chrons with duration less than 30 ky are omitted.

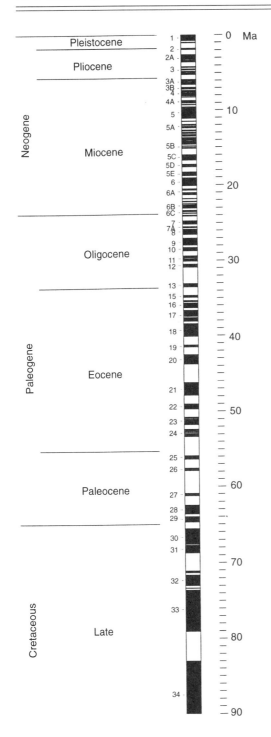

Table 7.2
Late Cretaceous Stage Boundaries

Stage boundary	Polarity chron boundary	Cande and Kent (1992a) (Ma)	Obradovich (1993) (Ma)	Herbert *et al.* (1995) (Ma)	Cande and Kent (1995) (Ma)
Cretaceous/Tertiary		*66.00*	65.4 ± 0.1		*65.00*
	Base C32r.1r	73.78			73.62
Campanian/Maastrichtian		*74.50*	71.3 ± 0.5		*74.50*
	Base C33n	78.78			79.08
	Base C33r	*83.00*			*83.00*
Santonian/Campanian		84.00	83.5 ± 0.5		84.00
Coniacian/Santonian			86.3 ± 0.5		
Turonian/Coniacian			88.7 ± 0.5		
Cenomanian/Turonian			93.3 ± 0.2		
Albian/Cenomanian			98.5 ± 0.5	99.5	
Aptian/Albian			112 ± 1	111.6	
Barremian/Aptian			121	122.0	

For Cande and Kent (1992a) and Cande and Kent (1995), italics indicate calibration data for spline fit.

Fischer, 1986), Herbert *et al.* (1995) have determined the durations of the Aptian, Albian, and Cenomanian stages on the basis of cyclostratigraphy in the pelagic sediments in central Italy. A consistent bundling of limestone–marl couplets in a nearly 5:1 ratio, observed in Barremian–Cenomanian Italian sections, first led De Boer (1982) and Schwarzacher and Fischer (1982) to propose that the carbonate rhythms represent precessional forcing grouped into 100 ky increments by the eccentricity envelope. Herbert *et al.* (1995) combined cyclostratigraphic estimates for the duration of the Aptian, Albian, and Cenomanian with Obradovich's (1993) estimate for the Cenomanian–Turonian boundary (93.5 ± 0.2 Ma) to calculate ages for the Barremian–Aptian, Aptian–Albian, and Albian–Cenomanian boundaries (Table 7.2).

One clear discrepancy concerns the Campanian/Maastrictian boundary (see Table 7.2). This stage boundary is generally considered to coincide with the top of the *Globotruncanita calcarata* foraminiferal zone in Italian pelagic limestone sections, and this zonal boundary lies in the upper part of C33n. Therefore, the Cande and Kent (1995) time scale implies an age >73.62 Ma (age of top of C33n) for the Campanian/Maastrichtian boundary, whereas the definition of the stage boundary in the Western Interior (at the base of the *Baculites eliasi* zone) yields an age of 71.3 Ma (Kennedy *et al.,* 1992; Obradovich, 1993). The younger age for this stage boundary appears to be consistent with Campanian foraminifera-bearing limestones from Mexico (Renné *et al.,* 1991). In the deep-sea fan deposits of the Point Loma Formation (San Diego, California), the Campanian/Maastrictian boundary lies in C33n and can be defined on the basis of ammonites and molluscs (Bannon *et al.,* 1989). Due to the endemic nature of the macrofauna, direct correlation from the Point Loma fauna to the Western Interior fauna is not possible.

8

Paleogene and Miocene Marine
Magnetic Stratigraphy

8.1 Miocene Magnetic Stratigraphy

Research on the magnetic stratigraphy of the Miocene began with the study of conventional piston cores with relatively low rates of sedimentation (Hays and Opdyke, 1967). In central Pacific cores, Foster and Opdyke (1970) obtained magnetic stratigraphies for sediments as old as Chron C5N (previously known as Chron 11) (Fig. 8.1). These studies pushed the limits of depth, and hence age of sediment, that could be reached by conventional piston cores. In the absence of suitable sections exposed on the continents, other methods of extending the polarity sequence and correlating it to the micropaleontological record had to be found. Consequently, studies were initiated on piston cores from the central Pacific in which hiatuses were present. These studies resulted in the correlation of siliceous fossil zones to sediments as old as Miocene (Opdyke et al., 1974; Theyer and Hammond, 1974).

The magnetic stratigraphy of the Miocene languished until the development of the hydraulic piston corer (HPC) and advanced piston corer (APC) by the Deep Sea Drilling Project (DSDP) and the Ocean Drilling Program (ODP). The magnetostratigraphic coverage of the Late Miocene is relatively complete and high quality data are available from land sections (Langereis et al., 1984; Krijgsman et al., 1994a; Hilgen et al., 1995) and HPC/APC cores from North and South Atlantic (Fig. 8.2, columns 1-2-3-4-5), Indian Ocean (column 7), and Pacific Ocean (columns 8-9). As a result, correlations for this time period are available for both high- and low-latitude sites. The magneto-biochronology for the late Miocene was established by Berggren et al. (1985b) and revised by Backman et al. (1990) and Berggren et al. (1995).

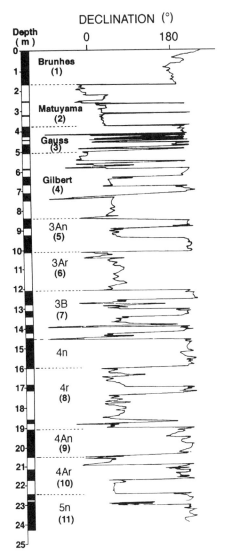

Figure 8.1 Magnetic declination in RC12-65 after alternating field demagnetization at peak fields of 5 mT. The black (normal) and white (reverse) bar diagram on the left indicates the proposed extension of the geomagnetic time scale based on this early study (after Foster and Opdyke, 1970). Old polarity chron nomenclature given in brackets.

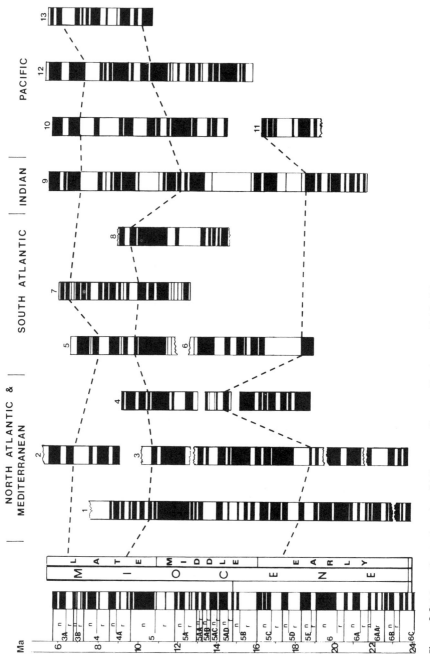

Figure 8.2 Magnetic stratigraphy of Miocene sediments. For key see Table 8.1.

Table 8.1
Miocene

	Rock unit	Lo Age Hi	Region	λ	Φ	NSE	NSI	NSA	M	D	RM	DM	AD	A	NMZ	NCh	%R	RT	F.C.T	Q	References
1	Site 608	Br-C6C	N. Atlantic	+42.84	−23.09	6	1.5	1	500	—	—	A	ZB	I	89	17	43	—	—	4	Clement and Robinson (1986)
2	Potamida-Kastello	Gil-C4r	Crete	+35.12	+25.20	7	325	3	75	—	—	T-A	ZB	D-I	26	3	48	—	—	4	Langereis et al. (1984)
3	Site 563	Br-C9	N. Atlantic	+33.64	43.77	1	.5 m	1	200	—	—	A	B	I	46	18	41	—	—	2	Khan et al. (1985)
	Site 563	Br-C9	N. Atlantic	+33.64	43.77	1	.5 m	1	200	—	—	A	B	I	46	18	41	—	—		Miller et al. (1985)
4	Site 643	Br-C5E	N. Atlantic	+67.28	+.0011	1	.3–.5 m	1	270	—	—	A	Z-V	I	145	16	24	—	—	4	Bleil (1989)
5	Site 519	Br-C5n	S. Atlantic	−26.14	−11.66	1	.1–4 m	1	300	—	—	A-T	ZB	I	34	9	34	R+	—	5	Tauxe et al. (1984)
6	Hole 521A	4A-5C	S. Atlantic	−26.07	−10.26	1	.07–1 m	1	37	—	1	A-T	ZB	VDI	20	9	54	—	—	5	Heller et al. (1984)
7	Site 704	Br-C5AN	S. Atlantic	+007.9	−47.1	1	.1–.05 m	1	430	—	1	A	ZB	I	48	13	50	—	—	5	Hailwood and Clement (1991)
8	Hole 696B	M-C17	S. Atlantic	−043	−61.8	1	.25 m	1	150	—	—	A	Z-V	I	47	22	45	—	—	4	Speiss (1990)
9	Site 710	Br-C6AA	Indian	−.04	+59.1	2	P	—	124	—	—	A	ZB	I	54	16	29	—	—	4	Schneider and Kent (1990b)
10	Site 588	Br-5AN	Pacific	−26.1	+161.22	2	1.2 m	1	310	—	V-K	A	ZB	I	42	13	62	—	—	6	Barton and Bloemendal (1986)
11	Site 575	Br-5D	E. Pacific	+5.85	−135.04	4	.5 m	1	130	—	—	A	FZ	D	28	10	58	—	—	5	Weinrich and Theyer (1985)
12	Site 578	Br-5B	Pacific	+33.93	+151.63	1	.1–2 m	1	160	—	—	A	B	I	71	15	43	—	—	3	Heath et al. (1985)
13	RC12-65	Br-5n	Pacific	+4.65	−144.97	1	.1 m	1	24.2	—	—	A	B	D	41	11	52	—	—	3	Foster and Opdyke (1970)
	Site 845	Gu-C5AB	E. Pacific	+9.58	−94.59	2	P	—	160	—	—	A	Z-B	D-I	48	9	49	—	—	5	Schneider (1995)
	Monterey Fm	C5R-C5B	USA	+35.17	−120.25	1	89	12	300	F-K	—	T	Z-P	F-V	16	7	49	R+	F+	8	Omarzi et al. (1993)
	Buff Bay	C5n-C5Ar		+18.75	−77.90	3	68	2	130	—	—	A	Z-P	—	3	2	57	—	—	5	Miller et al. (1994)
	Crete	C3AN-C4.2N	Crete	+35.12	+25.20	4	.3 m	1	60	Ar	K	T	Z	D-I	11	3	61	—	—	5	Krijgsman et al. (1994a)

Unquestionably, the best results from calcareous sediments of middle and early Miocene age were obtained by Clement and Robinson (1986) from the North Atlantic at DSDP Site 609. The magnetic stratigraphy from this site is the most complete available for the Miocene. As a result, ancillary studies such as $\delta^{18}O$ and $^{87}Sr/^{86}Sr$ stratigraphy (Miller *et al.*, 1991b) have been carried out. Unfortunately for Miocene magnetobiochronology, the microfauna and flora at this site are not representative of the tropical Miocene from which the standard foraminiferal and phytoplankton zonations are derived. Miller *et al.* (1991b) have, however, detected rare tropical foraminifers in this core by processing large volumes of sediment, enabling them to correlate the fauna to the low-latitude zonation.

The Miocene magnetobiochronology of Berggren *et al.* (1985b) relied heavily on South Atlantic cores from DSDP Leg 72 (Berggren *et al.*, 1983b). Unfortunately, magnetostratigraphic data were never published; only the interpretation was published as black and white bars. It is therefore impossible to assess the quality of the magnetostratigraphic data, which makes the magnetobiochronology suspect.

The Indian Ocean magnetostratigraphic record recovered at ODP Site 710 is of good quality, although hiatuses are present in the record (Schneider and Kent, 1990b). Correlation of low-latitude foraminiferal and nannoplankton zonations to the GPTS at this site has led to a revision of the Berggren *et al.* (1985a) magnetobiochronology (Backman *et al.*, 1990). These correlations have been confirmed by a recent study of Miocene sediment from the equatorial Pacific at ODP Holes 845A and 845B (Schneider, 1995). The Pacific magnetic stratigraphies were obtained using the ship-board pass-through cryogenic magnetometer at a 10-cm sampling interval after alternating field demagnetization at a peak field of 10 mT (Fig. 8.3). The correlation of the Miocene biozones to the GPTS is illustrated in Figure 8.8.

The Oligocene–Miocene boundary and its position with respect to the GPTS are matters of conjecture. This is due to the uncertainty in correlation of the type sections for the Aquitanian and Chattian stages, where an unconformity appears to coincide with the boundary. The problem was discussed extensively by Berggren *et al.* (1985b). The boundary was placed by them within C6Cn. The bioevent often used to denote the boundary is the first appearance of *G. kugleri*. Lowrie (1989) pointed out that the first appearance of this species appears to be diachronous between the Mediterranean and the South Atlantic. Backman *et al.* (1990) have suggested that the Oligocene–Miocene boundary be placed in Chron C6Cr; however, this placement is by extrapolation and not by direct correlation to the GPTS. Cande and Kent (1992a) have placed this boundary at the base of C6Cn.2 and this determination is followed here.

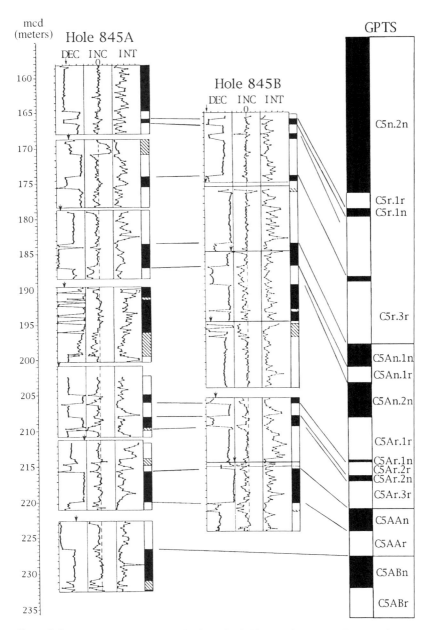

Figure 8.3 Magnetostratigraphic results from ODP Site 845 (after Schneider, 1995).

8.2 Paleogene Magnetic Stratigraphy

Unlike the record from the Miocene, which has depended heavily on marine cores, the correlation of the magnetic stratigraphy to the biostratigraphy of the marine Paleogene was first accomplished in a series of sections in north central Italy by Alvarez et al. (1977), Lowrie et al. (1982), and Napoleone et al. (1983), in a series of papers that have become classics. The pelagic limestones in these sections at Gubbio and Contessa have magnetization carried by magnetite and, in some cases, hematite. Both these magnetization components appear to have been acquired early in the history of the sediments. The antiquity of the magnetization has been verified by positive fold tests. Biostratigraphy in these sections required the use of new analytical techniques. Unlike most studies on deep-sea cores, the foraminifera in the indurated sediments at Gubbio had to be identified mainly in thin section (Premoli-Silva, 1977). Initially the nannofossils were considered too poorly preserved to be adequately identified (Premoli-Silva, 1977). It was not until the work of Monechi and Thierstein (1985) that a nannofossil biostratigraphy was available in these sections, although the poor nannofossil preservation in this area remains a hindrance to the precise correlation of nannofossil events to the GPTS. In the Contessa section, volcanic air fall ashes have been dated by the $^{40}Ar/^{39}Ar$ method (Montanari et al., 1988). The magnetic stratigraphy from the Contessa quarry section is shown in Figure 8.4. The sections of the region are well exposed in road cuts and could serve as type sections for integrated magnetobiochronology of Paleogene through Miocene time.

The highest frequency of reversals of the geomagnetic field in the last 100 My occurs in the Miocene, inhibiting correlation of Miocene magnetic stratigraphies to the GPTS. The reversal rate decreases going back in time from the Late Oligocene to the Paleocene, facilitating the correlation of magnetic stratigraphies to the GPTS. The better quality magnetic stratigraphies for the Paleogene are plotted in Figure 8.5. The polarity zones from the Gubbio section have been given letter designations such as N+, which correlates to Chron 26n. In the time interval covered by the Italian sections, 33 normal polarity intervals occur in the GPTS, whereas 28 normal polarity zones are observed in the Italian sections. The discrepancy of six normal intervals occurs in the Chron 15 to 18 interval. Further sampling in the Upper Eocene may reveal the fine structure of this part of the magnetostratigraphic sequence.

The magnetic stratigraphies from South Atlantic DSDP Sites 522, 523, and 524 (Tauxe et al., 1983c) cover most of the Paleogene except for a missing interval in the Eocene from Chron 21 to 23, representing approximately 6 My. The quality of the magnetic stratigraphy in these cores is

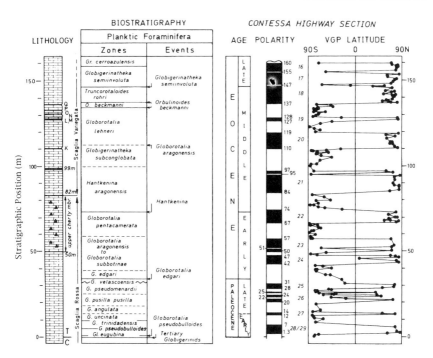

Figure 8.4 Magnetic stratigraphy and biostratigraphy of the Eocene part of the Contessa section (after Lowrie *et al.*, 1982).

exceptional, and their correlation to the Gubbio/Contessa sections and to the GPTS is unambiguous (Fig. 8.5). Farther to the south in Antarctic waters on the Maud Rise (ODP Leg 113), excellent core recovery at Holes 689 and 690 has facilitated the acquisition of high-quality magnetic stratigraphies (Speiss, 1990). Hiatuses hinder correlation of the magnetic stratigraphy to the GPTS in some intervals; however, excellent results were obtained for Late Oligocene to Middle Eocene time (Chron 8 to Chron 21 in Hole 689B). An excellent record was also obtained in Hole 690B from Late Paleocene (Chron 26) to Early Eocene time (Chron 24). A short magnetostratigraphic section has been reported from Hole 702B by Clement and Hailwood (1991) in sediments of Middle Eocene age (Chron 18–21). Other data are available but are often discontinuous (Fig. 8.5). It is interesting to note that almost all of these data come from the Atlantic Ocean and Mediterranean area. Data are lacking from the Pacific and Indian oceans. Data are available from the middle Eocene of the Atlantic margin (Miller *et al.*, 1990), southern England (Townsend and Hailwood, 1985), and from

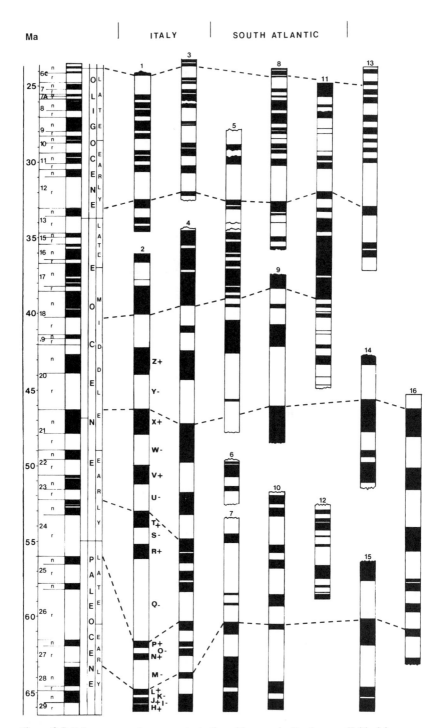

Figure 8.5 Paleogene marine magnetostratigraphic records. For key see Table 8.2.

134

Table 8.2
Paleogene

	Rock unit	Lo Age Hi	Region	λ	Φ	NSE	NSI	NSA	M	D	RM	DM	AD	A	NMZ	NCh	%R	RT	F.C.T	Q	References
1	Scaglia	C6-C20	Italy	+43.38	+12.56	1	.3–.6 m	1	140	A	I-K	T-A	Z-V	F-V	31	14	51	R+	F+	8	Lowrie *et al.* (1982)
2	Scaglia	C17-C29	Italy	+43.37	+12.58	1	202	1	150	—	I	T	Z-V	D-I	16	10	60	—	—	6	Napoleone *et al.* (1983)
3	Site 558	C5AD-C7	N. Atlantic	+37.77	+37.34	2	.25–.4 m	1	30	—	—	A	V	I	36	12	43	—	—	5	Khan *et al.* (1985)
4	Scaglia	C16-C29	Italy	+43.38	+12.56	3	.3–.6 m	1	160	A	I	T	Z-V	FV	33	13	51	R+	F+	9	Lowrie *et al.* (1982)
5	Site 523	C11-C20	S. Atlantic	-28.55	-2.25	1	.1–.4 m	1	130	—	—	A	Z-B	I	27	9	58	R+	—	5	Tauxe *et al.* (1984)
6	Site 524	C23-C31	S. Atlantic	-15.01	-3.51	1	1.4 m	1	290	—	—	A	Z-B	I	21	9	46	R+	—	5	Tauxe *et al.* (1984)
7	Site 524	C23-C31	S. Atlantic	-15.01	-3.51	1	1.4 m	1	290	—	—	A	Z-B	I	21	9	46	R+	—	5	Tauxe *et al.* (1984)
8	Site 522	C6-C16	S. Atlantic	-26.11	-.13	1	.1–.4 m	1	155	—	—	A	Z-B	I	63	13	54	R+	—	5	Tauxe *et al.* (1984)
9	Hole 702B	C18-C21	S. Atlantic	-50.95	-26.37	1	PTM	1	10	—	1	A	Z-V	I	9	5	57	—	—	6	Clement and Hailwood (1991)
10	Site 577	C23-C30	NW Pacific	+32.44	+157.72	2	.1–.5 m	1	69	—	—	A	Z-B	I	21	9	57	—	—	6	Bleil (1989)
11	Site 689	C7-C15	S. Atlantic	+004	-64	2	.25 m	1	150	—	—	A	Z-V	I	31	14	46	—	—	7	Speiss (1990)
12	Site 690	C9-C26	S. Atlantic	+003	-65	2	.25 m	1	144	—	—	A	Z-V	I	38	13	53	—	—	7	Speiss (1990)
13	Lincoln Creek	C6C-C17	USA	+47.00	-123.5	1	97	3	2900	K	—	T	B	V	27	23	68	—	—	4	Prothero and Armentrout (1985)
14	Atlantic Coastal Pl.	C8-C23	USA	+39.44	-74.72	1	62	1	100	S	—	A	V	I	7	4	50	—	—	4	Miller *et al.* (1990)
15	Alpine Scaglia	C24-C34	Italy	+46.00	+11.75	5	359	1	127	—	—	T	ZB	V	33	13	28	—	—	5	Channell and Medizza (1981)
16	London Basin	C21-C25	England	+51	000	10	152	2	—	—	—	A-T	V	—	28	5	71	—	—	2	Townsend and Hailwood (1985)
	Alabama Coastal Pl.	C11-C16	USA	-31.50	-88.00	3	153	1	76.2	O	1	A	V	I	14	6	62	—	—	6	Miller *et al.* (1993)

135

the late Eocene and Oligocene of coastal Oregon (Prothero and Armentrout, 1985).

The magnetobiochronology of the Paleocene has been discussed by Berggren et al. (1985a), Aubry et al. (1988), and Berggren et al. (1995). The Eocene/Oligocene boundary has been placed by the International Stratigraphic Commission in the Massignano section (Italy) at the level of the last appearance of the foraminifera *Hankenina* (Montanari et al., 1988). This position would place the stage boundary at C13r(0.14), close to the disappearance level of *Hankenina* at DSDP Site 522 (Poore et al., 1984), but above this bioevent in the Contessa Quarry section. From the original paleomagnetic investigation at Massignano, Bice and Montanari (1988) postulated the presence of a short normal polarity subchron immediately prior to the Eocene/Oligocene boundary. This subchron was not identified by Lowrie and Lanci (1994) and the present consensus is that the position of the boundary should be placed at C13r(0.14), as advocated by Cande and Kent (1992a).

The correlation of the Paleocene–Eocene boundary to C24r(0.6) (Cande and Kent, 1992a) is the same as that adopted by Berggren et al. (1985a) and close to that observed in the Umbrian sections by Lowrie et al. (1982). Berggren et al. (1995) give an uncertainty of about 1 My in the correlation of this stage boundary. The early/middle Eocene boundary is correlated to the GPTS in the Contessa sections and correlates to the top of C22n. The evidence for the placement of the middle/late Eocene boundary in the polarity sequence is not so clear, and this boundary is placed in late C18n in the Contessa Highway section. Berggren et al. (1995) choose to place this boundary in C17n on the basis of fossils from DSDP Site 523 in the South Atlantic (Poore et al., 1984). This placement is followed here; however, the correlation of this boundary to the GPTS depends on the paleontological criteria chosen to define it.

8.3 Integration of Chemostratigraphy and Magnetic Stratigraphy

In Plio–Pleistocene sediments, $\delta^{18}O$ stratigraphy provides a high-resolution stratigraphic correlation tool and a means of tuning the time scale and assigning absolute ages to polarity chrons (Ch. 6). The Cenozoic $\delta^{18}O$ record derived from Atlantic benthic foraminifera (Fig. 8.6) (Miller et al., 1987) may be interpreted as changes in the temperature of ocean bottom water or ice volume changes, or both. Shackleton and Opdyke (1973) have shown that ice volume changes dominated the Pleistocene benthic $\delta^{18}O$ record. Separating temperature change from ice volume for the early Cenozoic record is more

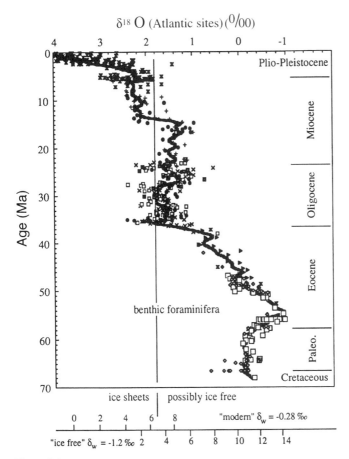

Figure 8.6 Composite benthic foraminiferal oxygen isotope record for Atlantic DSDP sites corrected to *Cibicidoides* and reported to Pee Dee Belemnite (PDB) standard. Chronostratigraphic subdivisions are drawn after Berggren *et al.* (1985a). The smoothed curve is obtained by linearly interpolating between data at 0.1-My intervals and smoothing with a 27-point Gaussian convolution filter, removing frequencies higher than 1.35/My. The vertical line is drawn through 1.8‰; values greater than this provide evidence for existence of significant ice sheets. The temperature scale is computed using the paleotemperature equation, assuming *Cibicidoides* are depleted relative to equilibrium by 0.64‰. The lower temperature scale assumes no significant ice sheets, and therefore $\partial w = 1.2‰$; the upper scale assumes ice volume equivalent to modern values, and therefore $\partial w = 0.28‰$ (after Miller *et al.*, 1988).

difficult. In the Paleocene, the $\delta^{18}O$ values vary about a value of 0‰, then in the late Paleocene $\delta^{18}O$ values begin to become more negative, peaking at $-1‰$ in the early Eocene (Fig. 8.6). The $\delta^{18}O$ values become steadily more

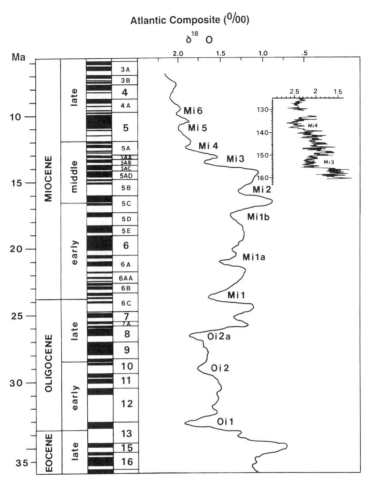

Figure 8.7 Composite record from the Atlantic indicating position of oxygen isotope events for Oligocene and Miocene time (after Miller *et al.*, 1991a). Inset shows high-resolution record for Mi3–Mi4 (after Woodruff and Savin, 1991). Time scale from Cande and Kent (1995).

positive throughout the Eocene, increasing sharply at the Eocene/Oligocene boundary to values of about +2.8‰. The $\delta^{18}O$ values then oscillate throughout the Oligocene and early Miocene, and a sharp change to more positive values occurs in the interval from 15 to 13 Ma, and the values then oscillate around +2.25‰ for the rest of the Miocene.

These changing values of $\delta^{18}O$ in benthic foraminifera have been interpreted by Miller *et al.* (1987, 1991a) in terms of bottom water temperature and ice volume changes. Miller *et al.* (1987) interpreted the changes ob-

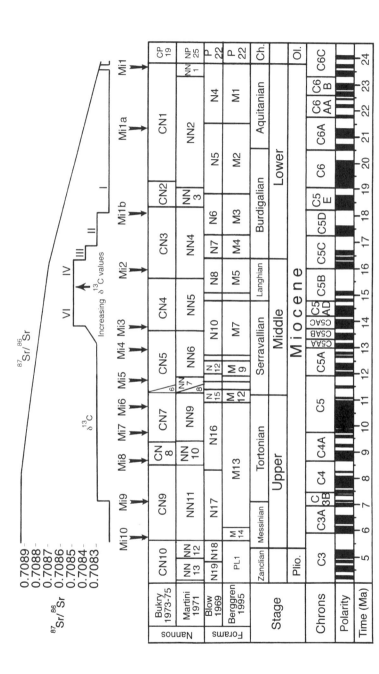

Figure 8.8 Polarity time scale for the Miocene (after Cande and Kent, 1995) correlated to foraminiferal zones and nannofossil zones (after Berggren *et al.*, 1995), to isotopic events recorded in $\delta^{18}O$ denoted by the prefix "Mi" (Miller *et al.*, 1991b), to $\delta^{13}C$ (after Woodruff and Savin, 1991), and $^{87}Sr/^{86}Sr$ (after Hodell and Woodruff, 1994).

served in Paleocene and Eocene sediments as largely due to changes in bottom water temperature. They argued that ice sheets became an important factor only after the Eocene/Oligocene boundary, and only since this time (~34 Ma) do oscillations observed in the $\delta^{18}O$ denote ice volume change. This is supported in some cases by the covariance of $\delta^{18}O$ values from benthic and planktic foraminifera. Times of sea level fall are presumed to be coincident with positive shifts on the post-Eocene $\delta^{18}O$ curve, with changes to more negative values representing deglaciation. It should be noted that because of the smoothing employed in the $\delta^{18}O$ record shown in Figure 8.6, it is not possible to see the high-frequency Milankovitch cycles characteristic of the Plio–Pleistocene record.

Miller *et al.* (1991a) have presented a new and updated $\delta^{18}O$ record for the Oligocene and Miocene (Fig. 8.7). They argue that continental glaciation began in Antarctica in the latest Eocene, at about 35 Ma. They have formally defined nine oxygen isotope zones in the Oligocene and Miocene which they have designated Oi1 and Oi2 and Mi1 to Mi7, ranging in age from Early Oligocene through early Late Miocene (Figs. 8.7 and 8.8). Miller *et al.* (1991a) interpret these isotope events in terms of glacial episodes. It is clear that this $\delta^{18}O$ stratigraphy does not have the resolution of the Plio–Pleistocene record and isotopic events have age uncertainties of 0.5 My using the sampling strategies employed by Miller *et al.* (1991a). The question arises as to whether early Neogene and Paleogene records can, in the future, be used to yield a high-resolution $\delta^{18}O$ stratigraphy. Woodruff and Savin (1991) have studied in detail the mid-Miocene $\delta^{18}O$ shift at DSDP Site 574, and it is possible to correlate isotope stage Mi3 to the section at Site 574 in the 150 to 156 mbsf interval (Fig. 8.7). It would appear that the potential exists for an extension of high-resolution $\delta^{18}O$ stratigraphy to sediments of Miocene age, and perhaps to the entire Cenozoic. Site 574 has no magnetic stratigraphy; however, the correlation to the Atlantic composite record is supported by the biostratigraphy.

Shifts in the $\delta^{13}C$ record have been utilized in sediment correlation (Berger, 1982; Haq *et al.*, 1980; Miller *et al.*, 1988, 1989). Changes in $\delta^{13}C$ are thought to be due to burial of organic carbon, erosion of carbonate sediments (reservoir exchanges), variations in biomass (climate change), or changes in productivity (Berger *et al.*, 1981). For the Pleistocene, in-phase changes of $\delta^{18}O$ and $\delta^{13}C$, with $\delta^{18}O$ becoming heavier and $\delta^{13}C$ becoming lighter during glacial episodes, are well known (Shackleton, 1977). Such in-phase changes are common throughout the Neogene beginning by Middle Miocene time (Woodruff and Savin, 1991). It is interesting to note that for two earlier $\delta^{18}O$ events thought to represent glacial events (Oi1 and Mi1), the $\delta^{13}C$ shift is synchronous with the $\delta^{18}O$ shift, but toward heavier values, not lighter ones.

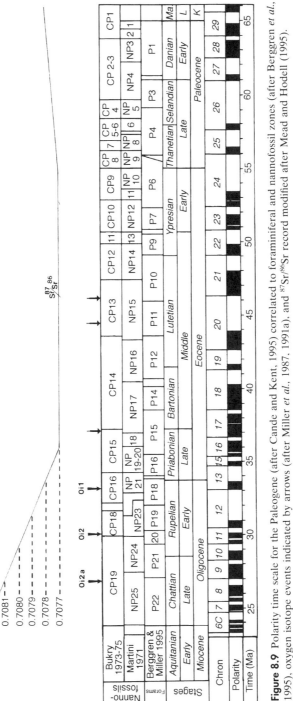

Figure 8.9 Polarity time scale for the Paleogene (after Cande and Kent, 1995) correlated to foraminiferal and nannofossil zones (after Berggren *et al.*, 1995), oxygen isotope events indicated by arrows (after Miller *et al.*, 1987, 1991a), and $^{87}Sr/^{86}Sr$ record modified after Mead and Hodell (1995).

The best documented $\delta^{13}C$ carbon shift is the prominant upper Miocene $\delta^{13}C$ shift of $-0.7‰$ (Keigwin, 1979) which coincides with Chron 3Bn at about 7 Ma (Haq et al., 1980) (Fig. 8.8). This carbon shift seems to occur not only in the marine record but also in ancient soils in the land record (Quade et al., 1989). The land and marine occurrences are both correlated to the GPTS. Recent studies of ODP cores off Greenland indicate that this event may correlate with the onset of glaciation on Greenland (Leg 152 shipboard party, 1994). This event is, therefore, an important link between the marine and nonmarine records. A second important $\delta^{13}C$ shift of about $+1.5‰$ close to the boundary between the lower and middle Miocene has been documented by Vincent et al. (1985). The middle Miocene is marked by the most positive $\delta^{13}C$ values of the entire Miocene and this interval has been called the Monterey event (Vincent and Berger, 1985). This $\delta^{13}C$ shift is hypothesized to be the result of organic carbon being locked up in sediments around the Pacific Rim. Woodruff and Savin (1991) have studied the $\delta^{13}C$ fluctuations in the middle Miocene in detail and have designated $\delta^{13}C$ isotope stages I to IX. The relative positions of the boundaries of carbon isotope stages I through VI are shown in Figure 8.8. The cores which record these stages do not have magnetic stratigraphies and the biostratigraphic zonation is nonstandard, which makes correlation difficult. Nevertheless, future studies will undoubtedly result in the $\delta^{13}C$ record being a useful correlation technique for Miocene time. Miller et al. (1988) have provided a $\delta^{18}O$ and $\delta^{13}C$ record for the Oligocene from DSDP Site 522, which has good magnetostratigraphic control. They suggest that this record serve as a type record for isotopic stages in the Paleogene. Two prominent $\delta^{13}C$ excursions are seen in this core which are correlative with the $\delta^{18}O$ events Oi1 and Mi1 (Fig. 8.7).

One of the most promising isotopic correlation techniques to emerge in the last decade is the use of the changing ratio of $^{87}Sr/^{86}Sr$ with time in ocean water (DePaolo and Ingram, 1985). It is particularly useful from the beginning of the Oligocene to the present, when $^{87}Sr/^{86}Sr$ in seawater changed from .7077 to .7090 (Figs. 8.8 and 8.9). High-resolution data are available for Miocene and younger sediments (Hodell et al., 1991). Miller et al. (1988, 1991b) have studied deep-sea cores with magnetic stratigraphy to create a record of Sr isotope changes for the Miocene that can be correlated to the GPTS. The study by Hodell et al. (1991) used a combination of magnetostratigraphically dated cores and data correlated to the GPTS indirectly through biochronology. For some time periods where there is rapid change in $^{87}Sr/^{86}Sr$, such as the Lower Miocene, it is possible to correlate magnetostratigraphic sections to the GPTS using the $^{87}Sr/^{86}Sr$ record.

The reason for the strontium isotopic ratio changing with time is a matter of debate. Strontium enters the ocean through the erosion of mountain belts, ancient shield areas, and oceanic ridge systems. Fortunately, it is not necessary to know the cause of the variation in order to use it for stratigraphic correlation. The integration of chemostratigraphy with biostratigraphy and magnetic stratigraphy can enhance the precision of stratigraphic correlation in the Cenozoic (Figs. 8.8 and 8.9). Berggren coined the term magnetobiochronology, which might now have to be amended to magnetobiochemochronology, which is a little awkward; but MBC chronology might be acceptable.

9

Cenozoic Terrestrial Magnetic Stratigraphy

9.1 Introduction

One of the long-standing problems of stratigraphy has been the correlation of nonmarine sediments containing vertebrate fossils to marine sediments containing invertebrate fossils. Due to correlation difficulties, the stratigraphy of marine and nonmarine sediments has developed more or less independently. As geomagnetic polarity reversals are globally synchronous, magnetic polarity stratigraphy offers the opportunity to correlate between these two contrasting sedimentary environments.

Magnetostratigraphic studies of terrestrial sequences began in the early 1970s in a Plio–Pleistocene mammal-bearing sequence of the San Pedro Valley, Arizona (N. M. Johnson *et al.,* 1975). This study, subsequently updated by Lindsay *et al.* (1990), demonstrated the potential of magnetic stratigraphy in terrestrial sequences. The authors were able to correlate widely separated fossil localities and place them in stratigraphic sequence. Correlation of polarity zones to the GPTS provided the time frame. The sediments of the basin were mapped using polarity zones and the rates of basin subsidence determined. In the 20 years since the publication of this paper, magnetostratigraphic studies of nonmarine Cenozoic sediments have become commonplace and have been used to solve problems in vertebrate paleontology, faunal migration, sedimentology, basin subsidence, and tectonic history.

Correlation of Plio–Pleistocene land mammal sequences to the GPTS has been accomplished in North America, Western Europe, southern Russia, China, the Indian subcontinent, South America, and Africa (Fig. 9.1, Table 9.1). First and last appearances of mammal taxa have been correlated to the GPTS in many individual sections, some of which have radiometric age control. Magnetic stratigraphy of Plio–Pleistocene lake and loess sequences has been motivated by the desire to understand continental climatic response to Plio–Pleistocene glaciation. Good quality Plio–Pleistocene lake sediment magnetic stratigraphies exist for North America, Japan, Australia, and Europe (Table 9.2). The climatic record from lakes is often provided by downcore changes in pollen type and abundance. For loess, good quality magnetic stratigraphies have been obtained from Europe, Central Asia, North America, South America, and, most importantly, China (Table 9.3).

9.2 North American Neogene and Quaternary

For North America, magnetic stratigraphy of mammal-bearing sequences covers almost the entire Cenozoic and extends into the Cretaceous. The terrestrial mammal-bearing stratigraphy in North America is divided into 19 land mammal ages (LMAs) based on stage of evolution and the appearance of immigrant mammal taxa from other continents, usually Eurasia, or at the end of the Cenozoic, from South America. Magnetic stratigraphy offers an independent means of ordering events in mammalian biochronology and thereby enhancing studies of tempo and mode in evolution.

The Plio–Pleistocene nonmarine magnetostratigraphic record is well known for most continents and the correlation of the mammalian zones to the GPTS is reasonably well established (Fig. 9.1). In North America the record is more or less complete, except for the Late Pleistocene and the earliest Pliocene. The boundary between Blancan and Irvingtonian fauna occurs at the top of the Olduvai subchron, with the first appearance of *Mammuthus* in North America. The base of the Blancan LMA is tentatively placed in the Sidufjall subchron at ~4.8 Ma. Figure 9.2 shows the correlation of first and last appearances of mammal taxa to the GPTS in North America from Late Miocene time. The first appearances may be evolutionary first appearances, such as that of *Equus* at Hagermann in Idaho, or migrationary first appearances such as that of *Mammuthus* (elephant) just above the Olduvai subchron at Anza-Borrego in southern California.

Considerable progress has been made in correlating the Miocene LMAs to the GPTS in North America (Fig. 9.3, Table 9.4). Initial studies on Miocene sediments by MacFadden (1977) and Barghorn (1981) have been followed by recent studies by Whistler and Burbank (1992) and MacFadden *et al.* (1990b) that extended this correlation to sediments containing Clarendonian and Barstovian fossils of middle and early Miocene age. Whistler and Burbank (1992) placed the boundary between the Clarendonian and Hemphillian in early C4Ar and the preceeding boundary between the Barstovian and Clarendonian at the base of C5Br. MacFadden *et al.* (1990b) placed the Hemingfordian–Barstovian boundary within chron C5Br. The boundary between the Arikareen and Hemingfordian has been placed within C5Er by B. J. MacFadden (personal communication, 1995).

9.3 Eurasian Neogene

Eurasia has relatively long Neogene terrestrial sequences with abundant fossil content. The most successful studies have been concentrated along the southern deformed margin of Eurasia where Cenozoic sediments have been exposed by Alpine–Himalayan orogeny. These areas extend from Pakistan and India, where much work has been done on the Siwalik sediments particularly in Pakistan, through the Caucasus and Turkey to the Mediterranean and Spain. Unfortunately, many of the areas of Western Europe where classical faunas were initially described are poorly suited to magnetostratigraphic study because of limited exposure and, in some cases, lack of knowledge of the precise location of fossil finds.

The European early Pliocene land mammal age, the Ruscinian (Fig. 9.1), is younger than the flooding of the Mediterranean at the Miocene–Pliocene boundary, which is now well-dated magnetostratigraphically and astronomically (Ch. 6). Magnetostratigraphic studies in Spain (Opdyke *et al.*, 1989, 1996) place the Turolian–Ruscinian boundary (=Mn 13/14 boundary) in the early Gilbert, at about the level of the lower boundary of the Sidufjall subchron.

The Vallesian–Turolian boundary (Fig. 9.3) is placed at 9 Ma by Berggren *et al.* (1985b). This is close to the placement of the Clarendonian–Hemphillian at 8.8 Ma (Chron 4A) (Tedford *et al.*, 1987; Whistler and Burbank, 1992). The position of the Astaracian–Vallesian boundary has been a subject of controversy because the base of the Vallesian is defined by the first appearance of *Hipparion,* several species of which migrated to Europe from North America. The appearance of *Hipparion* in the Siwaliks of Pakistan is well dated magnetostratigraphically at 10.8 Ma (Johnson *et al.*, 1982). On the other hand, Berggren *et al.* (1985b) preferred an earlier age of 12.5 Ma based on radiometric ages from Germany that correlate with

Figure 9.1 Selected Pliocene and Pleistocene terrestrial records from North America, Europe, Pakistan, China, Africa, and South America. Numbers refer to individual studies (see Table 9.1). Time scale from Cande and Kent (1995).

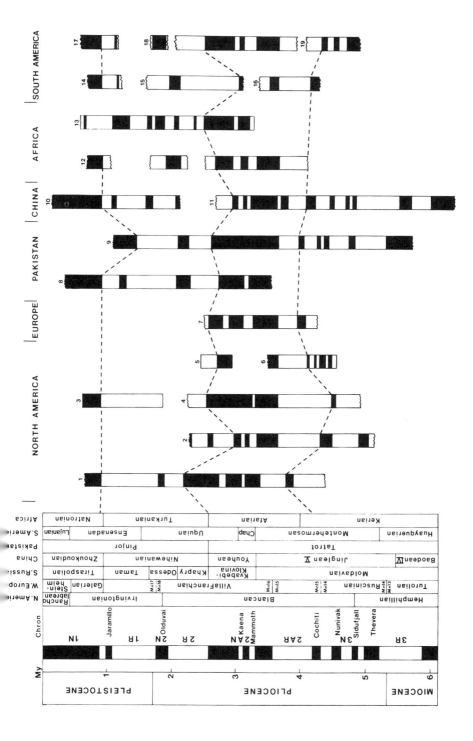

Table 9.1
Plio–Pleistocene Mammals

	Rock unit	Lo Age Hi	Region	λ	Φ	NSE	NSI	NSA	M	D	RM	DM	AD	A	NMZ	NCh	%R	RT	F.C.T.	Q	References
1	St. David Fm.	B-Gil	USA (A)	+32	−110.2	6	149	3	400	K	J-K	A	B	F	12	3	62	R+	—	7	N. M. Johnson et al. (1975)
2	Vallecito-Fish Crk	M-Gil	USA (Cal)	+33.15	−116.35	3	150	3	4000	—	J-I	A	B	D	12	3	56	R−	—	5	Opdyke et al. (1977)
	Vallecito-Fish Crk	M-Gil	USA (Cal)	+33.15	−111.35	3	150	3	4000	F	—	T	B	F	12	3	56	R+	—	7	N. M. Johnson et al. (1983)
3	Jaw Face	B-M	Canada	−50.66	−107.87	1	357	1	11	F	—	A	Z-P	F	2	2	78	R+	—	7	Barendregt et al. (1991)
4	Glen's Ferry	M-Gil	USA (Idaho)	+43	−115.29	9	176	3	500	K	J-I	A	B	F D-I	5	3	85	R+	—	7	Neville et al. (1979)
5	111 Ranchbeds	M-Ga.	USA (Ar)	+32.75	−109.5	4	35	3	100	K	—	A-T	B	Z-V	2	2	50	—	—	5	Galusha et al. (1984)
6	Verde Fm.	G-4n	USA (Ar)	+34.55	−112	9	164	3	450	K	V	A	B	V	32	5	46	R−	—	6	Bressler and Butler (1978)
7	Teruel Basin	M-Bil	Spain	+40.62	+.001	3	78	3	120	—	T	Z-P	V-F	16	3	50	R+	—	7	Opdyke et al. (1996)
8	Upper Siwalik	B-Ga	Pakistan	+32.93	+73.73	3	113	3-5	1100	F	—	A	B	F	9	3	50	R+	—	5	Keller et al. (1977)
9	Upper Siwalik	M-3A	Pakistan	+33.00	+73.5	8	132	3	1850	F	J	A	B	F-V	38	6	49	R+	F+	8	Opdyke et al. (1979)
10	Louchuan loess	B-M	China	+36	+109.23	1	.3 m	1	150	—	J-I-K	T		D-I	7	3	52	—	—	4	Heller and Liu (1984)
11	Yushe Basin	M-4N	China	+37	+113	3	280	3	810	—	—	T	Z-P	V	20	4	58	—	—	6	Tedford et al. (1991)
12	Koobe Fora	B-Gil	Kenya	+04	+36.3	9	—	—	290	A	J-I	A-T	Z-P	V	12	4	46	—	—	8	Hillhouse et al. (1986)
13	Shungura	M-Ga	Ethiopia	+5	+36	25	310	3	770	A	X	A	V	F	17	2	50	R+	—	7	Brown et al. (1978)
14	Tarija Fm.	B-M	Bolivia	−21.5	−64.75	4	100	3	650	K	—	T-A	Z-P	E.V.	4	2	47	R+	—	7	MacFadden et al. (1983)
15	Uquia	M-Ga	Argentina	−21.32	−65.35	1	20	1	200	K	—	—	B	V	4	2	85	—	—	4	Marshall et al. (1982)
16	Inchasi Beds	Gil	Bolivia	−19.75	−65.5	5	54	3	120	—	1	T-A	Z-P	F-V	4	1	82	R+	—	8	MacFadden et al. (1983)
17	Ensenadense	B-M	Argentina	−38	−57.57	3	129	1	14.5	K	—	A	B	D-I	3	3	46	R+	—	3	Nabel and Valencio (1981)
18	LaPaz-Calvorio	M-Gil	Bolivia	−16.5	−68.0	4	100	3	260	K	I-H	T	Z-B	F-DI	11	3	47	R+	—		Thouveny and Servant (1985)
19	Corral Quemado	Gil-4A	Argentina	−27.33	−66.9	2	29	3	900	K	J-I	T-A	Z-B	F-V	8	1	57	R+	F+	9	Butler et al. (1984)
	Victoria	B-Gil	Australia	−38	+144	12	80	3	24	K	—	A-T	PCA	V	11	4	—	—	—	6	Whitelaw (1991a)
	Buenos Aires	B-Ga	Argentina	−38.33	−57.66	2	.2 m	1	26	K	—	A-T	B	F-V	17	3	56	—	—	4	Orguira (1990)

Table 9.2
Lake Sediments

Rock unit	Lo Age Hi	Region	λ	Φ	NSE	NSI	NSA	TM	D	RM	DM	AD	A	NMZ	NCh	%R	RT	F.C.T.	Q	References
Lake George	B–Ga	Aust. (NSW)	−35.08	+149.42	1	89	3	36	—	—	A	V	1	9	3	35	—	—	4	Singh et al. (1981)
Tule Lake	B–Ga	USA (CA)	+42.0	−121.5	3	384	1	370	K	K-A	A	B	1	17	3	41	—	—	5	Rieck et al. (1992)
Bonneville Basin	B–Ga	USA (Utah)	+40.8	−112.54	1	144	1	111	K	—	—	—	F	3	2	7	—	—	1	Eardley et al. (1973)
Lake Tyrell	B–Ga	Aust (Vic)	35.34	+142.81	3	112	1	22	—	—	A-T	V-Z	D-I	8	4	47	—	—	5	An et al. (1986)
Lake Biwa	B—	Japan	+35.25	+136.27	1	~8000	1	200	K	—	A	B	D-I	1	1	0	—	—	4	Yaskawa et al. (1973)
Searles Lake	B–Ga		+35.72	−117.34	1	120	3	700	—	—	A-T	B	1	23	3	47	—	—	4	Liddicoat et al. (1980)

Table 9.3
Loess Records

Rock unit	Age	Region	λ	Φ	N.S.E.	NSI	NSA	M	D	RM	DM	AD	A	NMZ	NCh	%R	RT	F.C.T.	Q	References
Baoji	Ga-Gil	Shaanxi-China	+34.33	+107.01	1	576	1	190	—	K	T	B	D-I	12	4	51	—	—	5	Evans et al. (1991)
Lanzhou	B-M	Gansu-China	+35.2	+103.16	1	40	3	330	—	—	T	B	V	4	2	38	—	—	3	Burbank and Li (1985)
Fairbanks	B-Ga	Alaska-USA	+64.51	−147.45	6	283	1	32	F	—	A	B	I	10	3	49	—	—	5	Westgate et al. (1990)
Xifeng	B-Ga	Gansu-China	+35.7	+107.60	1	1848	1	200	—	K	T	Z-B	D-I	21	4	46	—	—	5	Liu et al. (1988)
Luochuan	B-M	Shaanxi-China	+35.8	+109.2	2	488	1	150	—	K-I	T	Z-V	D-I-F	7	3	49	R+	—	7	Heller et al. (1984)
Tashkent	B-M	Tadzhikistan	+41.20	+69.18	6	1500	1	90	—	K	T	B	D-I	3	2	25	—	—	5	Lazarenko et al. (1981)
Balneario	B-Ga	Argentina	−38.18	−57.64	1	180	2	20	—	K	A	Z-V	D-I	7	3	38	—	—	5	Ruocco (1989)
Lantian	B-Gil	China	+34.2	+109.2	2	500	1	195	—	—	T	Z-V	D-I	21	4	50	—	—	4	Zheng et al. (1992)
Shan Xian-Xifeng	B-M	Henan / Gansu-China	+34.3 / +35.70	+111.9 / +107.8	2	226	1	135	—	—	T	Z-B	V	6	2	48	—	—	5	Yue et al. (1984)
Baoji	B-Ga	Shaanxi-China	+34.33	+107.1	1	576	1	186	—	K	T	Z-B	D-I	16	3	49	—	—	5	Rutter et al. (1990)
Tajik	B-M	Tajikistan	+38.33	+68.48	7	250	1	100	—	K	T-A	Z-P	D-I	2	2	16	—	—	6	Forster and Heller (1994)

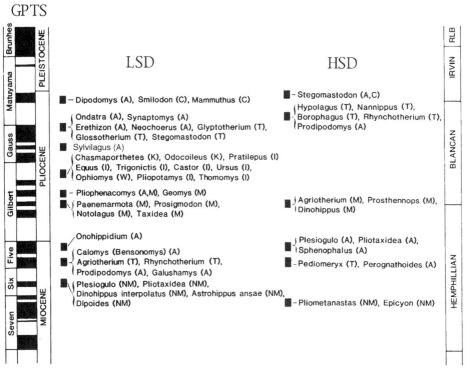

Figure 9.2 Lowest stratigraphic datum (LSD) and highest stratigraphic datum (HSD) of North American Plio–Pleistocene mammals determined from magnetostratigraphic studies (after Lindsay *et al.*, 1984). Letters in brackets refer to location of fossil finds: A = Arizona, T = Texas, M = Mexico, NM = New Mexico, C = California, I = Idaho.

early *Hipparion* fauna. Sen (1989) has suggested that *Hipparion* appears in the Mediterranean at approximately 11.5 Ma within Chron C5r. This age, which should correlate to the base of the Vallesian, is the same as the age given by Tedford *et al.* (1987) for the base of the Clarendonian. Whistler and Burbank (1992) have placed the base of the Clarendonian in C5Ar.3 at 12.8 Ma, however immigrant taxa were not used to define the boundary. If this placement of the boundary is accepted, then the base of the Vallesian is about 1 My younger than the base of the Clarendonian (Fig. 9.3).

The base of the Astaracian, the interval preceding Vallesian in the Europe land mammal zonation (Fig. 9.3), is placed by Krijgsman *et al.* (1994b) in C5ACn at about 13 Ma, much younger than previous estimates, and very different from the base of the Barstovian, which has been correlated to C5Br by MacFadden *et al.* (1990b). In North America, the boundary between the

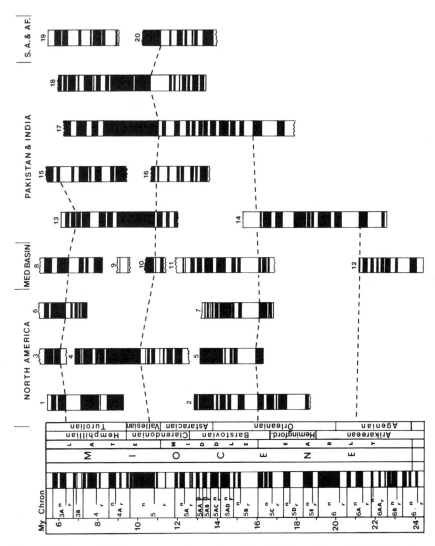

Figure 9.3 Terrestrial Miocene magnetic stratigraphies. Numbers refer to individual studies (see Table 9.4).

early and late Barstovian is placed at 14.5 Ma, coincident with severe climatic cooling in the middle Miocene and with the newly determined Orleanian/Astaracian boundary in Spain. Lindsay (1995) has recently placed the base of Barstovian land mammal age to the early part of chron C5Cn.3 (approximately 16.8 Ma) based on the appearance of the immigrant *Copemys*.

Table 9.4
Miocene Mammals

	Rock unit	Lo Age Hi	Region	λ	Φ	NSE	NSI	NSA	M	D	RM	DM	AD	A	NMZ	NCh	%R	RT	F.C.T.	Q	References
1	Chamita Fm.	Gilbert-C4A	USA, NM	+36.01	−106.05	1	135	3	500	F	J	A	B	F-DI	11	3	36	R+	—	6	MacFadden (1977)
2	Barstow Fm.	C5AB-C5C	USA, Colo	+35	−117	7	272	3	965	A	I	T-A	Z-P	F-V	22	6	49	R+	F+	10	MacFadden et al. (1990b)
3	Big Sandy Fm.	Gil-C3C	USA, Ari	+34.5	−113.58	4	54	3	75	F	—	A	B	V	6	2	43	R+	—	6	MacFadden et al. (1979)
4	Ricardo Gp.	C4-C5AB	USA, Colo	+35.3	−118	1	83	3–4	1500	F	—	T	Z-B	V	26	11	39	R+	—	7	Loomis and Burbank (1988)
5	Hepburn's Mesa	C5AD-C5AD	USA, Mont.	+45.3	−111.8	2	62	3	100	F	—	T	Z-V	F-V	12	3	39	R+	—	7	Burbank and Barnosky (1990)
6	Verde Fm.	Gauss-C4	USA, Ari	+34.5	−111.75	9	164	3	300	K	V	A	B	V	21	5	48	R−	—	6	Bressler and Butler (1978)
7	Tesuque Fm.	C5A-C5C	USA, NM	+36	−106	7	230	3	650	F	I-H	A	B	F-D	26	6	56	R+	—	7	Barghorn (1981)
8	Cabriel Basin	Gil-C4	Spain	+39.5	−1.33	2	42	3	210	—	—	T	Z-P	V	10	3	50	—	—	6	Opdyke et al. (1989)
9	Kastellios Hill	C4A	Crete	+35.18	+25.1	1	77	3	90	—	I-X	T	Z-P	V	5	1	96	—	—	6	Sen et al. (1986)
10	Bou Hanifia	Mid Miocene	Algeria	+36	+1	1	37	—	100	K	—	T	P	V	8	2	70	—	—	5	Sen (1989)
11	Aragon	5A-5C	Spain	+41.21	−1.38	2	357	2	280	—	I-J	T	Z-P	D-I	40	6	50	—	—	7	Krijgsman et al. (1994b)
12	Molasse	C6A-C6C	France	+47.57	+5.53	2	69	3	275	F	—	T	Z-V	F-V	20	4	59	R+	—	6	Burbank et al. (1992a)
13	Salt Range	Gil-C5A	Pakistan	+32.75	+73	6	360	3	1700	F	—	T	B	F-V	127	15	50	R+	F+	8	Opdyke et al. (1982); Johnson et al. (1982)
14	Dera Gazni Khan	C5B-C6AA	Pakistan	+30.1	+70.75	3	127	3	700	—	O	T	Z-P	V	19	7	57	R−	—	6	Friedman et al. (1992)
15	Haritalyangar	C3A-C5N	India	+31.55	+76.6	1	77	3	950	—	—	T	Z-B	F-V	15	3	43	R+	—	5	G. J. Johnson et al. (1983)
16	Huch Nala	C5-C5AC	Pakistan	+32.65	+72.2	1	71	3	700	—	—	T	B	V	16	5	55	R+	F+	5	Johnson et al. (1988)
17	Chinji	C4-C5D	Pakistan	+32.67	+72.5	2	159	3	1900	F	—	T	Z-B	V	33	11	4	R+	F+	8	Johnson et al. (1985)
18	Khaur	C6-C15	Pakistan	+33.16	+72.52	11	272	3	2600	F	—	T	Z-B	F-V	89	10	44	—	C+ F+	7	Tauxe et al. (1980); Tauxe and Opdyke (1982)
19	Corral Quemado	Gil-C7	Argentina	−27	−67.9	1	99	3	2300	K	J-I	A	Z-B	F-V	20	4	53	R+	—	7	Butler et al. (1984)
20	Ngorora Fm.	C5-C5AA	Kenya	+.66	+36	2	91	3	800	K	—	T	Z-V	F-V	14	4	46	R+	—	6	Tauxe et al. (1985)
	Ngorora Fm.	C5-C5AA	Kenya	+.66	+35.9	2	104	3	350	A	—	T	Z-B	F-V	18	4	44	R+	—	6	Deino et al. (1990)
	Honda Gp.	C5A-C5AA	Bolivia	−21.95	−65.42	3	106	3	300	K	1	T-A	Z-P	F-V	26	2	54	R+	—	9	MacFadden et al. (1990a)
	Hyargus Nuur	Gil-C3A	Mongolia	+49.26	+93.08	2	218	3	80	—	—	T	B	V	12	3	53	—	—	3	Pevzner et al. (1983)
	Gemerek	C5B	Turkey	+39.13	+36.05	3	114	1	170	K	—	T	P	D-I	3	1	59	—	—	6	Langereis et al. (1989)

The magnetic stratigraphy of the Oligocene/Miocene boundary in Europe based on fossil mammals has been reported by Burbank *et al.* (1992a). The section studied covers 280 m in the lower freshwater molasse in the Haute-Savoie (France) and includes the last two micro-mammal zones of the Paleogene (MP29 and MP30) and the first mammal zone (MN1) of the Neogene in stratigraphic superposition. Unfortunately, a 50-m gap separates probable MP30 faunas from faunas assigned to MN1. The polarity zones are designated R1 to N8 and the gap falls between N4 and N2. The authors, however, choose to place the Oligocene/Miocene boundary at the base of N2, which they correlate to C6Cn.3n. The magnetostratigraphic pattern, however, is ambiguous (Fig. 9.3, column 12). Another possible correlation to the GPTS would be N3 and N4 to C6Bn; and N5, N6, and N7 to C6AA. This would place the MP30/MN1 boundary at the base of C6Cn.1n (Burbank *et al.*, 1992a) and would place the base of the Agenian at this position. Barbera *et al.* (1994) have placed the Oligocene/Miocene boundary at the base of C6Cn.2n in a section in the Ebro basin of northern Spain.

Unquestionably the best studied area of continental Neogene vertebrate-bearing sediments is the Siwaliks of Pakistan and India. These studies were begun in the mid-1970s and have continued to the present. The region is wonderfully suited for magnetic stratigraphy because of well-exposed long sections of very fossiliferous sediments that became magnetized early in their history. In this region it is possible to begin in Brunhes age sediments and systematically work back in time to the onset of sedimentation in the basin (Keller *et al.*, 1977; Opdyke *et al.*, 1979, 1982; Johnson *et al.*, 1982, 1985; Tauxe and Opdyke, 1982). In many respects the area has served as a laboratory for the application of magnetic stratigraphy to (1) vertebrate evolution and migration (Flynn *et al.*, 1984), (2) sedimentary processes (Raynolds and Johnson, 1985; Behrensmeyer and Tauxe, 1982; McRae, 1990b), (3) tectonics (Burbank and Raynolds, 1988), and (4) basin development (Cerveny *et al.*, 1988). Almost all of these applications of magnetic stratigraphy to terrestrial sediments were anticipated by N. M. Johnson *et al.* (1975). In the Siwaliks, the magnetic stratigraphy is now so well known from multiple sections that vertebrate paleontologists usually correlate new fossil finds directly to a nearby magnetic stratigraphy and hence to the GPTS.

The correlation in China of the mammal-bearing sequences to the GPTS is known in considerable detail from the Late Miocene to the Pleistocene (Tedford *et al.*, 1991; Shi, 1994; An *et al.*, 1987, Heller and Liu, 1982, 1984). The importance of the Chinese record since the early Mutuyama is enhanced by the fact that the Chinese loess preserves a climatic record that can be correlated to Plio–Pleistocene climatic cycles (Kukla *et al.*,

1988; Kukla and An, 1989). A mandible and a cranium of *Homo erectus* have been correlated to loess magnetic stratigraphy (An and Ho, 1989). Other data exist from Asiatic Russia but unfortunately they are not easily accessible to Western workers (see Pevzner *et al.*, 1983).

9.4 African and South American Neogene

Terrestrial magnetic stratigraphy in Africa stems from the great interest in dating the evolution of man. Long well-exposed fossiliferous sequences are available in central Africa, and magnetostratigraphic data exist for the Gauss to Recent interval in the Omo area (Hillhouse *et al.*, 1986), in the Shungura sequence (Brown *et al.*, 1978), and in Afar (Renné *et al.*, 1993). Other studies of importance are by Tauxe *et al.* (1985) and Deino *et al.* (1990) in sediments from the Baringo basin correlative to the C5AA to C5N interval. The magnetic stratigraphy of Late Oligocene fossiliferous sediments from the Fayum has been reported by Kappelman *et al.* (1992). This important study is the first from the terrestrial Paleogene of Africa.

Magnetostratigraphic studies on South American mammal-bearing sediments have been concentrated mostly in the Neogene or Upper Paleogene in a series of studies by Marshall *et al.* (1979, 1982, 1986), Butler *et al.* (1984), and MacFadden *et al.* (1983, 1985, 1990a, 1993). South America separated from Africa in the mid-Cretaceous, allowing the South American mammalian faunas to evolve in isolation. During the Neogene, however, North America and South America approached one another and eventually became joined through the isthmus of Panama, allowing the mammals to move from one continent to the other. The timing of the exchanges has been revealed by the magnetostratigraphic dating of sediments that contain the oldest record of the exotic mammal groups. In South America, the oldest of the immigrants from North America, the *Procyonidae* (raccoons), appear at 7 to 7.5 Ma in lower Chron 7 (Butler *et al.*, 1984). In this study, the boundary between Huayquerian and Montehermosan land mammal ages was placed at 6.4 Ma. The second wave of North American immigrants occurs in sediments of the Uquian land mammal age dated by Marshall *et al.* (1982) from late Gauss to above the Olduvai. In a study of the type Chapadmalal, Orguira (1990) correlated this important migration to the late Gilbert; however, the reverse, normal, reverse (R-N-R) magnetostratigraphic sequence is more likely to correlate to the Kaena and Mammoth subchrons, which would place this migration event in the middle Gauss. In North America, the first appearance of South American forms occurs at the beginning of the Hemphillian; however, the most important migration event took place at the Gauss/Matuyama boundary (Galusha *et al.*, 1984).

It would seem that there is an assymmetry between migrations across the isthmus of Panama, with the most important immigrants such as *Equus* arriving in South America during the middle Gauss, while the South American migrants arrive in North America near the Gauss/Matuyama boundary. These studies illustrate the power of the magnetostratigraphic method in resolving problems of intercontinental faunal exchange.

An earlier intrusion of fauna into South America has also been recognized and recorded in sediments of the Deseadan land mammal age. This exotic fauna is of considerable interest since it contains caviomorph rodents and the first South American monkeys. This is a most curious interchange since it is believed that monkeys originated in Africa, and it is not clear how they managed to make the journey to South America. Early Oligocene dates on the Deseadan from South America compounded the problem, because the descendant species in South America appeared to be older than proposed ancestors in Africa. In a study of the Salla beds of Bolivia, MacFadden *et al.* (1985) were able to redate the Deseadan land mammal age and demonstrate that it was originally dated 10 My too old. They were able to place the first appearance of monkeys at 27 Ma and to demonstrate that rodents appear in South America prior to 28.5 Ma. This study addresses the ancestor problem but does not address the problem of how the animals managed to travel from Africa to South America.

9.5 North American and Eurasian Paleogene

Figure 9.4 shows the magnetostratigraphic studies of the Paleogene and Late Cretaceous of North America and Europe ranked in time and identified with author and year of publication (Table 9.5). The boundaries of the land mammal ages in the Paleocene have been updated by studies of Butler and Lindsay (1985) and Butler *et al.* (1987), and their placement of Paleocene mammalian boundaries is used here. In cases where individual sequences have been restudied and reanalyzed, the most up-to-date reference is given (Table 9.5).

Mammalian chronology for the Late Eocene and Oligocene has been reviewed and revised by Swisher and Prothero (1990) and Prothero and Swisher (1992). Our correlation of Duchesnian to Arikareen North American land mammal ages follows Prothero and Swisher (1992) except for some minor revisions. The boundary between the Duchesnian and Chadronian is placed within C17n (Fig. 9.4). This placement satisfies the magnetic stratigraphy and new radiometric results from the Vieja section of west Texas (Testamarte and Gose, 1979). Unfortunately, the magnetic stratigraphy for this section is poorly defined. Prothero and Swisher (1992) placed

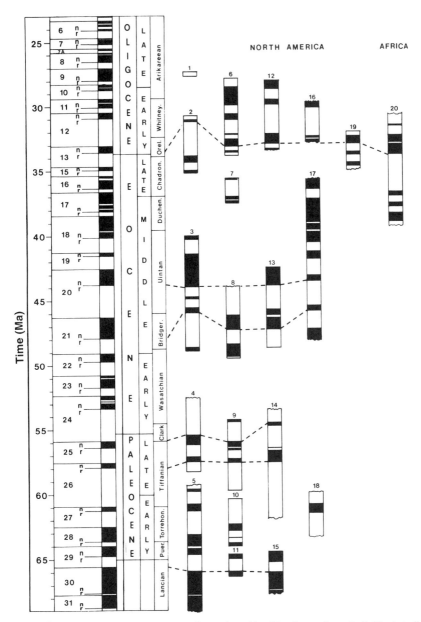

Figure 9.4 Terrestrial Paleogene magnetic stratigraphies. Numbers refer to individual studies (see Table 9.5).

Table 9.5
Paleogene Mammals

	Rock unit	Lo Age Hi	Region	λ	Φ	NSE	NSI	NSA	M	D	RM	DM	AD	A	NMZ	NCh	%R	RT	FCT	Q	References
1	Otay Fm.	Arikreean	USA (CA)	+32.75	−117.25	2	19	3	83	A	–	T	Z-V	F-V	1	1	100	–	–	6	Prothero (1991)
2	Flagstaff Rim	C12-C15	USA (Wy)	+42.6	−106.69	1	114	3	220	A	–	T	–	I	8	3	78	–	–	4	Prothero and Swisher (1992)
3	Washakie Basin	C19-C22	USA (Wy)	+41.5	−109	3	58		920	K	J-X	T	Z-V	F-V	9	3	56	R+	–	8	Flynn (1986)
4	Clarks Fork Basin	C24-C26	USA (Wy)	+44.8	−109	3	273	3	2012	–	–	A	Z-V	V		3	68	–	–	5	Butler et al. (1981a)
5	San Juan Basin	C26-C31	USA (NM)	+36.4	−108	4	44	3	750	F	I-V J-X	T	Z-V	V	7	3	47	R+	–	7	Butler and Lindsay (1985)
6	Toadstool Park	C11-C13	USA (Neb.)	+42.9	−103.63	1	123	3	225	A	K	A-T	Z-V	V	11	4	51	–	–	5	Prothero et al. (1983)
7	Vieja Gp.	C16-C17	USA (Tex.)	+30.28	−104.63	1	–		–	–	–	–	–	–		–	–	–	–	1	Testamarte and Gosé (1979)
8	East Fork Basin	C20-C22	USA (Wy.)	+43.65	−109.7	2	61	3	610	K	J-X	T	Z-V	F-V	7	2	74	R+	–	8	Flynn (1986)
9	Black Peak Fm.	C24-C26	USA (Tex.)	+29.40	−103.1	2	53	3	170	–	–	T	Z-B	F-V	8	3	79	R+	–	6	Rapp et al. (1983)
10	Dragon Canyon	C27-C29	USA (Utah)	+39.19	−111.3	2	35	3	110	A	O	A	Z-B	V	9	3	65	–	–	6	Tomida and Butler (1980)
11	Hell Creek	C29-C30	USA (Mont.)	+47.6	−107	4	39	3	100	–	O	A	Z-B	V	8	2	59	–	–	6	Archibald et al. (1982)
12	Big Badlands	C11-C13	USA (SD)	+43.71	−102.48	6	18	1	250	K	–	A-T	V	–	27	3	66	–	–	4	Prothero et al. (1983)
13	San Diego Area	C20-C21	USA (CA)	+32.75	−117.25	9	49	3	500	–	J-X	T	Z-V	F-V	6	2	58	R+	–	7	Flynn (1986)
14	Eureka Sound	C24-C26	Canada	+78.82	−82.45	3	101	3	2000	–	–	AT	Z-B	F-I	7	2	76	R+	F+	7	Tauxe and Clark (1987)
15	Red Deer Valley	C29-C31	Canada	+50.66	−110.72	1	170	3-4	135	K	–	A	B	F-V	19	3	32	R–	–	4	Lerbekmo et al. (1979)
16	Scott's Bluff	C12-C13	USA (Neb.)	+41.9	−103.82	1	81	3	135	A	–	T	–	V	11	3	65	–	–	4	Prothero and Swisher (1992)
17	Devil's Graveyard Fm.	C18-C22	USA (Tex.)	+29.47	−103.73	12	1-1.5 m	2-4	152	A	I-J	A-T	Z-V	I	16	5	49	–	–	7	Walton (1992)
18	Clarks For Basin	C26-C27	USA (Mont.)	+35.12	−108.95	1	20	3	150	A	J-I	A	Z-V	F-V	3	2	72	–	–	6	Butler et al. (1987)
19	Dilts Ranch	C12-C16	USA (Wy.)	+43.00	−105.5	3	106	3	125	–	–	T	Z-B	V	7	3	62	–	–	5	Prothero (1985)
20	Fayum	C12-C16	Egypt	+29.62	+30.55	2	53	3	340	A	I	T	Z-P	F-V	18	4	59	R+	–	9	Kappelman et al. (1992)
	Ebro Basin	C6-C8	Spain	+41.32	+.21	1	108	3	300	–	–	T	Z-P	F-V	11	4	43	R–	–	5	Barbera et al. (1994)
	Willwood Fm.	C23-C24	USA (Wy.)	+44.66	−108.42	1	37	3	1480	A	J-TdI	A	Z-P	F,D-I	8	3	62	R+	–	8	Clyde et al. (1994)
	Willwood Fm.	C23-C24	USA (Wy.)	+44.2	−108.50	1	90	3	775	A	–	T-A	Z-P	F-V	14	2	46	R+	–	7	Tauxe et al. (1994)

the boundary within C16n on the basis of the radiometric dates and an early version of the Cande and Kent (1992a) time scale. The Whitneyan/ Arikareen boundary is placed at the C11/C10 boundary based on magnetic stratigraphy from the Toadstool Park–Roundtop section in Nebraska (Prothero and Swisher, 1992).

The correlation of the magnetic stratigraphies to the GPTS (Fig. 9.4, Table 9.5) is supported in most cases by radiometric ages. An important magnetic stratigraphy from the Devil's Graveyard Fm of west Texas was correlated to C20 (Walton, 1992). A more reasonable correlation, supported by the Bridgerian to Duchesnian fauna, would be to correlate the section to the C21n to C17 interval.

One of the problems that can be addressed by magnetic stratigraphy is the heterochroneity (time transgression) of faunal datums. The land mammal ages in North America are based on stage of evolution and appearance of immigrant taxa. Magnetic stratigraphy can test this hypothesis because the evolution and dispersal of mammals and reversals of the geomagnetic field are independent. Two studies in particular have raised the possibility that faunal datums in North America of Paleocene age may be time transgressive. The first of these studies is an extensive investigation of Cretaceous and Paleocene sediments in the San Juan Basin of New Mexico (Butler *et al.*, 1977). This region has a record of the turnover from the dinosaur fauna of the Cretaceous to mammal faunas of Paleogene age. The original interpretation correlated the Torrejonian faunas of the San Juan Basin to C26. The same authors (Butler *et al.*, 1981a) correlated the Tiffanian of the Clarks Fork basin, Wyoming, to C26. This led to the anomalous situation in which two distinctly different mammal faunas, thought to be sequential in time, appeared to be coeval in the western United States. Further study by Butler and Lindsay (1985) resolved the problem by showing that one of the normal polarity zones originally proposed in the San Juan Basin was caused by a viscous magnetic overprint. When this spurious polarity zone was eliminated, agreement was achieved between the two studies. The corrected stratigraphy leads to the placement of the Cretaceous– Tertiary (K/T) boundary in reversely magnetized sediments of C29r. All other K/T boundaries occur in reverse polarity sediments considered to correlate to C29r (Archibald *et al.*, 1982; Lerbekmo *et al.* (1979), except for a core in New Mexico's Raton Basin which was studied by Payne *et al.* (1983) and was said to be normally magnetized. Tauxe and Butler (1987) cite a contrary opinion (as a personal communication by E. M. Shoemaker) that the core is entirely reverse polarity. One hopes these data are published so that the results can be properly evaluated. Within the resolving power of the paleomagnetic method the K/T boundary appears to be synchronous, between marine and nonmarine sediments.

A second study which initially seemed to indicate faunal diachroneity was a study of fauna and flora in Arctic sediments by Hickey *et al.* (1983). These authors originally claimed that plants and animals appeared in Arctic sediments several million years prior to their appearance in temperate North America. The Eureka Sound Formation, which yielded these results, has been restudied paleomagnetically by Tauxe and Clark (1987), who showed conclusively that the original study was flawed because of the incomplete removal of magnetic overprints and came to the conclusion that no diachroneity is apparent. The results of these studies confirm that land mammal ages are, indeed, useful chronostratigraphic units (Flynn *et al.*, 1984; Clyde *et al.*, 1994; Tauxe *et al.*, 1994).

9.6 Mammal Dispersal in the Northern Hemisphere

The fact that mammals disperse from continent to continent is well known, and modern continental faunas are greatly influenced by these processes. The interchange of faunas between the continents has been of great utility in the construction of mammalian biochronologies within the different continents, particularly in North America and Europe. The ability of animals to move from one continent to another is affected by the availability of corridors that allow animals to pass. Plate tectonics has had a first-order effect on the interplay of faunas, since the different plates are in constant motion, changing geographic relationships between continents (McKenna, 1975). The Age of Mammals began at the inception of the Cenozoic, at 65 Ma. Plate tectonic history since that time has been largely one of continental dispersal, which had led to the partial isolation of large continental areas such as Australia and South America. The other continents have a history of intermittent contact and subsequent isolation. Tectonic activity resulting from collision will tend to create links between continental blocks. For example, collision of India with Eurasia during the Cenozoic and the continuing collision of the African plate with Europe provided opportunities for faunal interchange (McKenna, 1975).

Faunal dispersal is affected not only by changing paleogeography but also by paleoclimatology, which can erect or destroy ecological boundaries, either expediting or impeding the migration of animals confined to specific habitats. In some cases faunas that appear in midcontinental North America and Europe may have been displaced climatically from the northern part of the continents, as they sometimes have no close relatives in either North America or Eurasia. Paleoclimatology can also create or destroy dispersal routes by glacioeustatic changes in sea level. These changes can occur rapidly, as has been demonstrated from the marine $\delta^{18}O$ record of Pleistocene ice volume change, which implies rapid sea level falls of up to 120 m (Shackleton and Opdyke, 1973). This climatic record can be correlated to the terrestrial faunal record by magnetic stratigraphy.

Barry *et al.* (1985) claimed that evolutionary events and dispersal first appearances in the Siwaliks could be directly related to the $\delta^{18}O$ record of the middle Miocene. It is interesting to speculate whether this correlation can be established outside the Indian subcontinent. The change in $\delta^{18}O$ for benthic foraminifera from the Atlantic Ocean (Fig. 9.5) (Miller *et al.*, 1991a) shows the overall climatic trend for late Eocene through Miocene time. Mammal migrations have occurred between North America and Eurasia throughout the Cenozoic (Fig. 9.5), and indeed the geochronologic units (land mammal ages) were established using immigrant taxa as part of their definitions (Wood *et al.*, 1941; Woodburne, 1987). It might be expected that the beginning of each new land mammal age would correlate with sea level drops because lowering of sea level would facilitate dispersal. It is unlikely, however, that all dispersals correlate with sea level change, since, in some cases, the corridors might stand so high above sea level as to be unaffected by that change. This is probably the case in the early Eocene when North America and Europe were connected (in the region of the present Norwegian–Greenland Sea), allowing unhampered migration between the two continents.

Opdyke (1990) argued that intercontinental mammal migration has resulted from sea level drops following the development of continental ice sheets. Since Neogene land mammal ages are often defined on the basis of immigrant taxa, it was suggested that Neogene land mammal ages in Europe and North America should be closely correlated if both were defined on the basis of immigrant taxa. Miller *et al.* (1991a) have presented a new and updated $\delta^{18}O$ record for the Oligocene and Miocene and argue that continental glaciation began in Antarctica in the latest Eocene (~35 Ma). These authors have formally defined nine oxygen isotope zones in the Oligocene and Miocene which they have designated Oi1 and Oi2 and Mi1 through Mi7, ranging in age from the early Oligocene through the early upper Miocene (Fig. 9.5). Miller *et al.* (1991a) argued that the pre-Pleistocene glacial stages may have been associated with glacioeustatic falls in sea level of 55 to 180 m; however, they argue that the actual maximum drop in sea level in unlikely to have been as large as Pleistocene values of 120m.

The fall in sea level postulated above, will, of course, cause the development of unconformities on passive continental margins (Haq *et al.*, 1987; Christie-Blick *et al.*, 1990). Pitman (1978) has argued that the rate of sea level change also controls the position of passive margin unconformities.

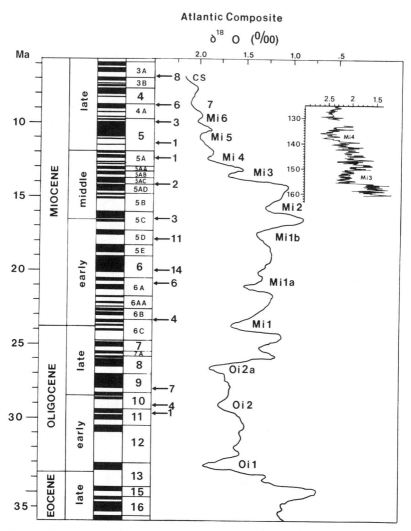

Figure 9.5 Correlation of mammal migrations to the GPTS (Cande and Kent, 1995) and oxygen isotope events (Miller *et al.*, 1991a). Positions in time of major mammal migrations are indicated by arrows with the number of immigrant taxa.

Christie-Blick *et al.* (1990) have suggested that the times of most rapid drop in sea level should coincide with the formation of unconformities. These passive margin unconformities may be detected using seismic reflection data (Vail *et al.*, 1977; Haq *et al.*, 1987). Evidence for sea level falls, combined with physical stratigraphy, has been formally organized into sequence stratigraphy. If there is a causal relationship between ice volume changes and sequence stratigraphy, then a correlation should exist between the isotopic record and sequence stratigraphy. Miller *et al.* (1991a) maintain that such a correlation does, in fact, exist in the Oligocene and Miocene. However, problems in correlating sequence stratigraphy to pelagic isotopic records will remain until the magnetic stratigraphy of continental margin sediments is rigorously documented.

Webb and Opdyke (1995) have reexamined the question of migratory events in North America and divide the migratory first appearances into first-order (>8 taxa) and second-order (>5 taxa) migratory events. Cenozoic faunal migration to and from North America and Eurasia is essentially divided into two segments, the first taking place in the Paleocene and Eocene and the second being Miocene and younger. McKenna (1975) and Opdyke (1990) have pointed out that tectonics and paleogeography are the primary controls on mammal migrations whereas climate-driven glacial eustasy is a secondary effect made possible by changes in paleogeography. The large first-order migratory patterns are undoubtedly caused by changes in paleogeography. It is probable that throughout the Paleogene, mammals could pass between North America and Europe over the Thulean land bridge across the Norwegian–Greenland Sea. The route probably remained open until the rifting apart of Svalbard and Greenland, probably at the end of the Eocene (Rowley and Lottes, 1988). The major drop in the temperature of bottom water which occurs in the world ocean at the beginning of the Oligocene (Oi1, Fig. 9.5) is not accompanied by any important mammal migration. The reason is undoubtedly that migration across the Thulean land bridge was no longer possible.

The other possible route from Eurasia to North America, the Bering land bridge, presented no apparent barrier to mammal migration during the Paleogene. However, at the end of the Mesozoic the Bering region was at paleolatitudes above 80°N, almost on the North Pole. The Bering land bridge region has been slowly moving into lower latitudes during the Cenozoic. It seems possible that even though the region was moving toward lower latitudes (70°N), the climate was continuing to deteriorate at the end of Eocene and early Oligocene time as indicated by ice-rafted debris around the Antarctic continent. It seems reasonable that the barrier that impeded mammal migration across the Bering land bridge was climatic. By Miocene time, the Bering region probably reached sufficiently low latitudes to allow mammals to cross at times of sea level drop caused by glacial eustacy.

The late Eocene "Duchesnian" interchange at about 40 Ma correlates with an unconformity on the continental margin (Miller *et al.*, 1987) and with a major offlap event in the Vail curve (TA 4.1). In addition, this event has also been correlated with an increase in isotopic values in both planktic and benthic foraminifera (Miller *et al.*, 1987) and with an estimated sea level drop of about 80 m (Prentice and Matthews, 1991). The next major migration event occurs in the middle Oligocene in C9r (R. H. Tedford, personal communication, 1994) where seven new immigrant species first appear coincident with, and following, isotopic event Oi2 (Fig. 9.5). The third major migratory event takes place in the late Arikareen and early Hemingfordian from 21 to 18 Ma. This is the most important period of immigration in the Cenozoic. These migratory events correlate to isotope events Mi1a and Mi1b (Miller *et al.*, 1991a). At least four positive oxygen isotope changes in the 18–21 Ma interval suggest that the lows in sea level at this time were comparable to those in the late Pleistocene (Prentice and Matthews, 1991). Two sequence boundaries occur in this interval (Haq *et al.*, 1987), a type 1 sequence boundary at 21 Ma and type 2 sequence boundary at 17.5 Ma, and it is probable that this migratory event is coincident with eustatic sea level fall.

The faunal migrations within the late Miocene occur within the Hemphillian LMA. The Hemphillian is characterized by three second-order migratory events at ~9.7 to ~6 Ma, involving 6 or 7 taxa (Tedford *et al.*, 1987). The first of these occurs near the beginning of the Hemphillian in C4Ar.2 at about 9.5 Ma, correlating with $\delta^{18}O$ isotope event Mi7 (Wright and Miller, 1992) and with a type 2 sequence boundary in C4A (Haq *et al.*, 1987). The second of the migratory events occurs at 7 Ma, roughly coeval with the Late Miocene carbon shift and a well-defined $\delta^{18}O$ event (Hodell *et al.*, 1994).

The final migratory event within the Hemphillian is believed to have occurred at about 6 Ma, which is close in time to the position of Gilbert oxygen isotope stages, TG20 and TG22, astronomically dated at 5.7 and 5.8 Ma (Fig. 6.7) (Shackleton *et al.*, 1995b). Lindsay *et al.* (1984) have placed the Hemphillian/Blancan boundary, which coincides with a first-order migratory event, between the Sidufjall and Nunivak subchrons. Tedford *et al.* (1987) placed the event at about 4.9 Ma, coincident with isotope stages Si_4 and Si_6, within the Sidufjall subchron (Fig. 6.7).

The final episode of faunal migration is well known to be at, or close to, the Gauss/Matuyama boundary at 2.5 Ma, and is clearly associated with the onset of the northern hemisphere glaciation (Fig. 6.7) (Shackleton *et al.*, 1984). At this time the horse, *Equus*, migrated to Eurasia and spread rapidly. Lindsay *et al.* (1980) proposed this correlation from the Siwaliks of Pakistan (Opdyke *et al.*, 1979), and further research in Eurasia has tended to support the appearance of *Equus* at or near the Gauss-Matuyama boundary. *Elephas* appears in sediments of middle Gauss age in the Siwaliks,

Table 9.6
Terrestrial Tectonics

	Rock unit	Lo Age Hi	Region	λ	Φ	NSE	NSI	NSA	M	D	RM	DM	AD	A	NMZ	NCh	%R	RT	F.C.T.	Q	References
1	Surai Khola	Mio-Pleis	Nepal	+28	−82.51	1	436	1	5000	−	I	T	Z-P	F-D-I	47	17	49	−	−	6	Appel et al. (1991)
2	Quebreda del Cura	Mio	Argentina	−30.25	68.5	2	112	3	1150	−	I	T	B	F-V	26	3	40	R+	−	5	Beer (1990)
3	Karewa Beds	Pleis-Plio	India	+34	74.43	9	200	3	800	F	−	A	Z-B	V	27	3	65	R+	F+	7	Burbank and Johnson (1983)
4	Pyrenean foreland	C18-C24	Spain	+42.25	−1.87	4	300	3	8000	−	−	T	Z-V	F-V	36	9	49	R±	F−	6	Burbank et al. (1992a)
5	Pschauar Beds	Pleis-Plio	Pakistan	+34	72	7	50	3	310	F	−	A-T	Z-B	V	15	3	69	R+	−	7	Burbank and Taberkheli (1985)
6	Oliana-Peramola	C13-C18	Spain	+42.07	−1.32	3	100	3	1300	−	K	T	Z-V	F-V	18	5	63	R±	F+	7	Burbank et al. (1992c)
7	Ricardo Gp.	Miocene	USA (CA)	+35.3	−118	1	84	3	1500	F	I	B	B	V	26	6	30	−	−	4	Burbank and Whistler (1987)
8	Mecca Hills	Pleis-Plio	USA (CA)	+33.61	−115.9	2	52	1	500	−	I	T-A	Z-P	F-D	5	1	89	−	F+	7	Chang et al. (1987)
9	Morales Fm.	Pleis-Plio	USA (CA)	+35.00	−119.5	6	49	3	2000	−	−	T-A	Z-P	F	13	2	63	R+	−	7	Ellis et al. (1993)
10	Rio Azul	Mio	Argentina	−30.69	68.75	9	114	3	2000	K	−	A-T	Z-B	F-V	29	3	58	R+	−	7	Jordan et al. (1990)
11	Siwalik	Mio	Pakistan	+32.71	72.46	3	216	3	200	F	−	T	B	F	8	2	33	R+	−	6	McRae (1990a)
12	Salla Beds	Mio-Oligo	Bolivia	−17.17	67.65	3	65	3	70	K	−	T	B	F-V	6	2	45	R+	−	6	McRae (1990b)
13	Manix Basin	Pleis-Plio	USA (CA)	+35	−116.5	6	140	1	80	K	J-A	FA	Z-P	F-V	5	1	76	R+	−	7	Pluhar et al. (1991)
14	Confidence Hills	Pleis-Plio	USA (CA)	+35.73	−116.56	1	156	1	190	K	−	T-A	Z-P	F-V	5	1	66	R+	−	8	Pluhar et al. (1992)
15	Axhandle Basin	Eo-K C21-C33	USA Utah	+39.42	−111.67	3	275	3	1050	−	I	F-A	Z-V	V	24	13	63	R+	−	8	Talling et al. (1994)
16	Upper Siwaliks	Pleis-Mio	Pakistan	+33	73.5	15	332	3	1500	F	−	T	Z-B	V	111	5	56	R+	F+	7	Raynolds and Johnson (1985)

as well as in Eurasia and China (Opdyke *et al.*, 1979; Tedford *et al.*, 1991). Faunal exchange at this time is very important between South and North America, and the culmination of movement of fauna from South America to North America is correlated to the Gauss-Matuyama boundary in North America (Galusha *et al.*, 1984).

Magnetic stratigraphy has proved to be of great utility in dating and correlating the sedimentary record of tectonic events (Table 9.6). One of the best examples is the study of Burbank and Reynolds (1984, 1988) in which they were able to date the tectonic development of the Himalayan

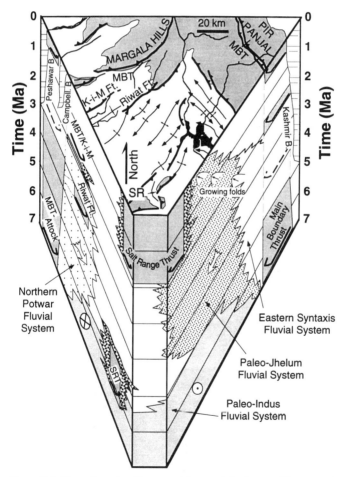

Figure 9.6 Tectonic and sedimentary history of the Potwar Plateau region (Pakistan) (after Burbank and Reynolds, 1988).

front using the magnetic stratigraphy of the sediments of the Potwar basin (Fig. 9.6). This type of study has expanded to other parts of the world such as the Pyrenees (Burbank *et al.*, 1992b,c) and Argentina (Jordan *et al.*, 1990). Magnetic stratigraphy has also been very important in studies of basin formation and subsidence (Burbank, 1983; Burbank and Johnson, 1983) and for dating of uplift episodes in mountain ranges (Cerveny *et al.*, 1988). Magnetic stratigraphy has also enabled scientists to study the sedimentation process itself and to address problems of stratigraphic completeness (Johnson *et al.*, 1988; McRae, 1990a,b; Badgley and Tauxe, 1990; Friend *et al.*, 1989).

10

Jurassic–Early Cretaceous GPTS

10.1 Oceanic Magnetic Anomaly Record

When the Heirtzler *et al.* (1968) geomagnetic polarity time scale (GPTS) was constructed, it was noted that magnetic anomalies older than anomaly 32 have lower amplitude and are poorly lineated, in both the Atlantic and Pacific Oceans. The lack of lineated magnetic anomalies in this so-called Cretaceous quiet zone implied either that there were no reversals of the Earth's magnetic field for a considerable length of time or that the magnetic properties of the sea floor were such that field reversals were not recorded. The weight of evidence which became available from magnetostratigraphic studies of Cretaceous rocks on land (Ch. 11) demonstrated that the quiet zone in the oceans corresponds to a time of few or no reversals of the geomagnetic field. In the Atlantic, a Mesozoic sequence (M-sequence) of correlatable anomalies prior to this "quiet zone" was identified and named the Keathly sequence by Vogt *et al.* (1971) (Fig. 10.1). Larson and Chase (1972) succeeded in correlating the Japanese, Hawaiian, and Pheonix M-sequence lineations, and Larson and Pitman (1972) were able to correlate these with the Keathley sequence in the Atlantic. The key to this correlation was the recognition of the fact that the Pacific anomalies had been formed in the southern hemisphere and transported into the northern hemisphere by plate motion. These correlations led to the extension of the GPTS into the Middle Jurassic.

The anomalies in the mid-Mesozoic M-sequence were initially numbered from M1 to M25. Unlike the Late Cretaceous–Cenozoic part of the GPTS, anomaly numbers were mainly assigned to reverse polarity anomalies (Fig. 10.2); however, the labels M2 and M4 were assigned to normally

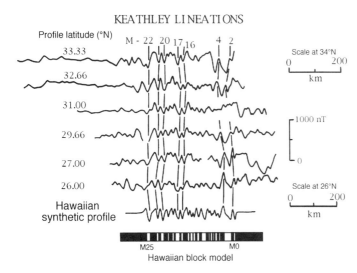

Figure 10.1 The Atlantic Keathley lineations correlated to the Pacific (Hawaiian) block model (after Vogt *et al.*, 1971; Larson and Pitman, 1972).

magnetized anomalies, resulting in a rather confusing situation. The magnetic anomaly between M10 and M11 is labeled M10N. This anomaly was originally omitted from M-sequence anomalies (Larson and Pitman, 1972) and was included by Larson and Hilde (1975) (Fig. 10.2) and designated M10N after Fred Naugler, co-chief scientist of the NOAA Western Pacific Geotraverse Project. Jurassic anomalies older than M25 were subsequently identified and the anomaly sequence was extended to M29 (Cande *et al.*, 1978a). Recent accumulation of oceanic magnetic anomaly data in the M0–M29 interval from the Pacific has led to the development of new block models for the Japanese, Hawaiian, and Phoenix lineations (Fig. 10.3) (Nakanishi *et al.*, 1989, 1992; Channell *et al.*, 1995a).

Difficulty in correlation and low amplitude of oceanic magnetic anomalies older than M29 led to the concept of a "Jurassic quiet zone"; however, aeromagnetic profiles from the western Pacific have succeeded in identifying lineated anomalies older than anomaly M29 and, as a result, the GPTS has been extended to anomaly M38 (Handschumaker *et al.*, 1988). The magnetic anomaly profiles from the western Pacific imply that the so-called Jurassic quiet zone is an interval of high reversal frequency. These magnetic anomalies are, however, of much lower amplitude than younger magnetic anomalies, and the reason for this remains unclear.

The M-sequence marine magnetic anomalies recorded by Larson and Hilde (1975) (Fig. 10.2) in the Hawaiian lineations have provided the means

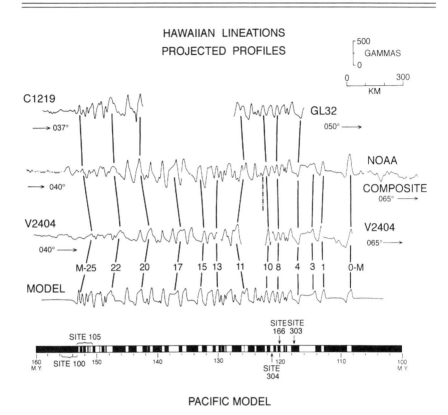

Figure 10.2 Magnetic anomaly profiles used to construct the Larson and Hilde (1975) Hawaiian block model (after Larson and Hilde, 1975).

of interpolation between age estimates for most Oxfordian to Aptian time scales. The time scales of Kent and Gradstein (1985) (KG85) and Harland *et al.* (1990) (GTS89) imply constant spreading rate in the M0–M25 Hawaiian oceanic anomaly block model of Larson and Hilde (1975) (LH75). The KG85 time scale uses the constant spreading rate assumption to interpolate between 119 Ma for the Barremian/Aptian boundary (base CM0) and 156 Ma for the Oxfordian–Kimmeridgian boundary (CM25n). The GTS89 time scale is based on chronogram ages for the Late Jurassic–Early Cretaceous stage boundaries. These ages were used to make a linear recalibration of the KG85 time scale, thereby inheriting the constant spreading rate assumption. The chronograms of GTS89 for this interval are poorly constrained and therefore do not validate the constant spreading rate assumption.

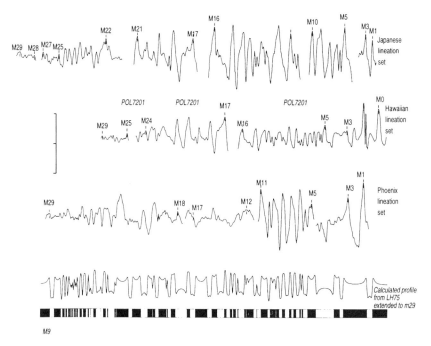

Figure 10.3 Profiles used to construct the Pacific (Japanese, Hawaiian, Phoenix) block models in Channell *et al.* (1995a). The calculated profile is the stimulation from the Larson and Hilde (1975) block model extended to M29.

10.2 Numerical Age Control

Since the KG85 and GTS89 time scales were published several numerical estimates for the age of M-sequence chrons have become available. Mahoney *et al.* (1993) obtained a $^{40}Ar/^{39}Ar$ age of 122.3 ± 1 Ma for the basaltic basement of ODP Site 807 on the Ontong Java Plateau. Tarduno *et al.* (1991) have argued that Ontong Java volcanism is constrained to a short interval in the Early Aptian; therefore this age estimate should be applicable to the Early Aptian. This interpretation was based on the observation that sediments overlying basaltic basement, and volcanoclastic sediments at distal sites, belonged to the *Globigerinelloides blowi* planktonic foraminiferal zone, which was considered to be confined to the Early Aptian (Sliter, 1992). Lower Cretaceous planktonic foraminiferal biostratigraphy in sections dated with ammonites and magnetic stratigraphy (Coccioni *et al.,* 1992) has, however, indicated that the *G. blowi* zone extends into the

Barremian. The *G. blowi* biozone does not, therefore, restrict Ontong Java volcanism to the Early Aptian.

At Site 878 on MIT Guyot, the oldest dated sediments (at 25 mm above basement) are from the lower part of the *C. litterarius* nannofossil zone, based on the first occurrence (FO) of *R. irregularis* and their location below the "nannoconid crisis." Pringle *et al.* (1993) obtained a $^{40}Ar/^{39}Ar$ age of 123.5 ± 0.5 Ma (converted for standard Mmhb-1 at 520.4 Ma) from these basalts. The remanent magnetization in the basaltic basement indicates a reverse polarity zone overlying a normal polarity zone. The reverse polarity zone is probably correlative to CM1.

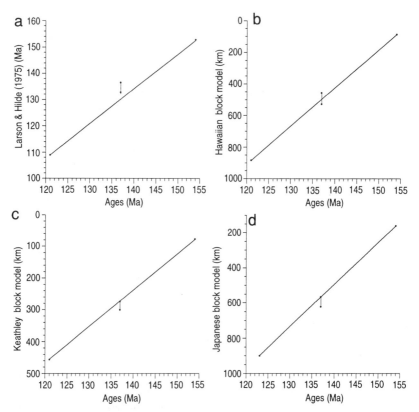

Figure 10.4 The selected radiometric age determinations for M0 (121 Ma), M16-M15 (137 Ma), and M25 (154 Ma) plotted against the (a) Larson and Hilde (1975) block model, (b) Hawaiian block model, (c) Keathley block model, and (d) Japanese block model. M0 is not present in the Japanese lineations and for (d) the young end of M1 is assigned an age of 123.2 Ma (after Channell *et al.*, 1995a).

Coleman and Bralower (1993) obtained a U-Pb zircon age of 122 ± 0.3 Ma from a bentonite in the Great Valley Group (northern California). This level has been correlated to the *C. litterarius* zone based on the FO of *C. litterarius*, however *R. irregularis* (a more useful marker for the Barremian–Aptian boundary, see Fig. 11.6) has not been found in the section (T. J. Bralower, personal communication, 1993). *C. litterarius* has been reported from several Barremian sections and the 122 ± 0.3 Ma bentonite layer from the Great Valley Group may be Barremian.

The $^{40}Ar/^{39}Ar$ age of 124 ± 1 Ma obtained from reversely magnetized granitic plutons from Quebec (Foland *et al.*, 1986) may be correlative to CM3, or to the shorter duration CM0 or CM1. Ten major igneous complexes with $^{40}Ar/^{39}Ar$ ages tightly bunched around 124 Ma record only reverse polarity (Foster and Symons, 1979), and it seems likely that the reverse polarity chron is CM3, which has more than three times the duration of

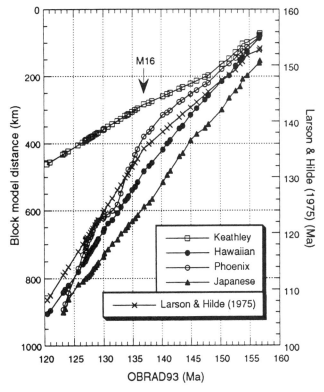

Figure 10.5 OBRAD93 time scale (Obradovich, 1993) compared to oceanic magnetic anomaly block models.

any reverse polarity chron in the younger part of the M-sequence. The duration of the interval between CM0 and CM3 is about 3.5 My (Herbert, 1992), leading to an age for the base of CM0 of about 120.5 Ma.

A U-Pb zircon age estimate of 137.1 (+1.6/−0.6) Ma from the Great Valley Sequence of northern California has been correlated to polarity chron CM16 or CM16n (Bralower *et al.*, 1990). This is an indirect correlation to the polarity time scale using nannofossil stratigraphy. The interval containing the two dated tuff layers was attributed to the Upper Berriasian *Assipetra infracretacea* subzone based on three nannofossil events: (1) the first occurrence of *Cretarhabdus angustiforatus,* which was used to define the base of the *A. infracretacea* subzone, (2) the occurrence of *Rhagodiscus nebulosus,* and (3) the absence of *Percivalia fenestrata*. Close inspection of the ranges of the three marker species (Fig. 11.6) suggests that the correlation of the two

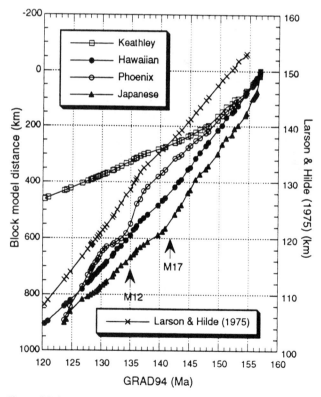

Figure 10.6 GRAD94 time scale (Gradstein *et al.,* 1994) compared to oceanic magnetic anomaly block models.

volcanic layers to CM16 or CM16n is too restrictive and that the CM16-CM15 interval may be more appropriate (see Channell *et al.*, 1995a).

The age of 154 Ma for the young end of CM25 is based on the correlation of this polarity chron boundary to the Kimmeridgian–Oxfordian boundary (Ogg *et al.*, 1984) and to age estimates for this stage boundary of about 154 Ma from California (Schweickert *et al.*, 1984) and Oregon (Pessagno and Blome, 1990). In the California Sierra Nevada, the ammonite-bearing Mariposa Formation is a synorogenic flysch supposed to contain the Oxfordian–Kimmeridgian boundary (Imlay, 1961) and is affected by Nevadan orogeny. K-Ar hornblende ages and U-Pb zircon ages on dykes and plutons appear to constrain the age of the Mariposa Formation and the Nevadan orogeny to the 152–159 Ma interval (see Schweickert *et al.*, 1984); however, many of the age determinations are not of high quality by modern standards. More recent ^{40}Ar/^{39}Ar and concordant U/Pb zircon ages from the Klamath Mountains apparently constrain the Oxfordian/Kimmeridgian boundary to the 150–157 Ma interval (see Pessagno and Blome, 1990). The definition of the Oxfordian/Kimmeridgian boundary in the Klamath Mountains is based largely on the overlapping occurrence of the radiolaria *Mirifusus* and *Xiphostylus*. This is a highly controversial definition as *Mirifusus* and *Xiphostylus* are *both* present from Aelenian time in Mediterranean sections (see Baumgartner, 1987).

10.3 Oxfordian–Aptian Time Scales

The age estimates mentioned above for the base of CM0 (121 Ma), CM16-CM15 (137 Ma), and the top CM25 (154-156 Ma) are not consistent with constant spreading rate in LH75 (Fig. 10.4a). Obradovich (1993) used the LH75 block model to interpolate between the same three age estimates: CM0 (121 Ma), CM16/16n (137 Ma), CM25n (156 Ma). The resulting time scale (OBRAD93) implies an abrupt spreading rate change at CM16 in LH75 and other Pacific block models (Fig. 10.5). Gradstein *et al.* (1994) have presented an integrated time scale (GRAD94) for Triassic to Cretaceous time. For the Late Jurassic–Early Cretaceous interval, constant spreading rate was assumed for most of the LH75 block model with an increase and then decrease in spreading rate in Valanginian and Berriasian time in order to satisfy the three principal age constraints. Synchronous apparent spreading rate changes in different block models may be an artifact of GRAD94 (Fig. 10.6).

A new Hawaiian block model (Nakanishi *et al.*, 1989, 1992; Channell *et al.*, 1995a) may represent an improved estimate of a constant spreading rate record for the M0–M29 interval. The three popular ages for CM0,

CM16-15, and CM25 are consistent with constant spreading in the new Hawaiian block model (Fig. 10.4b) and the resulting time scale (CENT94) does not imply abrupt or synchronous changes in spreading rate in the Pacific block models (Fig. 10.7). Figure 10.8 illustrates the relationship between CENT94 and other popular time scales. For numerical ages of Late Jurassic/Early Cretaceous polarity chrons and geologic stage boundaries, according to the various time scales, see Tables 10.1 and 10.2.

10.4 Hettangian–Oxfordian Time Scales

The Kent and Gradstein (1985) time scale for the Hettangian–Oxfordian intervals was constructed by interpolation between 208 Ma for the base of the Hettangian (Armstrong, 1982) and 156 Ma for the top of the

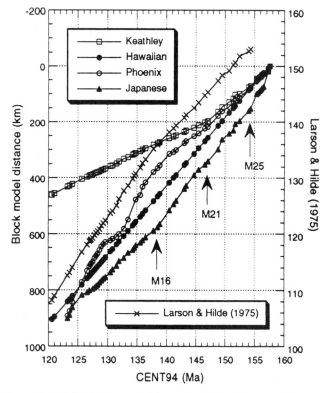

Figure 10.7 CENT94 time scale (Channell *et al.*, 1995a) compared to oceanic magnetic anomaly block models.

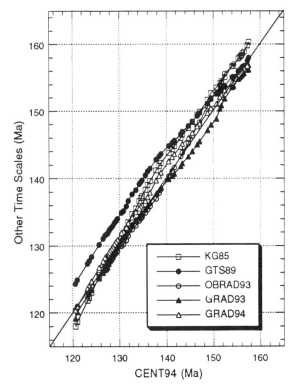

Figure 10.8 Plot of various time scales against CENT94. LH75 (Larson and Hilde, 1975), KG85 (Kent and Gradstein, 1985), GTS89 (Harland *et al.*, 1990), GRAD93 (Gradstein *et al.*, 1995) and GRAD94 (Gradstein *et al.*, 1994), CENT94 (Channell *et al.*, 1995a).

Oxfordian, assuming equal duration of the intervening 50 ammonite zones from Hallam (1975) (Table 10.2). In this interval, the GTS89 time scale (Harland *et al.*, 1990) utilizes the same age tie points (156 Ma for the top of the Oxfordian, and 208 Ma for the base of the Hettangian) and the same strategy of equal duration of ammonite zones; however, the use of a different ammonite zonation (Cope *et al.*, 1980a,b) leads to slightly different stage boundary age estimates. In GTS89, the Bathonian and younger Jurassic stage boundaries were also determined by interpolation from the magnetic anomaly record. The stage boundary age estimates derived assuming equal duration of ammonite zones, from magnetic anomalies, and from chronograms are listed in Table 10.2. For the Hettangian–Oxfordian interval in the time scale of Gradstein *et al.* (1994), maximum likelihood stage boundary age estimates were derived

Table 10.1
Comparison of M Sequence Time Scales

Reversed chron	LH75 (Ma)	KG85 (Ma)	GTS89 (Ma)	GRAD93 (Ma)	GRAD94 (Ma)	CENT94 (Ma)
M0 (top)	108.19	118.00	124.32	119.15	120.38	120.60
(base)	109.01	118.70	124.88	120.10	120.98	121.00
M1	112.62	121.81	127.35	122.56	123.67	123.19
	113.14	122.25	127.70	122.88	124.04	123.55
M3	114.05	123.03	128.32	123.45	124.72	124.05
	116.75	125.36	130.17	125.15	126.73	125.67
M5	118.03	126.46	131.05	125.96	127.68	126.57
	118.72	127.05	131.51	126.39	128.19	126.91
M6	118.91	127.21	131.64		128.33	127.11
	119.06	127.34	131.74	126.60	128.44	127.23
M7	119.27	127.52	131.89	126.80	128.59	127.49
	119.79	127.97	132.25	127.25	128.98	127.79
M8	120.21	128.33	132.53	127.68	129.29	128.07
	120.52	128.60	132.75	127.98	129.53	128.34
M9	120.88	128.91	132.99	128.32	129.79	128.62
	121.49	129.43	133.41	128.88	130.24	128.93
M10	121.94	129.82	133.72	129.31	130.58	129.25
	122.37	130.19	134.01	129.71	130.90	129.63
M10Nn-1	122.82	130.57	134.31	130.13	131.23	129.91
	122.88	130.63	134.36	130.20	131.28	129.95
M10Nn-2	123.31	131.00	134.65	130.60	131.60	130.22
	123.34	131.02	134.67	130.62	131.62	130.24
M10N	123.73	131.36	134.94	130.99	131.91	130.49
	124.07	131.65	135.17	131.26	132.10	130.84
M11	125.10	132.53	135.87	131.81	132.70	131.50
	125.68	133.03	136.27	132.12	133.05	131.71
M11	125.74	133.08	136.31	132.15	133.08	131.73
	126.22	133.50	136.64	132.42	133.37	131.91
M11An-1				132.66	133.64	
				132.68	133.66	132.40
M11A	127.16	134.01	137.30	132.92	133.93	132.47
	127.29	134.42	137.37	132.99	134.00	132.55
M12.1	127.68	134.75	137.63	133.19	134.23	132.76
	128.62	135.56	138.28	133.70	134.79	133.51
M12.2	128.74	135.66	138.36	133.75	134.85	133.58
	128.99	135.88	138.53	133.90	135.01	133.73
M12A	129.41	136.24	138.82	134.12	135.25	133.99
	129.56	136.37	138.92	134.20	135.34	134.08
M13	129.87	136.64	139.14	134.37	135.53	134.27
	130.41	137.10	139.50	134.66	135.84	134.53
M14	130.75	137.39	139.73	134.84	136.04	134.81
	131.81	138.30	140.46	135.41	136.67	135.57
M15	132.63	139.01	141.02	135.90	137.21	135.96
	133.30	139.58	141.47	136.38	137.89	136.49
M16	135.18	141.20	142.76	137.72		137.85
	135.94	141.85	143.28	138.26	140.40	138.50

Table 10.1 *continued*

Reversed chron	LH75 (Ma)	KG85 (Ma)	GTS89 (Ma)	GRAD93 (Ma)	GRAD94 (Ma)	CENT94 (Ma)
M17	136.42	142.27	143.61	138.51	140.86	138.89
	138.16	143.76	144.80	139.86	142.51	140.51
M18	138.82	144.33	145.25	140.33	143.14	141.22
	139.31	144.75	145.58	140.68	143.60	141.63
M19n-1	139.46	144.88	145.69	140.80	143.72	141.78
	139.55	144.96	145.75	140.88	143.79	141.88
M19	140.74	145.98	146.56	141.82	144.68	143.07
	141.28	146.44	146.93	142.25	145.08	143.36
M20n-1	141.64	146.75	147.17	142.54	145.35	143.77
	141.71	146.81	147.22	142.60	145.41	143.84
M20	142.47	147.47	147.75	143.21	145.99	144.70
	143.47	148.33	148.43	144.01	146.74	145.52
M21	144.74	149.42	149.30	145.02	147.69	146.56
	145.29	149.89	149.67	145.46	148.11	147.06
M22n-1	147.11	151.46	150.92	146.92	149.48	148.57
	147.17	151.51	150.96	146.95	149.52	148.62
M22n-2	147.23	151.56	151.00	147.01	149.57	148.67
	147.29	151.61	151.04	147.06	149.61	148.72
M22	147.38	151.69	151.10	147.13	149.68	148.79
	148.36	152.53	151.77	147.91	150.42	149.49
M22A	148.51	152.66	151.87	148.03	150.53	149.72
	148.72	152.84	152.01	148.20	150.69	150.04
M23	149.15	153.21	152.31	148.80	151.09	150.69
	149.48	153.49	152.53	149.24	151.39	150.91
M23	149.51	153.52	152.56	149.30	151.42	150.93
	150.24	154.15	153.06	150.31	152.10	151.40
M24	150.63	154.58	153.32	150.85	152.46	151.72
	151.06	154.85	153.61	151.44	152.86	151.98
M24	151.09	154.88	153.64	151.49	152.89	152.00
	151.33	155.08	153.80	151.82	153.11	152.15
M24A	151.48	155.21	153.90	152.02	153.25	152.24
	151.79	155.48	154.11	152.46	153.54	152.43
M24B	152.21	155.84	154.40	153.04	153.93	153.13
	152.39	156.00	154.53	153.30	154.10	153.43
M25	152.73	156.29	154.76	153.52	154.31	154.00
	153.03	156.55	154.96	153.72	154.49	154.31
M26					155.51	155.32
					155.69	155.55
M27					155.83	155.80
					156.00	156.05
M28					156.14	156.19
					156.29	156.51
M29					156.77	157.27
					156.85	157.53

Key: LH75, Larson and Hilde (1975); KG85, Kent and Gradstein (1985); GTS89, Harland *et al.* (1990); GRAD93, Gradstein *et al.* (1995); GRAD94, Gradstein *et al.* (1994); and CENT94, Channell *et al.* (1995a).

Table 10.2
Jurassic–Early Cretaceous Time Scales

Stage boundary	Kent and Gradstein (1985) (Ma)	Harland et al. (1990) Ma — Mag. anom.	Harland et al. (1990) Ma — Amm. zones	Harland et al. (1990) — Chronogram	Obradovich (1993) (Ma)	Gradstein et al. (1994) (Ma)	Channell et al. (1995a) (Ma)
Barremian/Aptian	119	*124.5*		125.5	121	121.0	121.0
Hauterivian/Barremian	124	*131.8*		131.0	127	127.0	126.0
Valanginian/Hauterivian	131	*135.0*		135.5	130	132.0	131.0
Berriasian/Valanginian	138	*140.7*	145.0	137.5	135	137.0	135.8
Tithonian/Berriasian (J/K)	144	*145.6*	153.5	148.0	142	144.2	141.6
Kimmeridgian/Tithonian	152	*152.1*	156.0	151.0		150.7	149.0
Oxfordian/Kimmeridgian	156	*154.7*	159.8	156.0		154.1	154.0
Callovian/Oxfordian	163	*157.1*	162.7	156.3		159.4	
Bathonian/Callovian	169	*161.3*	166.6	159.0		164.4	
Bajocian/Bathonian	176	*166.1*	168.2	159.2		169.2	
Aalenian/Bajocian	183		*173.5*	177.0		176.5	
Toarcian/Aalenian	187		*178*	178.0		180.1	
Pliensbachian/Toarcian	193		*187*	182.0		189.6	
Sinemurian/Pliensbachian	198		*194.5*	189.2		195.3	
Hettangian/Sinemurian	204		*203.5*	203.5		201.9	
Rhaetian/Hettangian (Tr/J)	208		*208*	210.5		205.7	

For Harland et al. (1990): italics indicate ages adopted in their GTS89 time scale.

from selected radiometric ages. The maximum likelihood stage boundary ages were then smoothed by means of a cubic spline fit, with the dependent variable being the unequally spaced maximum likelihood age estimates and the independent variable being the equal duration of ammonite subzones (Table 10.2).

11

Jurassic and Cretaceous Magnetic Stratigraphy

11.1 Cretaceous Magnetic Stratigraphy

The majority of Cretaceous magnetostratigraphic studies have been carried out on pelagic limestones in the Mediterranean region, following the pioneering work of Lowrie and Alvarez (1977a,b) on the Gubbio (Bottacione) section in Italy. The principal feature of Cretaceous geomagnetic polarity is the so-called Cretaceous long normal superchron, an interval of prolonged normal polarity lasting about 38 My, beginning just above the Barremian/Aptian boundary and ending close to the Campanian/Santonian boundary (Fig. 11.1, Table 11.1).

a. K/T Boundary Magnetic Stratigraphy

The position of the Cretaceous–Tertiary (K/T) boundary in the GPTS has always been of great interest because of the associated mass extinctions. The position of this boundary in magnetostratigraphic section was first determined at Gubbio (Bottacione) (Alvarez *et al.*, 1977). Since then, the position of this boundary has been accurately determined in more than 14 magnetostratigraphic sections both in outcrop and deep-sea cores. The average position of the K/T boundary in these studies implies a correlation to C29r (0.75) (75% up from base of the reverse chron). However, the location of the boundary within the polarity zone correlative to C29r is not a good estimate of the position of boundary *in time* due to a substantially reduced sedimentation rate in the Paleocene part of C29r. Cyclostratigraphy

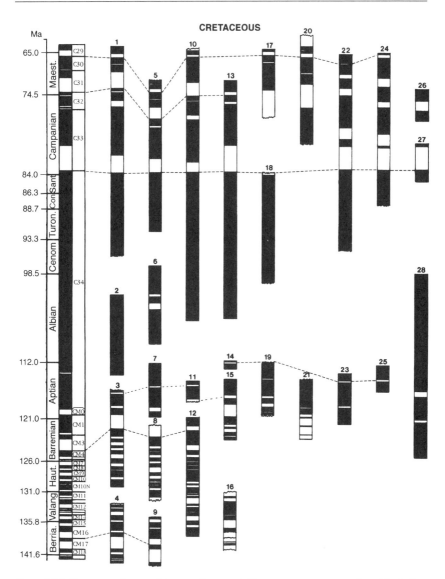

Figure 11.1 Summary of some of the more important Cretaceous magnetostratigraphic studies. For key, see Table 11.1.

Table 11.1
Cretaceous

	Rock unit	Lo Age Hi	Region	λ	Φ	NSE	NSI	NSA	M	D	RM	DM	AD	A	NMZ	NCh	%R	RT	F.C.T.	Q	References
1	Scaglia	C29N-C34N	Italy, Gubbio	+43.37	+12.58	1	5–1 m	1	250	—	J-1	A	V	V	13	6	22	R+	F+	6	Lowrie and Alvarez (1977b)
2	Fucoid Marls	34N	Italy, Poggio	+14.51	+12.58	1	52	3–4	78	—	1	A	V	F-V	1	1	—	—	F+	6	Lowrie et al. (1980b)
3	Maiolica	CM0-CM10	Italy, Gorgo a. Cerbera	+43.56	+12.53	2	.75–1 m	1	180	—	1	T	Z-V	F-V	24	11	47	—	F+	7	Lowrie et al. (1980a); Lowrie and Alvarez (1984)
4	Maiolica	CM13-CM19	Italy, Bosso	+43.56	+12.53	1	.8 m	1	120	—	—	T	Z-V	F-V	17	7	47	—	—	4	Lowrie and Channell (1984)
5	Scaglia	C24-C34	Italy, Belluno	+46.00	+11.75	5	359	1	127	—	—	T	V	V	33	13	28	—	—	5	Channell and Medizza (1981)
6	Fucoid Marls	C34	Italy, Albian. Contessa	+43.36	+12.56	1	72	1	30	—	Td1	T-A	Z-P	F-V	15	1	16	R-	—	6	Tarduno et al. (1992)
7	Maiolica	CM0-CM1	Italy Pie'Dosso	+45.55	+10.11	3	1.5 m	1	200	—	1	TA	Z-P	F-V	34	11	41	R+	F+	9	Channell and Erba (1992)
8	Maiolica	CM0-CM11	Italy, Polveno	+45.63	+10.05	3	1.5 m	1	200	—	1	TA	Z-P	F-V	34	11	41	R+	F+	9	Channell and Erba (1992)
9	Maiolica	CM15-CM19	Italy, Fonte Del Giordano	+43.30	+12.54	1	104	1	275	—	1	T	Z-B	F-V	21	9	43	—	F+	7	Cirilli et al. (1984)
10	Scaglia Rossa	C29-C34	Italy Moria	+43.50	+12.54	1	250	1	340	—	—	A	V	F-V	12	6	18	R+	F+	5	Alvarez and Lowrie (1978)
11	Maiolica	CM0-CM1	Italy Frontale	+43.56	+12.53	1	.75–1 m	1	130	—	1	T	Z-V	V	15	6	33	—	—	6	Lowrie and Alvarez (1984)
12	Maiolica	CM0-CM16	Italy, Capriolo	+45.60	+9.89	1	.3 m	1	167	—	—	T-A	Z-B	V	41	15	42	—	—	5	Channell et al. (1987)
13	Scaglia	C30-C34	Italy	+43.64	+12.71	1	450	1	300	—	1	T-A	Z-V	F-V	27	9	20	—	—	7	Alvarez and Lowrie (1984)
14	Fucoid Marls	C34N	Italy Valdorbia	+14.43	+12.71	3	24	4	30	—	1	A	V	F-V	3	1	06	—	F+	5	Lowrie et al. (1980b)

No.	Formation	Location	Chron	Lat	Long				N												Reference
15	Maiolica	Italy Presale	M0-M8	+43.56	+12.53	1	.75–1 m	1	94	—	1	T	Z-V	F-V	15	9	47	—	—	6	Lowrie and Alvarez (1984)
16	Berriasian Lmst.	France	CM14-CM19	+44.38	+4.25	1	163	1	30	—	1	T-A	Z-V	F-V	13	5	47	R+	—	6	Galbrun (1985)
17	Sopelana	Spain	C29N-C31R	+43.4	−.003.00	1	259	1	150	—	1	T-A	Z-P	F-V	8	4	50	R−	—	5	Mary et al. (1991)
18	Scaglia	Italy Cismon	C33R-CM8	+46	+11.76	2	1	1	394	—	1	A	B	V	26		20	—	—	6	Channell et al. (1979)
19	Maiolica	Italy Cismon	C33R-CM8	+46	+11.76	2	1	1	394	—	1	A	B	V	26		20	—	—	6	Channell et al. (1979)
20	Hole 525A	S. Atlantic	C28-C32	−29.08	+02.99	4	.3–.6 m	1	133	—	—	A	Z-B	I	12	5	38	—	—	5	Chave (1984)
21	Site 167	N. Pacific	C34-CM5	+07.00	−177.00	1	88	1	120	—	—	AT	Z-P	F-I	10	6	24	—	—	5	Tarduno et al. (1989)
22	Hole 530A	S. Atlantic	C26-C34	−19.19	−9.39	1	.35 m	1	630	—	—	A	Z-B	I	11	4	15	—	—	4	Keating & Herrero-Bervera (1984)
23	Site 463	Pacific	C34-CMO	+21.36	+175	1	56	1	160	—	—	A	Z-P	F-I	5	2	01	—	—	5	Tarduno et al. (1989)
24	Site 516	S. Atlantic	C29-C34	−30.35	−35.18	1	180	1	289	—	—	A.T.	B	I	9	6	35	—	—	3	Hamilton et al. (1983)
25	Site 317	C. Pacific	C34	−15.2	−146.8	1	74	1	50	—	—	A	Z-P	I	3	1	01	—	—	5	Tarduno (1990)
26	Point Loma fm	USA (CA)	C32-C33	+32.78	−117.29	3	45	1	200	—	J	T	Z-V	V	3	2	24	R+	F+	8	Bannon et al. (1989)
27	Great Valley Gap	USA (CA)	C33-C34	+40.00	−121.5	9	252	3	3000	—	A-1	A-T	Z-V	V	3	2	46	—	—	6	Verosub et al. (1989)
28	Christopher Fm	Canada	C34-M1	+80	−92	6	76	5	1500	—	V-O	T-A	Z-V	F	5	3	05	R+	F+	7	Wynn et al. (1988)
	Hole 690C	S. Atlantic	C29-C33	−65	+2	2	.25 m	1	70	—	—	A	Z-V	I	11	7	30	—	—	4	Hamilton (1990)
	Cehegin	Spain	CM15-CM17	+38.06	−1.81	2	73	1	12	—	—	T	Z-P	F	4	2	72	R+	—	7	Ogg et al. (1988)
	Aix-en-Provence	France	CM28-CM34	+33.5	+5.5	33	110	1–4	1166	—	—	T	Z-V	F	13	6	25	—	—	5	Westphal and Durand (1990)
	Deccan Traps	India	C29-C30	+20	+75	3	27	4–10	—	K	—	T.A	Z-P	F	3	2	55	R+	—	6	Vandamme et al. (1991)
	Sierra Geral	Brazil	Early Cretaceous	−18 / −30	−56 / −48	20	3	500	1000	K	J	T-A	Z-V	F	4		35	—	—	7	Ernesto et al. (1990)
	Leg 103	Atlantic	Haut-Alb	+42.17	+12.17	6	1.5 m	383	300	—	1	T-A	Z-P	I	20	11		—	—	5	Ogg (1988)
	Holes 752-755	Indian Oc.	E. Eocene-Turonian	−31	+93.83	4	576	1	310	—	1-A	T-A	Z-P	I	26	19	40	R+	—	7	Gee et al. (1991)

based on carbonate content, color density, and magnetic susceptibility from South Atlantic cores and Spanish land sections was used to demonstrate that the boundary occurred *in time* almost exactly in the middle of C29r (Herbert and D'Hondt, 1990; Herbert *et al.*, 1995).

Nonmarine terrestrial sediments from the western United States (Butler *et al.*, 1977, 1981b) as well as from Europe (Galbrun *et al.*, 1993) have contributed to the debate concerning the faunal crisis at the end of the Cretaceous by demonstrating that extinctions of marine and nonmarine fauna were essentially synchronous (see Section 9.5).

b. Santonian–Maastrichtian Magnetic Stratigraphy

The Late Cretaceous polarity zones at the young end of the Cretaceous long normal interval were recorded and correlated to foraminiferal biostratigraphy in the original Alvarez *et al.* (1977) study of the Gubbio (Bottacione) section (Fig. 11.2) Subsequent nannofossil biostratigraphy at Gubbio (Monechi and Thierstein, 1985) has provided a correlation of these polarity chrons to nannofossil zones (Fig. 11.3). The correlations of polarity chrons to nannofossil and foraminiferal biozonations at Gubbio have been ratified and refined in the Southern Alps (Channell and Medizza, 1981), Spain (Mary *et al.*, 1991), and in cores from the NW Pacific (Monechi *et al.*, 1985) and South Atlantic (Hamilton *et al.*, 1983; Chave, 1984; Tauxe *et al.*, 1983c; Poore *et al.*, 1984; Keating and Herrero-Bervera, 1984). Due to the scarcity of ammonites in the pelagic limestone sections, polarity chrons are generally correlated to the ammonite-bearing stage stratotype sections through the micropaleontology. Direct correlation of the GPTS to European Upper Cretaceous ammonite zones which define stage boundaries has not been achieved, although the C33r/C33n polarity chron boundary has been correlated to ammonite biostratigraphy in Wyoming (Hicks *et al.*, 1995).

c. Cretaceous Long Normal

From paleomagnetic data available at the time, Helsley and Steiner (1969) postulated the existence of an interval of constant normal polarity from Late Aptian to middle Santonian which could be correlated to long smooth intervals in coeval oceanic magnetic anomaly records (Raff, 1966; Heirtzler *et al.*, 1968). The presence of this long interval of normal polarity was confirmed by Keating *et al.* (1975) from deep-sea cores obtained from the Deep Sea Drilling Project (DSDP). Subsequent studies in Italian pelagic limestones resulted in correlation of the top of the quiet zone (base of C34r) to the early Campanian, just above the base of the *Globotruncana elevata* foraminiferal zone (Alvarez *et al.*, 1977;

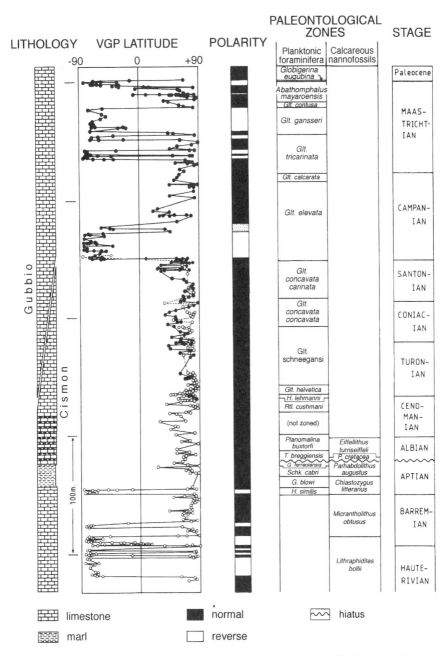

Figure 11.2 Summary of Cretaceous magnetostratigraphic studies at Gubbio and Cismon (Italy) (after Lowrie *et al.*, 1980b).

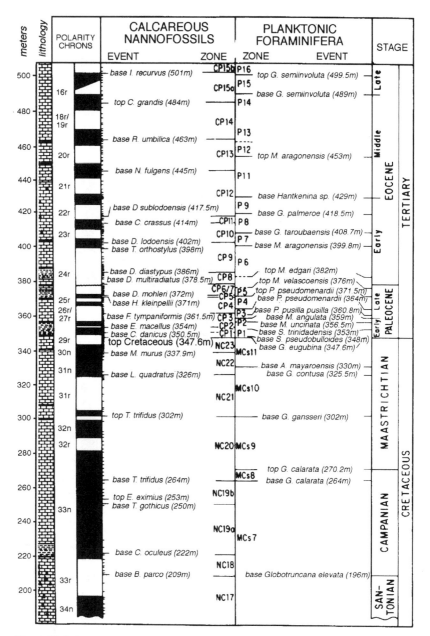

Figure 11.3 Correlation of Late Cretaceous–Eocene calcareous nannofossil and planktonic foraminiferal events/zones to polarity chrons at Gubbio, Italy (after Monechi and Thierstein, 1985).

Lowrie and Alvarez, 1977a,b; Channell *et al.*, 1979; Channell and Medizza, 1981) (Fig. 11.2). The entire Cretaceous long normal interval has been recorded in the Cismon section (northern Italy) and here the reverse polarity chron at its base (CM0) is close to the Barremian/Aptian boundary (Channell *et al.*, 1979) (Fig. 11.2). The base of CM0 has since been found to immediately postdate the first occurrence (FO) of nannofossil *R. irregularis* in two Italian land sections (Capriolo and Pie' del Dosso) (Channell and Erba, 1992) and at ODP Site 641 on the Galicia margin (Ogg, 1988). In the absence of the ammonite *Deshayesites,* which formally defines the Barremian/Aptian boundary, the FO of *R. irregularis* is considered the most reliable microfossil marker for this stage boundary. The base of CM0 is usually close to but slightly younger than this nannofossil event.

There are a number of magnetostratigraphic records of post-CM0 Aptian–Albian reverse polarity chrons. The first such observation in outcrop was by Lowrie *et al.* (1980a), who observed a reversely magnetized bed in the Late Aptian *Globigerinelloides algerianus* zone at Valdorbia (Umbria, central Italy). This so-called ISEA reverse polarity zone (Fig. 11.4), also known as polarity chron CM−1 (*minus* 1), was not observed at two other sections in Umbria where the same foraminiferal zone was present in comparable or greater stratigraphic thickness; however, Tarduno *et al.* (1989) observed two samples with reverse magnetizations just above the FO of *G. algerianus* at DSDP Site 463. The documentation of this short reverse polarity chron is strengthened by two samples with reverse magnetization spanning 43 cm in a core from ODP Site 765 (Ogg *et al.*, 1991b), where the reverse polarity zone postdates CM0 and is late Aptian in age.

Seven reverse polarity zones have been observed by Tarduno *et al.* (1992) in the middle Albian interval (Fig. 11.4) (*A. albianus* nannofossil zone and top *Ticinella primula* to base *Biticinella breggiensis* foraminiferal zones) at the Contessa section near Gubbio (Umbria, central Italy). In these reverse polarity zones, the magnetization is carried by hematite and the reddening of the sediment may be late diagenetic in origin. The reverse magnetization components in this interval are not antipodal to the normal magnetization components, and the reverse directions are offset toward directions consistent with the Late Cretaceous or Paleogene. The duration of the proposed middle Albian polarity zones is well known from lithologic cyclostratigraphy (see Herbert *et al.*, 1995, and references therein). Of the seven polarity zones recognized by Tarduno *et al.* (1992), the thickest (3.25 m) represents about 800 ky. This duration is greater than that estimated for reverse chron CM0 and about twice that estimated for CM1 (see Herbert, 1992), both of which are usually recognized by shipboard magnetic anomaly surveys. It is, therefore, doubtful that these middle Albian reverse

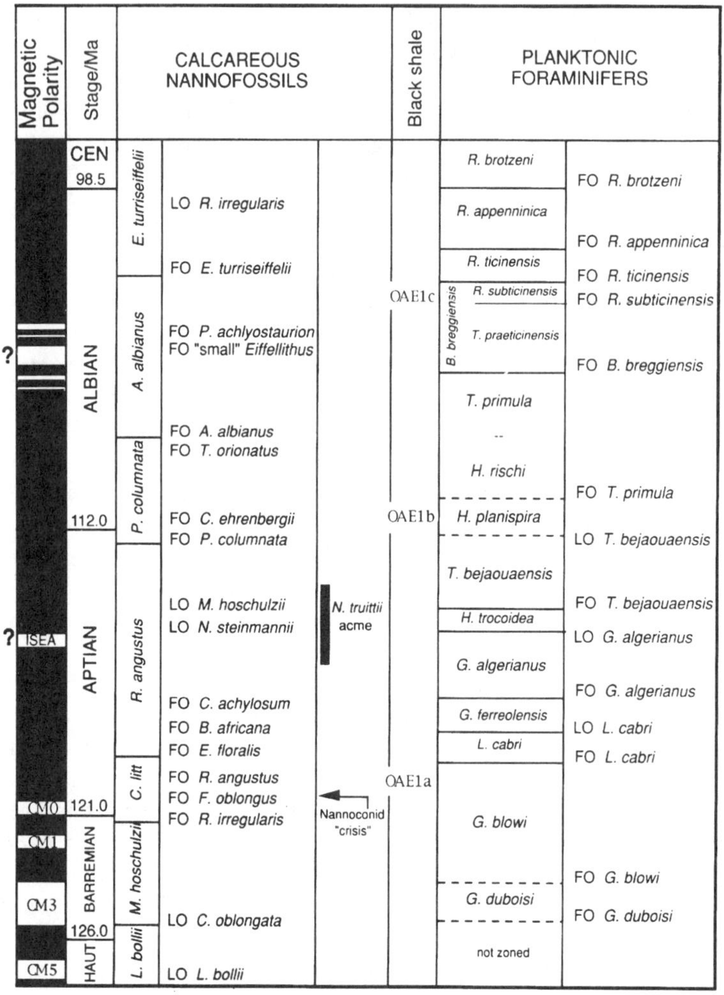

Figure 11.4 Correlation of mid-Cretaceous calcareous nannofossil and planktonic foraminiferal events to polarity chrons (after Larson *et al.*, 1993).

polarity zones represent the geomagnetic field at the time of deposition of the sediments. Until the middle Albian reverse polarity chrons are recognized elsewhere, we consider that they should not be incorporated into the geomagnetic polarity time scale (GPTS).

d. Berriasian–Aptian Magnetic Stratigraphy

"M-sequence" polarity chrons are numbered according to the correlative oceanic magnetic anomaly. These anomaly numbers, by tradition, correlate to reverse polarity chrons except for M2 and M4, which correlate to a normal polarity chrons. We use a prefix "C" to distinguish polarity chrons from magnetic anomalies and follow Harland *et al.* (1982) in labeling normal polarity chrons (other than CM2 and CM4) using the number of the next older polarity chron with the appendage "n." Using this nomenclature, CM9 denotes the reverse polarity chron correlative to magnetic anomaly M9 and CM9n denotes the normal polarity chron between CM9 and CM8. It should be noted the magnetic anomaly between M10 and M11 is labeled M10N. This anomaly was originally omitted from M-sequence anomalies (Larson and Pitman, 1972) and was included by Larson and Hilde (1975) and designated M10N after Fred Naugler, co-chief scientist of the NOAA Western Pacific Geotraverse Project.

Land section magnetic stratigraphy in the Mediterranean region has been the basis for the correlation of the M-sequence polarity chrons to biozonations and hence to geologic stage boundaries. The oceanic magnetic anomaly record from the Hawaiian lineations (Larson and Hilde, 1975) remains the template for the M-sequence polarity pattern. Some pelagic limestone sections in Italy have recorded substantial portions of the M-sequence pattern. Notable among these sections are Gorgo a Cerbara (CM0–CM9) (Lowrie and Alvarez, 1984), Polaveno (CM3–CM16) (Fig. 11.5) (Channell and Erba, 1992; Channell *et al.*, 1995b), Capriolo (CM8–CM16) (Channell *et al.*, 1987), and Bosso (CM14–CM19) (Lowrie and Channell, 1984) (see Fig. 11.1, Table 11.1). It is remarkable that the polarity pattern derived by Larson and Hilde (1975) from Hawaiian oceanic magnetic anomalies has been confirmed time and time again by these land section studies, particularly in the CM0–CM19 interval. The only polarity chron observed in land section which is not included in the Larson and Hilde (1975) polarity pattern is a second reverse subchron between CM11 and CM12. These two subchrons were recognized at Capriolo (Italy) (Channell *et al.*, 1987) and have also now been observed in the oceanic magnetic anomaly record (Tamaki and Larson, 1988).

The vast majority of the magnetostratigraphic studies in the CM0–CM19 interval have been carried out in the Maiolica Limestone Formation of Italy. This is a Tithonian to Aptian white/gray thin bedded magnetite-bearing pelagic limestone with fairly constant sedimentation rates (typically ~15 m/My), thereby aiding the recognition of polarity zone patterns. The biostratigraphic control in the Maiolica is mainly from calpionellids for the

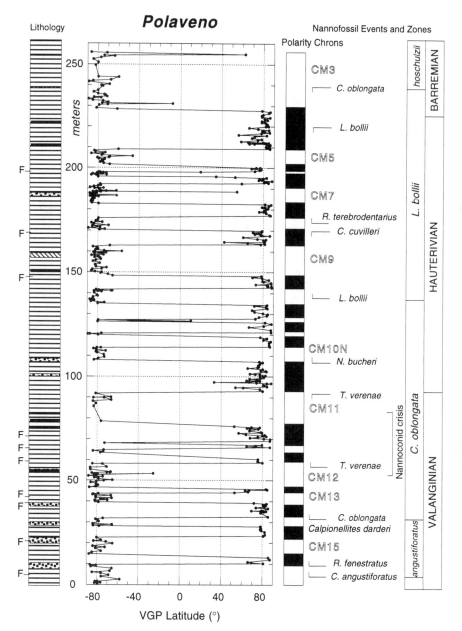

Figure 11.5 Virtual geomagnetic polar (VGP) latitudes from the Polaveno section (Italy), correlation to the GPTS, and correlation to nannofossil and calpionellid events (after Channell and Erba, 1992; Channell *et al.*, 1995b). In the lithologic log, F indicates small faults (with minimal offset), speckles indicate cherty intervals, diagonals indicate marly interval, and black bars indicate black shales. Horizontal thin bars indicate pelagic limestone, the dominant lithology.

Tithonian and Berriasian and from calcareous nannofossils for Tithonian to Aptian. It has been demonstrated in numerous Maiolica sections that polarity chrons correlate consistently to nannofossil events (Fig. 11.6) and calpionellid events (Fig. 11.7). For correlations to calpionellids, see Channell and Grandesso (1987) and Ogg *et al.* (1991c); for correlations to nannofossils see Bralower (1987), Channell *et al.* (1987, 1993), Bralower *et al.* (1989), Ogg *et al.* (1991b), and Channell and Erba (1992). The result of these studies is that the combination of nannofossil or calpionellid biostratigraphy and magnetic stratigraphy gives a very precise and useful integrated stratigraphic tool for pelagic limestones of this age. Even for short sections where polarity zone patterns are not distinctive, the nannofossil/calpionellid events indicate the approximate stratigraphic position relative to the GPTS, and the magnetic stratigraphy then gives precise stratigraphic control. Although the correlation of polarity chrons to nannofossils and calpionellids is firmly established, the correlation to stage boundaries requires correlation to ammonite zones. Although ammonites are very rare in the Maiolica limestones, recent finds place the Hauterivian/Barremian boundary in the upper (younger) part of CM4 (Cecca *et al.*, 1994) and the Valanginian/Hauterivian boundary at the young end of CM11 (Channell *et al.*, 1995b) (Fig. 11.6). Previous estimates of the correlation of these stage boundaries to polarity chrons relied on the correlation of nannofossil events to ammonite zones from Thierstein (1973, 1976). For the Berriasian–Valanginian boundary, the relevant ammonite zonal boundary has been correlated directly to CM15n at Cehegin (Spain) (Ogg *et al.*, 1988). For the Jurassic/Cretaceous (Tithonian/Berriasian) boundary, there are two alternative definitions with respect to ammonite zones, with no clear consensus. In the Tethyan realm, the boundary lies either at the *B. grandis/B. jacobi* zonal boundary or at the underlying *B. jacobi/Durangites* zonal boundary. At Carcabuey (Spain), the *B. jacobi/Durangites* zonal boundary lies in a normal polarity chron at the top of the section, which is interpreted as CM19n (Ogg *et al.*, 1984). Elsewhere, nannofossil and calpionellid events in the vicinity of the Jurassic/Cretaceous boundary have been consistently correlated to polarity chrons in land sections and deep-sea cores (Lowrie and Channell, 1984; Cirilli *et al.*, 1984; Channell and Grandesso, 1987; Bralower *et al.*, 1989; Ogg *et al.*, 1991c). These studies imply that the Jurassic/Cretaceous boundary lies in the upper part of CM19n or at the CM19n/CM18 polarity chron boundary, and it has been advocated that this polarity chron boundary be used to define the stage boundary (Ogg and Lowrie, 1986). At Berrias (France), the Berriasian stratotype section yielded a magnetic stratigraphy that is correlated to calpionellids, nannofossils, and ammonite zones (Galbrun, 1985; Bralower *et al.*, 1989); however, the base of the section is within the *B. grandis* zone and

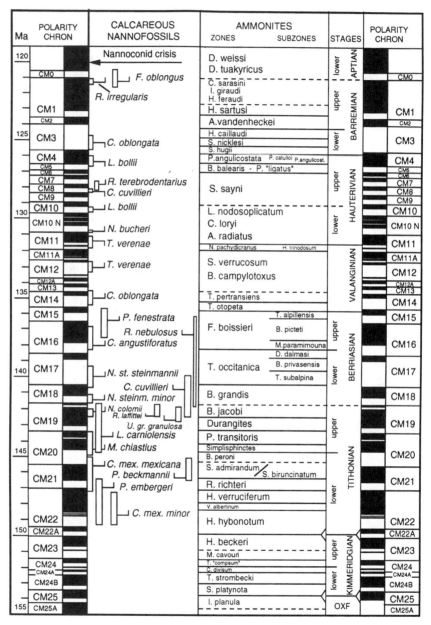

Figure 11.6 Summary of correlation of Oxfordian to Aptian polarity chrons to nannofossil events/zones and ammonite zones. Open bars indicate the range of the nannofossil events with respect to polarity chrons (from Channell *et al.*, 1995a).

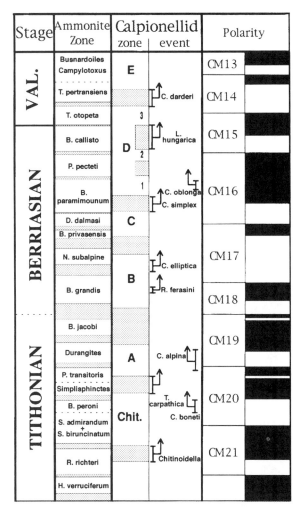

Figure 11.7 Correlation of calpionellid events to polarity chrons (after Channell and Grandesso, 1987; Ogg *et al.*, 1991c).

therefore the section does not include either of the two definitions of the Jurassic/Cretaceous boundary. Ogg *et al.* (1991c) have identified CM17 in the Purbeck Limestone of southern Britain, providing a first M-sequence magnetostratigraphic correlation out of the Tethyan into the Boreal Realm.

11.2 Jurassic Magnetic Stratigraphy

a. Kimmeridgian–Tithonian Magnetic Stratigraphy

As for the Early Cretaceous (CM0–CM18), the Larson and Hilde (1975) oceanic magnetic anomaly block model is the template for geomagnetic polarity in the Tithonian and Kimmeridgian stages of the Jurassic (CM18–CM25). Subsequent extension of the oceanic anomaly block model to M29 (Cande *et al.,* 1978), and to M38 (Handschumacher *et al.,* 1988) has extended the record into the middle Jurassic interval. Land section magnetic stratigraphies have been correlated to CM18–CM25; however, correlations to older M-sequence chrons and between-section pre-Kimmeridgian correlations have not been adequately achieved (Fig. 11.8, Table 11.2). The problem for pre-Kimmeridgian magnetic stratigraphy is twofold. First, pre-Kimmeridgian calcareous nannofossil biostratigraphy does not yet allow precise correlation among Tethyan sections; and second, pre-Kimmeridgian Jurassic facies in the Tethyan realm are often highly condensed and/or highly siliceous.

The oldest polarity chron recorded by the Maiolica Limestones of Italy is CM19. The Maiolica limestones are usually underlain either by siliceous limestones (Calcari Diasprigni) or by condensed nodular limestones (Ammonitico Rosso). The Sierra Gorda and Carcabuey sections of southern Spain are the key sections for the correlation of CM19–CM25 to ammonite zones (Ogg *et al.,* 1984). In these sections, the "Ammonitico Rosso"-type sediments have mean sedimentation rates of 2–3 m/My. The polarity patterns are distorted by variable sedimentation rates and it is not easy to correlate polarity zones to polarity chrons. Nonetheless, the correlations of Ogg *et al.* (1984) place the *H. beckeri/H. hybonatum* ammonite zonal boundary which defines the Kimmeridgian/Tithonian boundary in CM23n, and the *I. planula/S. platynota* zonal boundary which defines the Oxfordian/Kimmeridgian boundary in CM25 (Fig. 11.6). Ammonites are lacking in this interval in the Belluno Basin and Trento Plateau (Southern Alps, Italy), where the Oxfordian/Kimmeridgian and Kimmeridgian/Tithonian boundaries lie in the unsubdivided "*Saccocoma* Zone." The lack of calpionellid or other microfossil events in the Italian sections precludes the accurate definition of these boundaries; however, the polarity zone pattern in these sections is more readable than in the Spanish sections. Estimates of

Figure 11.8 Summary of some of the more important Jurassic magnetostratigraphic studies. For key, see Table 11.2.

JURASSIC

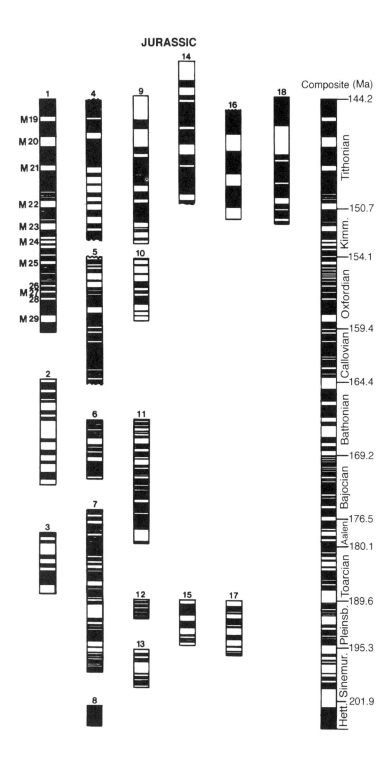

Composite (Ma)

M19
M20
M21
M22
M23
M24
M25
M26
M27
M28
M29

1
2
3
4
5
6
7
8
9
10
11
12
13
14
15
16
17
18

—144.2

Tithonian

—150.7

Kimm.

—154.1

Oxfordian

—159.4

Callovian

—164.4

Bathonian

—169.2

Bajocian

—176.5

Aalen.
—180.1

Toarcian

—189.6

Pleinsb.

—195.3

Sinemur.

—201.9

Hett.

Table 11.2
Jurassic

	Rock unit	Lo Age Hi	Region	λ	Φ	NSE	NSI	NSA	M	D	RM	DM	AD	A	NMZ	NCh	%R	RT	F.C.T	Q	References
1	Marine Magnetic Anomalies																				
2	Sierra Harana	Bathonian/Bajocian	Spain	+37.2	−3.7	1	106	1	13	—	—	T	Z-P	F-D-I	21	—	60	R−		6	Steiner et al. (1989)
3	Toarcian	Toarcian	France	+47.00	−.02	2	.1 m	1	5	—	1	T-A	Z-V	F-D-I	15	—	50	R+	—	7	Galbrun et al. (1988a)
4	Carcabuey	CM18-CM25	Spain	+37.5	−3.3	2	.1 m	1	10	—	—	T	Z-V	FV	23	6	40	R+	—	6	Ogg et al. (1984)
5	Ammonitico-Rosso	Ox-Callo.	Italy	+46	+11	3	.05 m	1	10	—	—	T	Z-P	V	55	—	32	—	—	6	Channell et al. (1990a)
6	La Fuente	Batho.-Bajo.	Spain	+37.5	−3.3	2	99	1	10	—	—	T	Z-P	F-D-I	19	—	42	R+	F+	8	Steiner et al. (1987)
7	Breggia	Baj.-Car.	Switzerland	+45.87	+9	2	.2 m	1	120	—	1	T-A	Z-V	F-V	82	—	47	—	—	7	Horner and Heller (1983)
8	Newark Supergroup	Het-Carnian	USA	+40.25	−75.25	3	4	5	4180	K	1	T	Z-P	FV	1	1	0	R+	F+	9	Witte et al. (1991)
9	Xausa	CM14-CM23	Italy	+45.7	+11.46	1	139	1	60	—	—	T-A	Z-B	V	25	9	52	—	F+	5	Channell et al. (1987)
10	Aquilon	L.Oxfordian M	Spain	+41.3	−1.00	5	105	1	13	—	—	T	Z-P	F-D-I	22	4	54	R−	F+	7	Steiner et al. (1985)
11	Carcabuey	Bath-Aal	Spain	+37.5	−3.3	2	.15 m	1	33	—	—	T	Z-P	F-D-I	51	—	41	—	F+	8	Steiner et al. (1987)
12	Alpe Turati	Ba-Car.	Switz	+45.87	+9	2	.2 m	1	120	—	1	T-A	Z-V	F-V	82	—	47	—	—	7	Horner and Heller (1983)
13	Kandelbach Grab.	Sin.-Het.	Austria	+4.7	+13.5	3	169	1	24	—	—	T	B	D-I	14	—	62	C+		6	Steiner and Ogg (1988)
14	Foza	Ber.-Kim	Italy	+45.54	+13.5	3	90	1	—	—	1	T-A	—	—	17	6	49	—	—	2	Channell et al. (1982b)
15	Bakonycsoke	Pliens	Hungary	+47.1	+18	1	.1 m	1	8.9	—	1	T	Z-V	V	11	—	55	R−	—	5	Marton et al. (1980)

No.	Name	Stage	Country	Lat.	Long.														Ref.	Reference
16	Quero	Titho.	Italy	+45.55	+11.56	1	74	1	50	—	—	T	Z-B	V	6	4	39	—	—	6	Channell and Grandesso (1987)
17	Cingoli	Pleins-Sinme	Italy	+43.33	+13.21	1	64	1	20	—	1	T	Z-V	F-V	16	—	58	—	—	6	Channell et al. (1984)
18	Frisoni	Titho-Kimm	Italy	+45.53	+11.33	1	132	1	80	—	—	T	Z-B	V	18	7	47	—	—	6	Channell and Grandesso (1987)
	Ammonitico-Rosso	L Toarcian E	Spain	+37.39	−3.49	1	.13	1	20	—	1	T	Z-P	F-D-I	21	—	53	R−	—	6	Galbrun et al. (1990)
	Kayenta Fm.	Pliensbachian	USA	+38.6	−109.6	2	101	1	100	—	—	T	-V	D-I	8	—	19	R+	—	5	Steiner and Helsley (1974)
	Krakow Uplands	M.Ox. E. Call.	Poland	+50.1	+19.6	5	204	1	107	—	—	T	Z-P	F	19	—	—	R+	—	7	Ogg et al. (1991a)
	La Luna	Sant. Sen	Venezuela	+9.9	−61.0	3	48	1	50	—	1	T-A	Z-P	F-D-I	4	—	63	—	—	7	Castillo et al. (1991)
	Lebombo Grp.	E. Jurassic	S. Africa	−24	+31.75	1	25	45	6000	K	A	A-T	B	F	4	—	30	R+	—	6	Henthorn (1981)
	Morrison Fm.	Kim-Ox	USA	+38.13	−108.21	1	215	1	85	—	—	T	V	D-I	5	—	78	—	—	4	Steiner and Helsley (1975a)
	Morrison Fm.	Kim-Tith	USA	+38.1	−108.2	1	.3	1	165	—	—	T	Z-V	F-D-I	13	—	80	—	—	4	Steiner and Helsley (1975b)
	Purbeck Ls.	Ber.-Tith	England	+50.6	−2.24	1	78	1	105	—	—	T	V	F-D-I	10	—	33	R(c) F(c)	—	5	Ogg et al. (1991c)
	Sierra Gerral		Brazil	—	—	—	—	—	—	—	—	—	—	—	—	—	—	—	—	—	Valencio et al. (1983)
	Sierra Paloma	Toarcian	Spain	+40.65	−1.1	1	47	1	75	—	—	T	Z-V	V	14	—	36	—	—	4	Galbrun et al. (1988b)
	Summerville & Curtis Fm.	Callovian	USA	+38.8	−111.1	2	457	1	120	—	—	T	V	F	9	—	92	R+	—	5	Steiner (1978)
	Umbria	Tithonian-Toar.	Italy	+43.33	+13	4	.5	1	200	—	1	T	Z-B	F-V	74	—	14	—	—	6	Channell et al. (1984)

the location of the Kimmeridgian/Tithonian boundary in the Italian sections place it in CM22 (Ogg *et al.*, 1984; Channell and Grandesso, 1987).

Early work on Jurassic magnetic stratigraphy was carried out on the Kimmeridgian–Oxfordian Morrison Formation in Colorado (Steiner and Helsley, 1975a) and the Callovian Summerville and Curtis formations in Utah (Steiner, 1978). Correlation to European sections is hindered by poor biostratigraphy in these terrestrial and near-shore sediments, complex magnetic behavior which compromises the fidelity of the magnetostratigraphic records, and variable sedimentation rates which distort the polarity zone pattern. Steiner *et al.* (1994) have made the case that magnetic stratigraphy can be used as a means of correlation in the Morrison Formation in Colorado and New Mexico.

b. Oxfordian–Callovian Magnetic Stratigraphy

The correlations of European land section magnetic stratigraphies to M0–M25 oceanic magnetic anomalies are fairly robust; however, magnetostratigraphic correlation to M26–M38 has not been well established. CM26 to CM30 have been correlated to Oxfordian ammonite zones in northern Spain (Steiner *et al.*, 1985/1986; Juárez *et al.*, 1994, 1995); however, the correlation between land sections and oceanic anomaly records remains somewhat tenuous. The difficulty in correlation is partly a result of discontinuous and low mean sedimentation rates (~1–3 m/My) in the condensed limestone facies. Similarly condensed Callovian–Oxfordian sediments in Monti Lessini (northern Italy) yield polarity reversals and sporadic ammonite control (Channell *et al.*, 1990a) but the polarity pattern cannot be correlated to the oceanic magnetic anomaly record. The existence of polarity reversals in the Callovian–Oxfordian is confirmed by studies of short sections from the Krakow Uplands (Poland) (Ogg *et al.*, 1991a); however, here again the lack of long, continuously deposited sedimentary sections does not allow a convincing correlation to the oceanic anomaly record. Although the magnetostratigraphic correlation to the M26–M38 oceanic magnetic anomaly record has not been achieved, it is clear that the Jurassic "quiet zone" in the central Atlantic magnetic anomaly record is not due to a prolonged interval of normal polarity, as had been implied by magnetostratigraphic study of the Valdorbia section (central Italy) (Channell *et al.*, 1984). It is now clear that the Callovian–Oxfordian interval, which correlates to the Jurassic quiet zone, was an interval of frequent polarity reversal (Fig. 11.8, Table 11.2).

c. Pre-Callovian Jurassic Magnetic Stratigraphy

For the Bajocian and Bathonian, several sections from southern Spain indicate a very high frequency of polarity reversal (Steiner *et al.*, 1987), an average reversal rate of at least 5.5 reversals/My for the Bajocian. The mean sedimentation rate in these sections is in the 1–4 m/My range. The high frequency of reversal and the lack of any discernable "fingerprint" in the polarity zone patterns have precluded clear correlation among land sections. The elucidation of the magnetic stratigraphy in this interval will be aided by studies of more expanded sections. Such sections occur in central Italy but they lack ammonites and therefore await refinement of nannofossil and radiolarian biostratigraphies in this interval.

One of the most important magnetostratigraphic studies in the Jurassic is that of Horner and Heller (1983) from the Breggia section (southern Switzerland) (Fig. 11.9). The sedimentation rates in the nodular limestones at Breggia are about 14 m/My for the Pliensbachian and 4–7 m/My of the Toarcian and Aalenian. These sedimentation rates are several times greater than for more typical Ammonitico Rosso-type limestones, and this results in a considerable improvement in the clarity of the magnetostratigraphic record. In addition, the ammonite biostratigraphy for the Pliensbachian–early Bajocian interval at Breggia has been determined in detail by Wiedenmayer (1980). Márton *et al.* (1980) acquired a magnetic stratigraphy from the

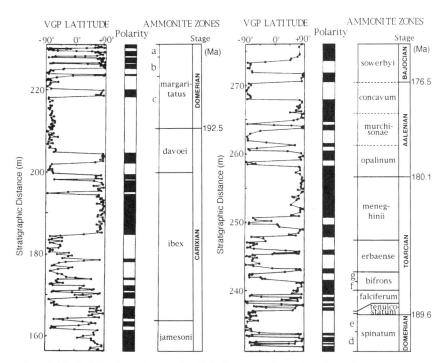

Figure 11.9 Carixian (Pliensbachian) to Bajocian magnetic stratigraphy and ammonite zones at Breggia (Switzerland) (after Horner and Heller, 1983).

9-m-thick Bakonycsernye section (Hungary). Here sedimentation rates in the Pliensbachian are about 1.5 m/My and 10 reversals are recognized in this stage. Pliensbachian sections in Italy with similar sedimentation rates yielded comparable numbers of reversals (Channell *et al.*, 1984). However, at Breggia, the sedimentation rate for the Pliensbachian is an order of magnitude greater (~14 m/My) and 39 reversals are recognized in the stage. Correlation between Breggia and Bakonycsernye shows that seven normal polarity zones at Breggia have been concatenated into one at Bakonycsernye. There is clearly a problem in resolving complete magnetic stratigraphies in the Ammonitico Rosso-type facies when sedimentation rates are a few meters/My, probably because of frequent lacunae even in sections where ammonite finds imply continuity of sedimentation. The Toarcian stratotype sections at Thouars and Airvault (France) are gray marls and limestones, and their magnetic stratigraphies have been resolved by Galbrun *et al.* (1988a). As the sections are about 5 m thick and the entire stage is represented, the mean sedimentation rates are less than 1 m/My. Nonetheless, a reasonable magnetostratigraphic correlation can be made from the stratotype sections to the Breggia section. The magnetic stratigraphy provides a means of correlation of the West European ammonite zonation in the stratotype sections to the Tethyan ammonite zonation at Breggia.

For the Hettangian and Sinemurian, three red nodular limestone sections in Austria (at Kendelbach Graben and Adnet) have been studied by Steiner and Ogg (1988). Correlations among sections are hampered by the condensed nature of the sedimentation and the very poor biostratigraphic control in these sediments. The same problems affect gray/white pelagic limestones of this age in central Italy (Channell *et al.*, 1984). The best quality Hettangian/Sinemurian magnetic stratigraphy is from a core drilled in the Paris Basin (Yang *et al.*, 1996) indicating high reversal frequency in the vicinity of the Hettangian–Sinemurian stage boundary.

11.3 Correlation of Late Jurassic–Cretaceous Stage Boundaries to the GPTS

Late Jurassic and Cretaceous chronostratigraphy is based on stage stratotypes defined by ammonite zones. However, due to absence or uneven distribution of ammonite faunas in land sections and oceanic cores, Cretaceous chronostratigraphy is often based on calcareous microplankton biostratigraphy and the supposed correlation of these events to ammonites. A review of calcareous nannofossil events in sections with ammonites, microplankton, or magnetic stratigraphy has revealed the uncertainties in correlation of nannofossil and calpionellid datum planes to stage boundaries (Figs.

11.6 and 11.7). All the nannofloral datums have been directly correlated to polarity chrons. Ammonite biozones have been directly correlated to magnetic stratigraphy in parts of the Oxfordian–lowermost Valanginian (see Ogg *et al.*, 1991c), Valanginian–Hauterivian (Channell *et al.*, 1995b), and uppermost Hauterivian–Barremian intervals (Cecca *et al.*, 1994).

The Oxfordian/Kimmeridgian boundary is correlative to the base of the *Sutneria platynota* ammonite zone, which has been correlated to CM25 in southern Spain (Ogg *et al.*, 1984). The Kimmeridgian/Tithonian boundary is correlative to the base of the *Hybonoticeras hybonotum* ammonite zone, also in southern Spain (Ogg *et al.*, 1984).

The Tithonian/Berriasian (Jurassic/Cretaceous) boundary does not have a universally accepted definition. Many of the candidate biomarkers have been correlated to the polarity chrons (Ogg and Lowrie, 1986; Channell and Grandesso, 1987; Bralower *et al.*, 1989; Ogg *et al.*, 1991c). The base of CM18 has become the generally accepted correlation to the Tithonian/Berriasian boundary.

The Berriasian/Valanginian boundary is defined by the base of the *T. otopeda* ammonite zone and falls within CM15n (Ogg *et al.*, 1988) and between the first occurrence (FO) of *Cretarhabdus angustiforatus* and the FO of *Calcicalathina oblongata* (Fig. 11.6). The Valanginian/Hauterivian boundary coincides with the base of the *A. radiatus* ammonite zone and is close to the FO of *Nannoconus bucheri* and at the base of CM11n (Channell *et al.*, 1995b). The base of the *S. hugii* ammonite zone defines the Hauterivian/Barremian boundary, which falls between the last occurrence (LO) of *Lihtraphidites bollii* and the LO of *Calcicalathina oblongata* and in the upper part of CM4 (Cecca *et al.*, 1994; Channell *et al.*, 1995b).

The Barremian/Aptian boundary was formally defined at the first occurrence of *Deshayesites*. None of the nannofossil events proposed by Thierstein (1973) to define this boundary have proved to be reliable. The FO of *Rucinolithus irregularis* is correlated to the upper part of the *C. sarasini* ammonite zone and is therefore slightly older than the Barremian/Aptian boundary (Channell and Erba, 1992). The base of CM0 coincides closely to this boundary, but direct correlation of this polarity chron with the ammonite biozone has not been documented.

The Santonian–Campanian is formally defined by the appearance of the ammonite *Placenticeras bidorsatum,* the index species of the oldest Campanian ammonite zone. Unfortunately, this species is extremely rare even in the type area (northwest Europe) and the Santonian–Campanian boundary is often informally defined on the basis of the FO of the nannofossil *Broinsonia parca* and/or the FOs of foraminifera *Globotruncana arca* and *Bolivinoides strigillatus.* The microfossil markers of the Santonian–Campanian boundary result in a correlation of this boundary to the basal part of C33r, close to the top of the Cretaceous normal superchron (e.g., Alvarez *et al.*, 1977; Channell *et al.*, 1979; Monechi and Thierstein, 1985).

The type area of the Campanian–Maastrictian boundary (northwest Europe) is characterized by a major hiatus, and hence there is more than average debate over the definition of this stage boundary. In the pelagic realm, the LO of the foraminifera *Globotruncanita calcarata* is often used as an informal definition, and this event usually appears in the top part of C33r (e.g., Alvarez *et al.*, 1977).

12

Triassic and Paleozoic Magnetic Stratigraphy

12.1 Introduction

The absence of preserved marine magnetic anomalies for times preceding the breakup of Pangea forces a change of strategy in the study of pre-Late Jurassic magnetic polarity stratigraphy. Irving and Pullaiah (1976) suggested the use of type sections to build the pre-Jurassic GPTS in the absence of the oceanic magnetic anomaly template. Up to now, type sections for magnetic stratigraphy have not been established; however, we believe that the type section approach would be helpful for pre-Late Jurassic time. Type sections for magnetic polarity sequences will have the same problems presently associated with type sections in classical stratigraphy, such as unconformities, poor age control, lack of diagnostic fossils, and inadequate descriptions. The ideal magnetostratigraphic type section would have not only a well-defined magnetic stratigraphy but also high-resolution biostratigraphy facilitating correlation to geologic stages. Many limestones with good biostratigraphic control have very weak magnetization intensities or are remagnetized. Red siltstones and sandstones, on the other hand, often carry a well-defined primary magnetization but are often difficult to correlate to the standard geologic stages due to lack of diagnostic fossils.

In the Soviet Union, a hierarchical system of hyperzones (megachrons), superzones (superchrons), and zones (chrons) was established for Cambrian to Permian time (Khramov and Rodionov, 1981). Unfortunately, the original data on which this scheme was based are not available to Western researchers, and descriptions of type localities were not given. The proposed GPTS (Fig. 12.1) is, therefore, difficult to evaluate. In intervals for which

Modified from KHRAMOV 1987

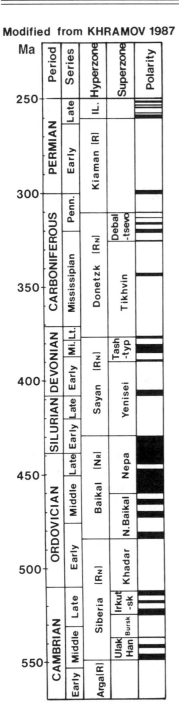

Figure 12.1 Geomagnetic polarity time scale derived from the Paleozoic of the former USSR. Normal polarity (black), reverse polarity (white) (after Khramov and Rodionov, 1981).

more recent data are available, such as the middle Carboniferous, many more polarity chrons are known to exist.

12.2 Triassic

Triassic magnetic stratigraphy was pioneered by Picard (1964) and Burek (1964, 1970) and expanded in clastic sediments from the western United States by C. Helsley and his students in the early 1970s (e.g., Helsley, 1969; Helsley and Steiner, 1974). Upper Triassic magnetic stratigraphy is best known from the lacustrine sediments of the Newark Basin. The initial magnetostratigraphic results from the Newark Basin were from industry boreholes and outcrops (McIntosh *et al.*, 1985). The upper part of the sequence above the Watchung Lava flows, as well as the flows themselves, are considered Early Jurassic (Hettangian) in age on the basis of pollen stratigraphy. These units are normally magnetized and were labeled the Upper Normal Interval (McIntosh *et al.*, 1985). This normal polarity zone extends down into the Upper Triassic (Rhaetian). Lower in the section the sediments were found to be characterized by mixed polarity over a stratigraphic thickness exceeding 2500m. Later studies by Witte and Kent (1989) and Witte *et al.* (1991), based on more complete demagnetization procedures and denser sampling, along with field tests for the age of the magnetization, led to a more complete documentation of the polarity sequence for sediments from the Newark Basin. From the Late Triassic Carnian stage to the earliest Jurassic Hettangian stage, at least six normal magnetozones were identified in the Newark Basin outcrops (Fig. 12.2, column 16).

The excellent results obtained from the Newark Basin outcrops led to a study of drill cores. A series of seven drill sites were located to obtain a complete record of the sedimentary history of the basin. Core recovery was almost 100% and overlapping cores, each about a kilometer in length, resulted in a complete magnetostratigraphic record of lacustrine beds covering the time interval from the early Carnian to the Triassic/Jurassic boundary. The magnetostratigraphic study of these sediments is enhaced by the presence of a well-defined cyclostratigraphy tuned to Milankovitch frequencies. The stratigraphy was divided into lithologic units, called McLaughlin cycles, which are believed to represent the 413-ky eccentricity cycle of the Earth's orbit. An age of 201 Ma based on ^{40}Ar/^{39}Ar and U/Pb techniques has been obtained for the lowest flow unit of Watchung basalts at the top of the sedimentary sequence (Sutter, 1988; Dunning and Hodych, 1990). It is, therefore, possible to date events within the drill cores by counting down from the Watchung basalts using the cyclostratigraphy.

The magnetic stratigraphy of these cores has been studied by Kent *et al.* (1995). Samples were taken from the drill cores at 2-m intervals. All samples were analyzed by progressive thermal demagnetization and oriented using the direction of the magnetic overprint. The magnetic stratigraphy from the Weston core (Fig. 12.3) illustrates the high quality of the data. The sedimentary rocks in the Newark Basin dip to the north and the drill sites were located so that the stratigraphies from each core overlap. This has resulted in overlapping patterns of polarity zones (Fig. 12.4). The borehole studies confirmed the previously determined outcrop studies, but the increased resolution led to an increase in the number of polarity zones from 12 to 42, an increase of almost a factor of 4. The Milankovitch cyclostratigraphy places the sediment sequence in a precise time frame and yields a reversal frequency of 2 My^{-1} over late Carnian and Norian time (Fig. 12.2, column 19). Witte *et al.* (1991) have suggested that the Newark Basin qualifies as the type locality for magnetic polarity stratigraphy of the late Carnian and Norian interval. The lack of a marine biostratigraphy is offset by the essentially complete magnetic stratigraphy and the precise cyclostratigraphic timeframe.

Middle and Late Triassic magnetic stratigraphies have been obtained from condensed marine "Hallstatt-type" reddish nodular limestones at two localities in southwest Turkey and from Austria (Gallet *et al.,* 1992, 1993, 1994) (Fig. 12.2, columns 6, 9, 10, 17). The sedimentary sections are not thick (~30 m), and unconformities are present. The quality of the magnetostratigraphic data is excellent and field and laboratory tests for stability have been carried out. There are, however, discrepancies in correlation between the two Turkish sections which the authors attribute to the sections originating in different hemispheres. The southernmost section has been correlated to a third section in Austria, the polarity of which is thought to be unambiguous as it must have originated in the northern hemisphere. The polarity sequence is not easily correlated to that of the Newark Supergroup.

Kent *et al.* (1995) have offered a partial correlation of the Turkish sections to the Newark Basin record. They correlate the Kavur Tepe magnetozones a+, e+, and g2+ (Fig. 12.2, column 9) to magnetozones d+, h+, and j+ (Fig. 12.2 columns 16, 19). They have also suggested a correlation of the Chinle section (Reeve and Helsley, 1972) to the Newark sequence. The Chinle section has four normal magnetozones (Fig. 12.2, column 1), the youngest of which (N4) is correlated to l+ of the Newark sequence (Fig. 12, column 16). N2 and N3 of the Chinle section could be correlated to j+ of the Newark sequence.

The Carnian/Norian boundary based on pollen in the Newark sequence occurs between the upper part of polarity zone b+ and the base of polarity zone d+. This boundary, based on conodonts, is present in the sequence at

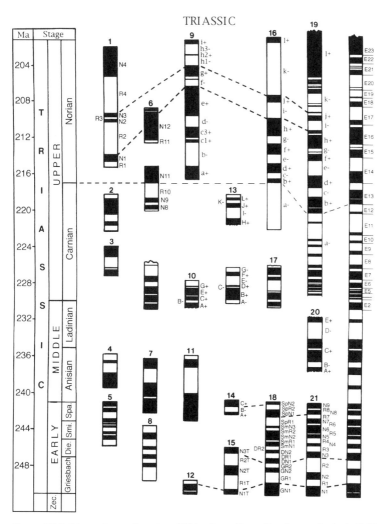

Figure 12.2 Magnetic stratigraphy of Triassic sections. Numbers refer to individual studies in Table 12.1. Right-hand column is composite section.

Bolucektasi Tepe just prior to a boundary from reverse to normal polarity, informally designated N11 (Fig. 12.2, column 6); prior to N8 in this column the quality of the magnetic stratigraphy deteriorates and the VGPs fall to low latitudes. Kent *et al.* (1995) tentatively correlated the base of polarity zone N11 to the base of b+ in the Newark magnetic stratigraphy; however, a secure pattern match is not possible. Unfortunately, the base of the

Table 12.1
Triassic

	Rock unit	Lo Age Hi	Region	λ	Φ	NSE	NSI	NSA	M	D	RM	DM	AD	A	NMZ	NCh	%R	RT	F:C:T	Q	References
1	Chinli	Norian	USA, NM	+35.03	−104.08	2	.5–.9 m 240	1	145	—	I-O	T	B	F	9	5	49	R+	—	5	Reeve and Helsley (1972)
2	Chinli	Carnian	USA, NM	+35	−106.5	3	20	5	55	—	I-O	T	Z-P	F	5	3	51	R−	C+ F+	7	Molino-Garza et al. (1991)
3	Fleming Fjord	Carnian	East Greenland	+71.73	−23.41	1	134	1	12.2	—	—	T	B	F-D-I	5	—	60	—	—	3	Reeve et al. (1974)
4	Moenkopi	Anisian	USA, NM	+35	−106.5	4	75	5	110	—	I-O	T	Z-P	F	19	6	51	R−	F+ C+	8	Molino-Garza et al. (1991)
5	Moenkopi	F. Triassic		+38.58	−108.93	3	.35 m	1	150	—	—	—	B	F-D-I	15	12	62	—	—	5	Helsey and Steiner (1974)
6	Bolücektasi Tepe	Norian-Carnian	Turkey	+37	+30	1	262	1	72	—	I-Tdl	A-T	Z-P	F-V	23	—	26	R+	—	7	Gallet et al. (1992)
7	Moenkopi	Ani. Spa.	USA, Ari	+35.8	−111.45	2	.3 m 800	1	118	—	—	T	B	F D-T	5	3	84	R+	F+ C+	6	Purucker et al. (1980)
8	Chugwater Fm.	E. Triassic	USA	+43	−107.00	5	563	1	179	—	J	T	B	F-D-I	23	5	80	R+	F+	7	Shive et al. (1984)
9	Kavur Tepe	Norian	Turkey	+38	+30.5	1	179	1	30	—	I-Tdl	T-A	Z-P	F-V	15	—	50	—	—	8	Gallet et al. (1993)
10	Mayerling	Carnian	Austria	+48.05	+16.13	2	67	1	25	—	1	T	Z-P	F-D-I	8	—	40	R+	—	8	Gallet et al. (1994)
11	Anton Chico	Anisian	USA, NM	+34.9	+105.7	2	87	1	23	—	O	T	Z-P	F-D-I	6	—	46	—	—	6	Steiner and Lucas (1992)
12	Feixianguan Fm.	Die-Greis.	China	+32	−105.5	1	200	1	130	—	1	T	Z-P	F-V	6	—	30	R−	—	7	Heller et al. (1988)
13	Mezartik	Carnian	Turkey	+36.53	+30.56	1	53	1	5	—	1	T	Z-P	F-D-I	14	—	47	R+	—	8	Gallet et al. (1994)

No.	Locality	Age	Country	Lat	Long																Ref	Reference
14	Chios	Anisian	Greece	+37.31	+22.82	4	125	1	13	—	I	—	T	Z-P	F-V	3	—	—	R+	F+	9	Muttoni et al. (1995)
15	Russia		Russia			2	150	1	1600	—	—	—	—	—	F-DI	7	—	57	R-	—	3	Gurevich and Slautsitays (1985)
16	Newark Supergroup	Het.-Carnian	USA	+40.25	-75.25	3	24	5	4180	K	1	—	T	Z-P	F-V	4	—	55	R+	F+	9	Witte et al. (1991)
17	Bolucektasi Tepe	Norian	Turkey	+37	+30	1	262	1	72	—	—	—	T	Z-P	F-V	23	—	26	R+	—		Gallet et al. (1992)
18	Arctic Archipelago	Spt.-Grie	Canada	+80.5	-95	3	342	1	1100	—	—	—	T	Z-P	F-DI	27	17	62	R+	F+	9	Ogg and Steiner (1991)
19	Newark Supergroup	Het.-Norian	USA	+40.75	-75.5	7	1695	1	4660	K	1	—	T	Z-P	F-I	45	12	51	R+	F+	10	Kent et al. (1995)
20	Hydra	Anisian-Ladinian	Greece	+37.3	+23.43	1	36	1	23	—	1	—	T	Z-P	F-V	8	—	52	R+(c)	—	7	Muttoni et al. (1994)
21	Feixianguan Fm.	Spat-Gries	China	+29.9	+106.3	4	501	1	960	—	—	—	T	Z-P	F-DI	16	12	65	R+	F+	9	Steiner et al. (1989)
22	Wombat Plateau	Norian-Carnian	Australia	-16.95	+115.61	2	356	1	400	—	—	—	A-T	Z-P	1	16	—	63	—	—	6	Galbrun et al. (1992)
23	Iberia	Zeoh-Carnian	Spain	+29.9	+115.61	4	200	1	900	—	1-O	—	T-A	Z-P	F	10	—	37	R+	—	4	Turner et al. (1989)

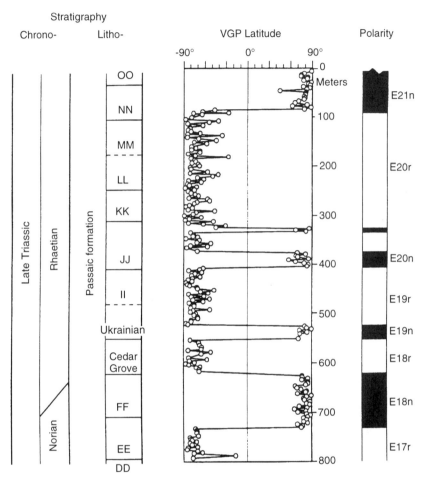

Figure 12.3 Magnetic stratigraphy of the Weston core, Newark Basin (after Kent *et al.*, 1995). Polarity chrons are labeled as normal and reverse pairs, with the reverse overlying normal.

Carnian is not identified in the Newark basin. In the Mayerling section (Austria) (Fig. 12.2, column 10), the Ladinian/Carnian boundary occurs in the normal polarity zone A+. At other sections, this boundary occurs at unconformities (Gallet *et al.*, 1994).

Middle Triassic magnetic stratigraphy is less well established than that of the Late Triassic. In the Middle Triassic Moenkopi Fm. of Arizona and New Mexico, the biostratigraphy must rely largely on rare vertebrate finds. The magnetic stratigraphy of the Moenkopi Formation is well known

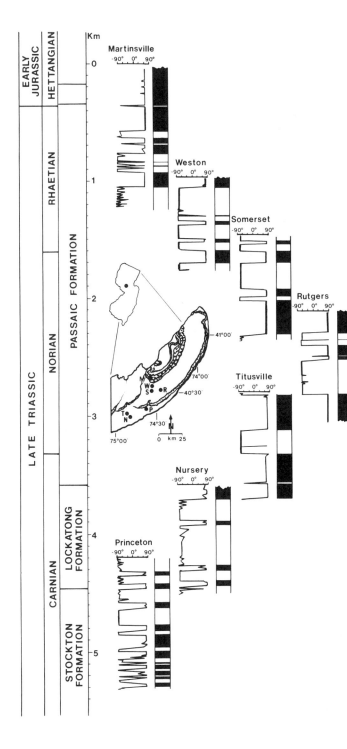

Figure 12.4 Correlation of Triassic cores from the Newark Basin (after Kent *et al.*, 1995). Inset indicates location of cores in Newark Basin. VGP latitudes are depicted by a line passing through the data points.

through the studies of Helsley (1969), Helsley and Steiner (1974), Purucker *et al.* (1980), and Molina-Garza *et al.* (1991). Steiner *et al.* (1993) place the upper part of this formation in Arizona and Colorado in the early Anisian. Molina-Garza *et al.* (1991), on the other hand, place a section of the Moen-kopi from New Mexico in the late Anisian/Ladinian, based on fossil verte-brates (Fig. 12.2, column 4). Steiner and Lucas (1992) have obtained a magnetic stratigraphy from the Anton Chico member of the Moenkopi Fm. which they place in the Anisian on the basis of vertebrate fossils (Fig. 12.2, column 11). The polarity zone patterns from the Anton Chico member given by the two studies from New Mexico are very similar and are said to correlate to the Holbrook member of the Moenkopi Fm. in Arizona.

The only studies of marine sections from the Middle Triassic are by Muttoni *et al.* (1994) and Muttoni and Kent (1994). The first of these studies delineates the magnetic stratigraphy across the Anisian/Ladinian boundary from a section on Hydra Island (Greece). The quality of the data is good and the stage boundary occurs within a normal polarity zone informally designated C+ (Fig. 12.2. column 20). The other study is on limestones from the Southern Alps and the section is in sediments of the latest Anisian age. The section studied is entirely normally magnetized and probably correlates with magnetozone C+ of the Hydra section. Magnetostrati-graphic results have been reported from the Middle Triassic Muschelkalk Fm. of Spain by Turner *et al.* (1989). The quality of the magnetostratigraphic data from this study is poor. A large proportion (85%) of the samples did not yield interpretable demagnetization data, and this magnetic stratigraphy is not included in Figure 12.2.

Lower Triassic results have been obtained by Steiner *et al.* (1989) from the Feixianguan and Jialingiang formations of Sichuan (China) from a 1-km-thick section which ranges in age from the Permo–Triassic boundary to the Spathian (upper part of Lower Triassic) (Fig. 12.2, column 21). In a recent compilation (Steiner *et al.,* 1993), stratigraphic gaps are shown in the Spath-ian, but how these are recognized is not clear. Ogg and Steiner (1991) have studied the stratotype sections of the Early Triassic from Ellesmere and Axel Heiberg islands in the Canadian Arctic. The data quality of the three sections is not as high as that from the Sichuan; however, the magnetic stratigraphy is readily interpretable (Fig. 12.2, column 18). If these two sections cover the entire Early Triassic then the correlation of polarity zones would seem to be straightforward, since nine normal polarity zones are present in each section (Fig. 12.2). More detailed biostratigraphy is needed to solve the problem of possible stratigraphic gaps in the late Early Triassic of Sichuan. Unfortu-nately the boreal fauna of the Arctic cannot be easily correlated to the tropical fauna of the Chinese sections, inhibiting unequivocal magnetostratigraphic correlation. The magnetic stratigraphy of the boundary of the Early/Middle

Triassic (Spathian/Anisian) has been studied by Muttoni *et al.* (1995) in fossiliferous limestones from Chios Island (Greece). The stage boundary, based on ammonites, falls within normal magnetozone C+ (Fig. 12.2, column 14). The authors correlate normal magnetozone C+ at Chios with normal polarity zone SpN2 in the Arctic sections and with N8 in the Sichuan section, although a correlation to N9 appears more likely.

Magnetostratigraphic studies across the Permo–Triassic boundary indicate that a polarity zone boundary from reverse to normal is present at, or shortly above, the Permo–Triassic boundary. Studies in the former Soviet Union and China (Steiner *et al.*, 1989; Heller *et al.*, 1988) indicate that the basal Triassic normal polarity zone corresponds to most of early Griesbachian time.

McElhinny (1973), in an analysis of polarity bias, concluded on the basis of data available at that time that the geomagnetic field during Triassic time was 75% normal polarity. In Figure 12.2, the right-hand column represents a composite magnetic stratigraphy, and it can be seen that the magnetic field is not biased toward normal polarity. The reversal sequence given here for the Triassic is most reliable for the Early Triassic, upper Carnian, and Norian times and least secure in the Middle Triassic.

12.3 Permian

Irving and Parry (1963) were the first to recognize the long period of reverse polarity spanning Late Carboniferous and Permian time. They named it the Kiaman Magnetic Interval, after the name of the town in New South Wales (Australia) where rocks showing reverse polarity were first reported by Mercanton (1926). The name Kiaman has been retained by Russian workers but was abandoned by Irving and Pullaiah (1976) and replaced by the Permo–Carboniferous Reverse Superchron (PCRS). The PCRS is known to end within the Late Permian. Irving and Parry (1963) originally placed the termination of the Kiaman at the beginning of the Narabeen Chocolate Shales of the Sidney Basin (Australia), which are Late Permian and Early Triassic in age. They called this polarity change the Illawara reversal. A restudy of this formation was carried out by Embleton and McDonnell (1981), who confirmed a reverse to normal polarity change in red claystones near the base of the Triassic. The older sediments in this sequence are unstably magnetized so a complete magnetic stratigraphy could not be established.

The Upper Permian and Lower Triassic sequences in North America, Europe, and in many of the Gondwanaland continents often contain important unconformities. Upper Permian marine sediments are present, however, along the northern margin of Gondwanaland (Pakistan and India) as well as in China. The best quality magnetic stratigraphy for later Permian time comes from the Wargal and Chidru Formations of the Salt Range in northern Pakistan (Haag and Heller, 1991) (Fig. 12.5, column 13). These sections comprise mainly carbonate sediments and yield an excellent fauna allowing correlation to other sections. The magnetostratigraphic record spans late Kazanian and Tartarian time, with six normal and five reverse polarity zones. The top of the PCRS was not determined as the lower part of the Wargal Formation is normally magnetized. Unfortunately, it has become apparent that there is an unconformity at the top of the Permian in this section; sediments equivalent to Changxingian time in the Chinese sequence are apparently missing (Wignall and Hallam, 1993). The biostratigraphic and magnetostratigraphic correlation of the Salt Range sequences to China (Fig. 12.5, columns 11 and 12) (Steiner *et al.*, 1989; Heller *et al.*, 1988) is therefore tenuous, although the correlation of the thick normal polarity zone near the top of the sequences in both China and Pakistan is reasonable (Fig. 12.5).

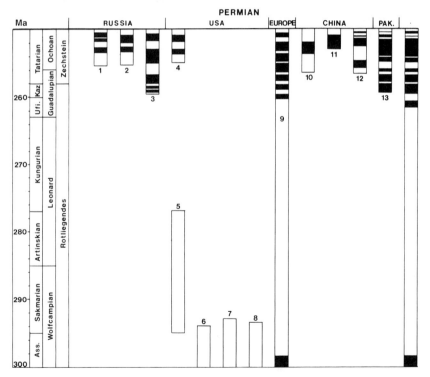

Figure 12.5 Magnetic stratigraphy of Permian sections. Numbers refer to individual studies in Table 12.2. Right-hand column is composite section.

Table 12.2
Permian

	Rock unit	Lo Age Hi	Region	λ	Φ	NSE	NSI	NSA	M	D	RM	DM	AD	A	NMZ	NCh	%R	RT	F.C.T	Q	References
1	Tartarian	Tartarian	E. USSR	—	—	—	—	—	—	—	—	—	—	—	—	—	—	—	—	—	—	Molostovsky (1992)
2	Tartarian	L Tri-Tar	USSR	—	—	2	2.5–5 m 760	1	1500	—	V	T	B	F	6	—	57	R–	—	5	Gurevich and Slautsitays (1985)
3	Tartarian	Tartarian-Kz	Urals	—	—	—	—	—	—	—	—	—	—	—	—	—	—	—	—	—	Khramov (1974)
4	Dewey Lake	Ochoan	USA, W. Tex.	+34.5	–101.5	3	19	5	52	K	I	T	Z-P	F	4	—	.34	R+	—	9	Molina-Garza et al. (1989)
5	Cutler Fm.	Permian	USA, Utah	+38.41	–109.58	1	.6 m	1	200	—	J	T	B	F DI	1	1	100	—	—	5	Gose and Helsley (1972)
6	Maroon-Minturn Fm.	E. Permian	USA, Colo	+39.73	–106.4	1	68	1–5	700	—	—	T	Z-P	F-D	1	1	100	—	F+	6	Miller and Opdyke (1985)
7	Casper Fm.	E. Permian-L. Carb.	USA, Wyo.	+41.45	–105.21	1	.37 m 549	1	190	—	J-I	T	Z-V	F	1	1	—	—	—	5	Diehl and Shive (1981)
8	Ingleside Fm.	E. Permian-L. Carb.	USA, Wyo.	+40.8	–105.2	1	243	1	70	—	J-I	T	Z-B	F	1	1	—	—	—	5	Diehl and Shive (1979)
9	Rotleigande	Ach-Rot.	Germany	+51	+10.30	40	—	3	4000	—	—	T-A	V	F	7	—	95	—	—	4	Menning et al. (1988)
10	Tarim	Tri-Permian	China, Tarim	+42.13	–83.35	1	21	5	870	—	—	T	Z-P-G	F	3	—	.72	R+	—	5	McFadden et al. (1988)
11	Dalong Fm.	Lopingian	China, Sich.	+32	+105.5	1	200	1	130	S	1	T	Z-P	F-V	6	—	29	R–	—	6	Heller et al. (1988)
12	Wujiaping	Changsing-Wujiaping	China, Sich.	+32.4	–106.4	2	85	1	120	—	1	T	Z-P	F-V	7	—	.58	R–	—	7	Steiner et al. (1989)
13	Wargal and Chidru	Zechstein	Pakistan	–32.6	+71.8	1	220	1	200	—	1	T	Z-P	F-V	10	4	38	—	—	6	Haag and Heller (1991)

The correlation of the Upper Permian marine sequence in Pakistan to nonmarine red beds which dominate the rock record in Russia, Western Europe, and North America and South America is not clear. A recent study of the Ochoan (Tartarian) Dewey Lake Formation in northern Texas (Molina-Garza *et al.,* 1989) shows at least two normal polarity zones in the Late Permian (Fig. 12.5 column 4). Khramov (1974) shows at least five normal polarity zones in sediments from the Upper Permian type section of the Ural Mountains (Fig. 12.5, column 3). The data quality appears to be high, and the results are internally consistent. Unfortunately, these sections have an unconformity in uppermost Permian and lowermost Trias sic, similar to the situation in the western United States.

A considerable amount of magnetostratigraphic work has been carried out on the Permian Rotliegende and Zechstein Formations of Germany, both in outcrop and bore holes (Menning *et al.,* 1988) (Fig. 12.5, column 9). There are normal polarity zones in the upper Rotliegende and through- out the Zechstein Formation. Menning *et al.* (1988) showed at least seven normal polarity zones in the Late Permian. This number of normal polarity zones is more or less consistent with the results from the former USSR and Pakistan, where five and six normal polarity zones have been recorded in this interval, respectively. In the composite polarity column for the Permian stage (Fig. 12.5, right column), eight normal magnetozones are given for Late Permian time, derived principally from the records from China, Pakistan, and central Europe.

The position of the top of the PCRS is older than the Tartarian of Russia, the Zechstein and upper Rotliegende Formations of central Europe, and the base of the Wargal Formation in the Salt Range of Pakistan. The Wargal Formation is Murgabian in age, correlative to the Guadalupian or Wordian of the North America and to the base of the Kazanian stage in Russia (Haag and Heller, 1991; Catalano *et al.,* 1991). The upper boundary of the PCRS in North America lies beneath Ochoan units of west Texas, which have an associated $^{40}Ar/^{39}Ar$ age of 251 ± 4 Ma on volcanic ash (Molina-Garza *et al.,* 1989). The boundary between the Permian and Triassic in China has been dated at 251 Ma by Claoué-Long *et al.* (1991) using the SHRIMP ion probe at the Australia National University. Taken at face value, the date from Texas would place the Ochoan magnetic stratigraphy very close to the Permian–Triassic boundary. Harland *et al.* (1990) have subdivided the Permian stage into the Rotliegende and Zechstein epochs in a bipartite division. The base of the Zechstein Epoch is given a numerical age of 256 Ma by Harland *et al.* (1990). Valencio *et al.* (1977) in Argentina have reported a normal polarity zone lying above an intrusion dated at 259 ± 7 Ma which they called the Q. Del Pimiento event. In the same region, Creer *et al.* (1971) also noted normally magnetized lavas at 258 ± 5

Ma. It is possible that these normal polarity zones may mark the end of the PCRS and are not subchrons within the PCRS. Further sampling in the Argentine sections may offer an opportunity to date the end of the PCRS. Due to the lack of reliable radiometric ages directly associated with sediments of Late Permian age, stage boundaries of the Late Permian are poorly constrained in time. The precise date of the end of the PCRS is therefore unknown. Haag and Heller (1991) believed it to be older than 261 Ma, whereas M. Menning (personal communication, 1992) gives an age of 265 Ma. In Figure 12.5, an age of 262 Ma is given for this boundary; however, it should be understood that nowhere is it directly dated.

Over the years, reports of short normal subchrons within the PCRS have been published in both Russian and Western literature. Helsley (1965) found normally magnetized sediments in the Dunkard Formation of West Virginia, which is believed to be Early Permian in age. The initial study was done with alternating field demagnetization, which is notoriously ineffective for removal of secondary overprints from hematite-bearing samples. Samples from the normal polarity horizon were subsequently subjected to thermal demagnetization by Gose and Helsley (1972) and were found to retain their normal polarity. Normal polarity magnetizations have been obtained from Lower Permian Pictou sediments of Prince Edward Island by Symonds (1990). The normal polarity zone in the Pictou beds may be correlative to that in the Dunkard Fm. of West Virginia. Menning *et al.* (1988), in their study of the Rotliegende Formation of eastern Germany, found normally magnetized sediments in the Early Permian part of this sequence which they placed at the Permian/Pennsylvanian boundary, implying a correlation with the Dunkard series of West Virginia. It does seem certain that one or more normal subchrons occur within lower Permian strata, although it is unclear whether the normal polarity zones recognized at different locations are correlative.

Several sections have been sampled across the Permo–Carboniferous boundary in the western United States (Diehl and Shive, 1979, 1981; Miller and Opdyke, 1985) and no normally magnetized rocks were detected. Sinito *et al.* (1979) gave evidence for a short subchron within the PCRS, positioned 100 m below a lava dated at 288 ± 5 Ma. This may be equivalent to the normal polarity subchron seen in the Dunkard Fm., although it was originally thought to be the base of the PCRS. Normally magnetized sites of Early Permian age have been reported in several tectonic studies (Halvorson *et al.*, 1989; Wynne *et al.*, 1983; Irving and Monger, 1987) and may be correlative with the Dunkard subchron. Although normal subchrons may occur within the PCRS, magnetostratigraphic studies by Diehl and Shive (1979, 1981), Miller and Opdyke (1985), and Magnus and Opdyke (1991) have shown that the PCRS is dominantly reverse polarity, as claimed by Irving and Pullaiah (1976) (Fig. 12.5, columns 6–9).

12.4 Carboniferous

The age of the base of the PCRS has not been resolved. In North America, McMahon and Strangway (1968) reported normal polarity zones in sediments of the Upper Demoinesian Minturn Formation of Colorado. The section was restudied by Miller and Opdyke (1985), who did not find any normal polarity zones in the section. The study was extended throughout the Minturn Formation by Magnus and Opdyke (1991), who determined that the reverse polarity zone (PCRS) extended to the base of the Minturn Fm., which is earliest Demoinesian or late Atokan in age.

Roy and Morris (1983) had shown that sediments with dual polarity are present in rocks as young as Westphalian A in the Maritime provinces of Canada. In the same region, DiVenere and Opdyke (1990, 1991b) documented an internally coherent set of polarity zones in the Maringouin, Shepody, and Claremont Formations of Namurian and Westphalian A age. At least seven normal polarity zones are present, giving a reversal rate of about 1 My^{-1} (Fig. 12.6). The age of magnetization is constrained to the Late Carboniferous by a positive fold test. An attempt was made to extend the magnetic stratigraphy to younger formations exposed along the shores of the Bay of Fundy. All samples were reversely magnetized; however, a data gap of 700 m occurs in the sequence (Fig. 12.6). These data yield the best and most complete reversal sequence now available for the early Late Carboniferous. The dual polarity for rocks of early Pennsylvanian age is supported by a study on red Morrowan paleosols from Arizona (Nick *et al.*, 1991).

The age of the beginning of the PCRS was originally determined to be 310 Ma based on the K/Ar age of the Paterson Volcanics in Australia. These volcanics were shown by Irving (1966) to be normally magnetized and were used to define as the base of the PCRS (Kiaman). Sites taken from the Seaham Formation, directly overlying the Paterson Volcanics, were reversely magnetized and were thought to be Late Carboniferous in age. Upper Carboniferous sites from the Currabubula and Lark Hill Formations were also reversely magnetized and passed the fold test. The base of the Kiaman was, therefore, thought to be firmly determined. However, a recent study of the Paterson Volcanics using single-crystal zircon dating by Claoué-Long *et al.* (1995) has yielded a date of 328 Ma, which places them within the Visean, much older than the normally magnetized sequences from the Canadian Maritimes (Fig. 12.6). Therefore, the Paterson

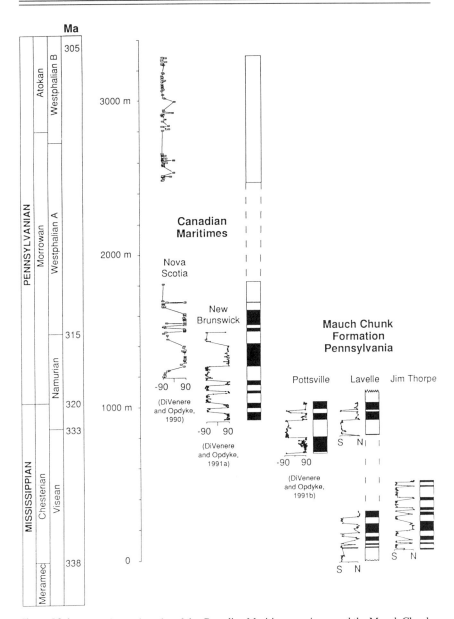

Figure 12.6 Magnetic stratigraphy of the Canadian Maritime provinces and the Mauch Chunk Fm. (Pennsylvania).

Volcanics cannot be used to fix the onset of the PCRS. Recent studies in .Australia (Opdyke *et al.*, 1960) have located the base of the Kiaman and dated it at approximately 320 Ma (Claoué-Long, personal communication).

The other significant data set that bears on this problem is the magnetic stratigraphy of the Middle Carboniferous strata of the Donetz Basin in the Ukraine. Khramov (1974) has presented a magnetic stratigraphy for the Upper Carboniferous which shows normal polarity zones as young as Upper Moscovian. Opdyke *et al.* (1993) restudied these sediments, and although it was possible to recover a prefolding magnetization which is probably Carboniferous in age, it was not possible to confirm Khramov's original magnetic stratigraphy. At this time, the only unimpeachable data that bear on the age of the base of the PCRS are those from Maritime Canada. The highest normal polarity zone in this sequence occurs within sediments of Namurian age and yields a duration for the PCRS of approximately 58 My (from 262 to 320 Ma).

In North America, the reversal pattern determined in the Lower Pennsylvanian has been extended into the Upper Mississippian (Visean) in sediments of the Mauch Chunk Formation of Pennsylvania (DiVenere and Opdyke, 1991a) (Fig. 12.6). Opdyke and DiVenere (1994) have reported a minimum of nine normal polarity zones within the Mauch Chunk Formation (Fig. 12.7, column 7) with a reversal frequency of 1.5 My^{-1}, similar to that of the early Pennsylvanian. The geomagnetic field appears to have been about 50% normal and 50% reverse in Middle Carboniferous time (Fig. 12.7).

Little is known of the polarity pattern for the rest of Early Carboniferous time; however, both polarities are present in sediments of Visean and Tournaisian age (Irving and Strong, 1985). The beginnings of a magnetic stratigraphy has been obtained from Lower Carboniferous lavas of the Midland Valley of Scotland of Torsvik *et al.* (1989) (Fig. 12.7, column 8). Turner *et al.* (1979) and Palmer *et al.* (1985) have reported magnetic stratigraphy from Carboniferous limestones from Great Britain; however, these limestones have been remagnetized (McCabe and Channell, 1994) and the magnetic stratigraphy is therefore unreliable.

12.5 Pre-Carboniferous

Middle and Early Paleozoic magnetic stratigraphies are largely restricted to the former Soviet Union. Lower Paleozoic data are mainly from the Siberian platform. In North America and Western Europe, severe remagnetization problems have resulted in very few reliable paleomagnetic pole positions and a paucity of magnetostratigraphic data.

A very condensed sequence across the Famennian/Frasnian boundary of the Late Devonian appears to record several reversals of the geomagnetic

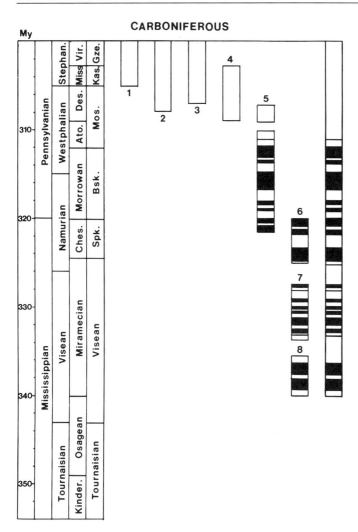

Figure 12.7 Magnetic stratigraphy of Carboniferous sections. Right-hand column is a composite. Numbers refer to individual studies in Table 12.3.

field (Hurley and Van der Voo, 1990). A reversal sequence of this age does not appear in the Russian GPTS. Douglas (1988) has studied an ~1 km thick section of the Ringerike Fm. of Norway which is considered to be Ludlow (Late Silurian) in age. Two formations were studied, the Sundvollen Fm. and Stubdal Fm. Most of the Sundvollen Fm. appears to be normally magnetized while the uppermost Sundvollen Fm. and overlying Stubdal

Table 12.3
Carboniferous

	Rock unit	Lo Age Hi	Region	λ	Φ	NSE	NSI	NSA	M	D	RM	DM	AD	A	MZ	NCh	%R	RT	F:C:T	Q	References
1	Supai Gp.	Virgillian	USA, Arizona	+35.2N	–113	3	145	1	48	—	—	T	Z-P	F	1	1	100	—	—	6	Steiner (1988)
2	Casper Fm.	E.P.-L.C.	USA, Wyoming	+41/45	–105.21	1	.37 549	1	190	—	J-I	T	Z-V	F	1	1	100	—	—	5	Diehl and Shive (1981)
3	Minturn Fm.	Vir-Des	USA, Colorado	+39.73	–106.4	11	68	1-5	700	—	—	T	Z-P	F-D	1	1	100	—	—	5	Miller and Opdyke (1985)
4	Minturn Fm.	Miss-Atoka	USA, Colorado	+38.2	–105.85	1	55	3	1280	—	—	T	Z-P	F	1	1	100	—	C+	6	Magnus and Opdyke (1991)
5	Cumberland Grp.	Na. West	N. Sco., Canada	+45.75	–64.38	1	177	1	2100	—	—	T	Z-P	F-V	9	—	78	R+	F+	7	DiVenere and Opdyke (1991)
	Cumberland Grp.	Na. West	Canada, N.B.	+45.75	–64.29	1	.5	1	600	—	—	T	Z-P	F-V	11	—	5	R+	F+	7	Divenere and Opdyke (1990)
6	Mauch Chunk Fm.	Late Miss.	USA, Penn.	+40.75	–76.42	1	105	1	340	—	—	T	Z-P	F-V	6	—	41	R+	F+	7	Divenere and Opdyke (1991a)
7	Mauch Chunk Fm.	Late Miss.	USA, Penn.			2	164	1	550	—	—	T	Z-P	F-V	22	—	53	R+	F+C+	7	Opdyke and DiVenere (1994)
8	Kinghorn	E. Carb.	U.K., Scotland	+56.2	–3.1	1	16	5	566	K	J-K	T-A	Z-P	F	4	—	46	R+	—	8	Torsvik et al. (1989)

Fm. are reversely magnetized. The normal polarity zone is well documented and the high unblocking temperature magnetization is most probably Silurian in age. Trench *et al.* (1993) have compiled Silurian paleomagnetic data from all continents. They identified 19 studies which they considered yield reliable polarity information. The compilation of this data set yields mixed polarity for early and later Silurian time; however, rocks of Wenlock (middle Silurian) age yield only normal polarities. This study provides an indication of reversal pattern but this type of analysis cannot take the place of magnetostratigraphic study of individual sections. It should also be noted that this compilation is in serious conflict with the Russian GPTS (Fig. 12.1).

For the Lower Paleozoic, the magnetic stratigraphy of the Ordovician is best known. Early work on Ordovician magnetic stratigraphy was reported by Rodionov (1966) and Khramov (1974) (Fig. 12.1). These early studies employed temperatures of 100°C and alternating peak fields of less than 10 mT. This low-level treatment is clearly inadequate by modern standards, and the reverse and normal directions are not 180° apart, indicating unresolved magnetic overprints. Nevertheless, it appears that the directions record two polarities and that the observed polarity zones are reproducible on a regional scale across the Siberian platform.

A second important study, but of more recent vintage, has been carried out on Ordovician (Llanvirn–Caradoc) limestones from Vastergotland, southern Sweden (Torsvik and Trench, 1991). The composite section is about 26.5 m in thickness and therefore very condensed. An internally consistent sequence of normal and reverse polarity zones has been obtained (Fig. 12.8, column 4). The top of the sequence, Caradoc in age, is normally magnetized. The Llandeilo is mainly reversely magnetized and is separated from the overlying and underlying parts of the sequence by gaps in data. The basal 9 m of section is Llanvirn in age and, although dominantly reversely magnetized, contains two normal polarity zones.

Trench *et al.* (1991) have reviewed the polarity of Ordovician paleomagnetic data worldwide. They point out that a study of the Upper Ordovician (Ashgill) Juniata Fm. of Pennsylvania by Miller and Kent (1989) indicates entirely normal polarity. Fourteen other studies which were designed to produce data for pole positions, rather than for magnetostratigraphic purposes, were judged by Trench *et al.* (1991) to possess reliable polarity information. These studies help to ratify the magnetic stratigraphy from the primary sections (Fig. 12.8, columns 1,4). An important result of these correlations is a reassessment of the biostratigraphic correlation between the Baltic Shield and the Siberian platform (Chugaeva, 1976; Trench *et al.*, 1991).

The Ordovician Period covers ~60 My and from data now available, it is certain that the magnetic polarity record for the Ordovician is much

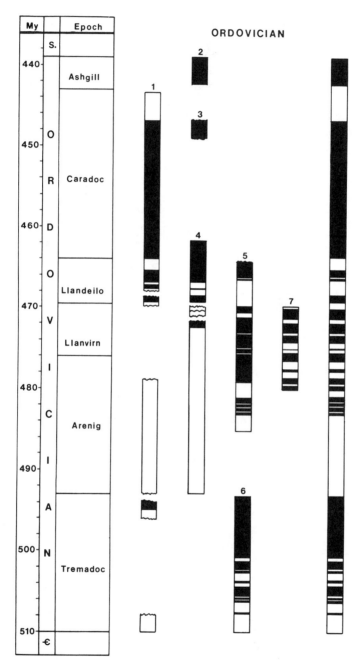

Figure 12.8 Magnetic stratigraphy of Ordovician sections. Right-hand column is a composite. Numbers refer to individual studies in Table 12.4.

Table 12.4
Cambro-Ordovician

#	Rock unit	Lo Age Hi	Region	λ	Φ	NSE	NSI	NSA	M	D	RM	DM	AD	A	MZ	NCh	%R	RT	F.C.T	Q	References
1	Lena River	Coradoc-Tremadoc	Russia	—	—	—	—	—	—	—	—	—	—	—	—	—	—	—	—	—	Khramov et al. (1965), cited in Trench et al. (1991)
2	Juanita Fm.	Ashgill	USA, Penn.	+40.5	−78	—	21	5	—	—	—	T	Z-P	F	1	—	—	—	F+	5	Miller and Kent (1989)
3	Llanbedrog and Mynytho Gps	Longvillian	N. Wales	+53.17	−3.50	—	69	—	—	K	J	A	V	F	1	—	—	—	F+	6	Thomas and Briden (1976)
4	Vustergotland	Caradoc-Llanvern	Sweden	+58.3	−13.9	3	43	1	20	O	J	T	Z-P	F-D	6	—	65	—	—	7	Torsvick and Trench (1991)
5	Lena River	Middle Ord.	Russia	—	—	1	.1	1	39	I	—	—	B	—	24	—	42	—	—	—	Khramov (1974)
6	Siberian Platform	Low. Ord.	Russia	—	—	1	231	1	90	—	—	—	—	DI	16	—	36	—	—	3	Metallova et al. (1984)
7	Everton Fm.	Ll-Arenig	USA, Arkansas	+36.2	−92.66	3	360	1-3	114.5	I	—	TA	Z-P	F	16	—	47	—	C+	8	Farr et al. (1993)
	Black Mtn.	E.O.-L.C.	Queensland, Aus.	+22.6	−14.03	1	169	1	1000	—	—	T	Z-P	F-D	13	—	67	R+	—	7	Ripperdan and Kirschvink (1992)

more complex than that proposed by Trench *et al.* (1991). A new study of Ordovician rocks from Arkansas (Farr *et al.,* 1993) has yielded 16 polarity zones spanning late Arenig and Llanvirn time (Fig. 12.8, column 7) and a section presented by Khramov for the Middle Ordovician of the Lena River also shows many (~22) polarity zones. Unfortunately, faunal information is lacking from the Lena River section. What seems to be certain, however, is that the Middle Ordovician is a time of relatively frequent reversal of the geomagnetic field (Fig. 12.8). A study of a Lower Ordovician section from the Lena River yields at least 16 polarity zones in Tremodoc sediments (Metallova *et al.,* 1984) (Fig. 12.9; Fig. 12.8, column 6). The base of the Early Ordovician appears to be reversely magnetized in this section. It is clear, when the original data are compared to the polarity scale presented by Khramov and Rodionov (1981) (Fig. 12.1), that their scale was greatly simplified.

The most important magnetic reversal studies in rocks of Cambrian age are from the Siberian Platform (Khramov, 1974; Kirschvink and Rozanov, 1984). The stratigraphy presented by Khramov spans the time interval from the middle Cambrian into the Ordovician. Data are now available from Australia from an expanded section at Black Mountain (Queensland) for Late Cambrian and Early Ordovician time (Ripperdan and Kirschvink, 1992) (Fig. 12.10). The section in Australia is 1 km thick, while the Siberian sections rarely exceed 100 m in thickness. An unambiguous correlation between these two regions is not possible, but there are some common features in the magnetostratigraphic records. Both studies show reverse polarity at the Cambro–Ordovician boundary and both show five normal polarity zones in the Late Cambrian. A more definitive correlation between the two regions will require further study of the Russian sequences and a more complete presentation of the biostratigraphy of the Siberian sections. Kirschvink and Rozanov (1984) have analyzed sediments from the lowermost Cambrian of the Siberian Platform, and a magnetic stratigraphy across the Tommotian–Atdabanian boundary was obtained. The magnetization components pass a fold test; however, they yield a pole position which is far away from the Siberian apparent polar wander path, making polarity designations ambiguous. Kirschvink *et al.* (1991) have inverted the initial polarity designations based on changing ideas of paleogeography. The correlation of this pattern to a similar age record from the Amadeus Basin of Australia (Kirschvink, 1978) is not possible, perhaps because of unconformities in the Australian record. A magnetostratigraphic record from this time interval given by Wu *et al.* (1989) is now thought by the authors to be the result of remagnetization.

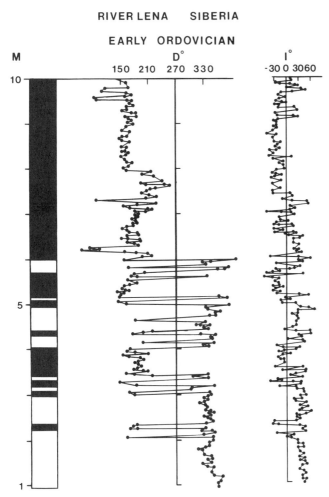

Figure 12.9 Magnetic stratigraphic study of Early Ordovician sediments from the Lena River, Siberia (after Metallova *et al.*, 1984). D and I denote declination and inclination, respectively. Polarity interpretation is reproduced in Fig. 12.8, column 6.

12.6 Polarity Bias in the Phanerozoic

The reversal record of the geomagnetic field from the Carboniferous to present is known well enough to allow us to examine the overall reversal pattern for the last 330 My of Earth history. It was previously thought that long segments of this record were characterized by polarity bias. Except

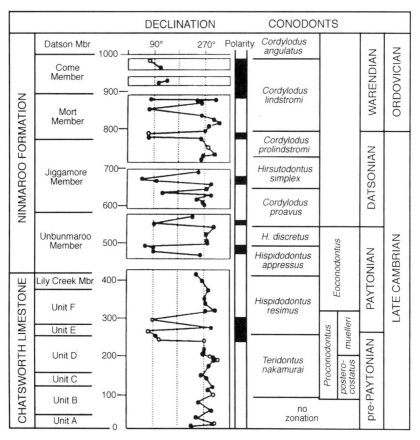

Figure 12.10 Black Mountain section (Australia) across Cambro–Ordovician boundary (after Ripperdan and Kirschvink, 1992).

for the Cretaceous quiet zone and PCRS (Kiaman), there is no evidence of substantial polarity bias in the Phanerozoic. At the present state of knowledge of the polarity structure of the paleomagnetic field, the duration of polarity chrons from Middle Carboniferous to the end of the Cretaceous is similar to the distribution of durations seen in the Cenozoic (Fig. 12.11). In the logarithmic histogram of polarity duration since 330 Ma (Fig. 12.12), polarity chron durations generally lie in the 0.1–1 My interval. The PCRS (Kiaman) and Cretaceous quiet zone (KQZ) fall well away from the bulk of the data. Polarity chrons with durations greater than 2 My are very rare. In Figure 12.13, the number of reversals was counted in 4-My bins and

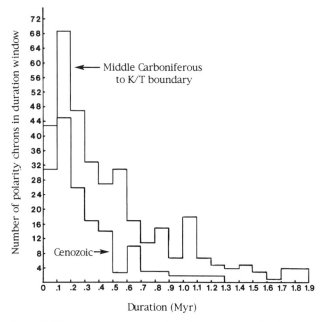

Figure 12.11 A comparison of duration of polarity chrons for the Cenozoic and the Carboniferous–Cretaceous interval.

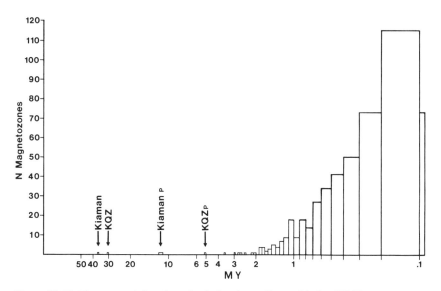

Figure 12.12 Histogram of duration of polarity chrons (log scale), for 330 Ma to present.

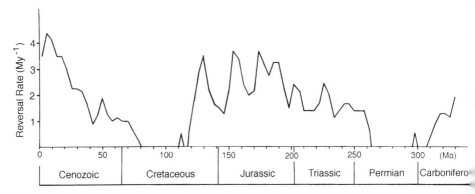

Figure 12.13 Reversal rate from 330 Ma to the present. The number of reversals was counted in 4-My bins and smoothed with a running mean.

smoothed with a running mean. The number of reversals per million years ranges from zero to about 4 My^{-1} in the late Cenozoic. The average reversal rate, excluding the two superchrons, appears to be about 2 My^{-1}. If the record is considered as a whole it would seem that the geomagnetic field has two states, one in which it reverses at a rate of two or more per million years and another in which it does not reverse at all. The change between these two apparent geomagnetic states has been correlated to activity of mantle plumes which rise rapidly from the D″ layer and manifest themselves as large igneous provinces such as Ontong Java Plateau and the Deccan Traps (Courtillot and Besse, 1987; Larson and Olson, 1991).

McElhinny (1973) analyzed the polarity record from paleomagnetic studies available at that time and concluded that the ratio of normal to reverse polarity indicated that for long periods of time the geomagnetic field seemed to favor one polarity. He recognized polarity bias in favor of normal polarity in Mesozoic rocks and in favor of reverse polarity in the Paleozoic. Intervals of polarity bias were postulated by Algeo (1996) for the Jurassic and Ordovician based on a global analysis of Phanerozoic paleomagnetic data. The analysis given here shows no such bias in the magnetostratigraphic record. The apparent bias is probably caused by unremoved secondary magnetizations acquired during the Kiaman (PCRS) or during the KQZ and by inadequacies in amount of high quality magnetostratigraphic data, particularly for pre-Late Jurassic time.

13

Secular Variation and Brunhes Chron Excursions

13.1 Introduction

"Magnetic stratigraphy" or "magnetostratigraphy" most commonly refers to polarity reversal stratigraphy. However, directional changes associated with secular variation of the geomagnetic field are an important means of correlation in recent sediments. The technique has been largely restricted to lake sediments due to their high sedimentation rates and lack of bioturbation. At typical deep-sea sedimentation rates of about 1 cm/ky, secular variations are averaged out by bioturbation and/or the remanence acquisition process. However, marine sediments from restricted basins (such as the Mediterranean and Black Sea) with sedimentation rates of several tens of cm/ky yielded some of the earliest secular variation records (e.g., Opdyke *et al.,* 1972; Creer, 1974). The first convincing demonstration that secular variation of the geomagnetic field can be recorded in lake sediments is attributed to Mackereth (1971), who used this technique to provide a time frame for sedimentation in Lake Windermere (UK). He correlated remanence declination changes to observatory records, particularly to the early 19th century westerly declination maximum, and showed that the chronology is consistent with the older ^{14}C ages but not with those influenced by detrital carbon input from recent burning of fossil fuels. This work paved the way for the development of the field of secular variation magnetic stratigraphy, which can provide a time frame in lake sediments and is also our main source of information on the nature of Holocene geomagnetic secular variation.

Whereas conventional (polarity reversal) magnetic stratigraphy depends on the recognition of directional changes of remanent magnetization of about 180°, secular variations generally do not exceed a few tens of degrees. Hence sampling and measurement techniques in secular variation studies are particularly critical. The development of a pneumatic coring device suitable for lake sampling (Mackereth, 1958) was an important step toward obtaining lake sediment cores that were both long enough and sufficiently undeformed for secular variation studies. In his pioneering study, Mackereth (1971) used an astatic magnetometer to measure subsamples sliced from the cores. The advent of pass-through fluxgate magnetometers (Molyneux et al., 1972) allowed the declination of remanence of the whole core to be rapidly measured without subsampling, and the more recently available 3-axis pass-through cryogenic magnetometers (Goree and Fuller, 1976; Weeks et al., 1993) allow both declination and inclination to be determined rapidly.

13.2 Sediment Records of Secular Variation

Although secular variation records can provide a means of precise correlation, only about 20% of all lakes studied have yielded secular variation records of sufficient resolution to define turning points in declination and inclination which can be used for correlation (Thompson and Oldfield, 1986). Some of the clearest and sharpest records are from Western Europe, such as those from Lake Windermere (Mackereth, 1971; Creer et al., 1972), Lough Neagh (Thompson, 1973), Lake Vuokonjarvi (Stober and Thompson, 1977), Lac de Joux (Creer et al., 1980a), and Loch Lomond (Turner and Thompson, 1979). The better dated British records have been merged to give a well-dated "master curve" of directional changes for the last 10,000 years (Fig. 13.1) (Turner and Thompson, 1981) which can be used as a calibration curve.

There is also a large body of data from North America which comes mainly from the Great Lakes (Creer et al., 1976; Mothersill, 1979, 1981; Creer and Tucholka, 1982), from Minnesota (Banerjee et al., 1979; Lund and Banerjee, 1985; Sprowl and Banerjee, 1989), from Oregon (Verosub et al., 1986), from California (Lund et al., 1988; Brandsma et al., 1989), and from British Columbia (Turner, 1987). The detailed 36–125 ka record from Mono Lake (California) (see Lund et al., 1988) records the Mono Lake excursion at about 28 ka and a distinctive waveform in the subsequent declination and inclination record which appears to recur every 2500–3500 years (Fig. 13.2). It was suggested that this distinctive waveform evolved out of the Mono Lake excursion and the persistence of the subsequent

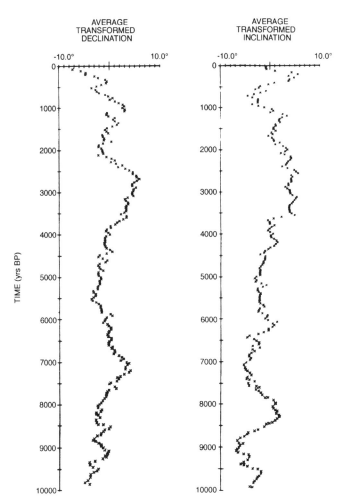

Figure 13.1 British master secular variation curve (after Turner and Thompson, 1981). Transformed declination and inclination averaged over ten cores for 0–7 ka and three cores for 7–10 ka.

waveform suggests a long-term memory in the core-dynamo process (Lund, 1989). This hypothesis is supported by the work of Negrini *et al.* (1994), who recorded a repetitive waveform in the secular variation in lake sediments from Oregon which immediately postdate the Pringle Falls excursion.

Comparison of the British Columbia (Mara Lake) record with the composite record from the Great Lakes indicates little correlation for the

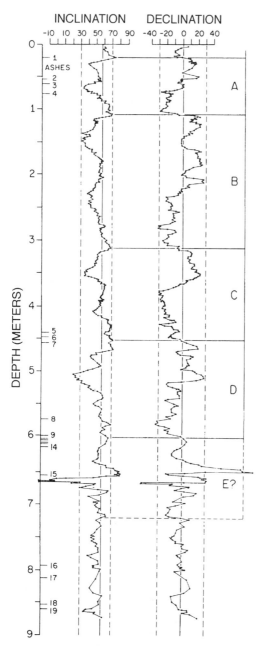

Figure 13.2 Secular variation record from Mono Lake sediments. The Mono Lake excursion occurs in the 6–7 m interval and is followed by several proposed repetitions of the secular variation waveform (after Lund *et al.*, 1988; Lund, 1989).

past 1500 years; however, the correlation between 5000 and 2000 yr B.P. (Fig. 13.3) has been interpreted in terms of a lag of 400 years over the 30 degree longitude difference between the two sites, implying a westward drift of the nondipole field of 0.0785 degree/yr (Turner, 1987). Such interpretations are highly dependent on the precision of available (radiocarbon) age control. The correlation of the record from Fish Lake, Oregon (Verosub *et al.*, 1986) with that from British Columbia is problematic (Fig. 13.3), with an apparent 800-yr difference in the age of corresponding features at the two sites (Turner, 1987). This may be explained by radiocarbon ages that are systematically 800 yr too old at the Oregon site (due to the introduction of radiogenically old detrital carbon) or by a time lag between deposition and stabilization of the postdepositional remanence at this site (Turner,

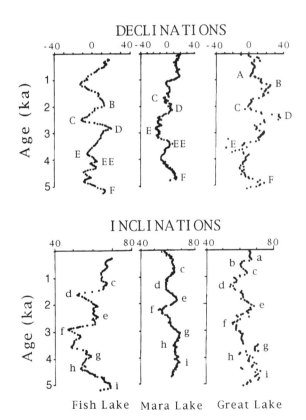

Figure 13.3 Declination and inclination records from Fish Lake, Oregon (Verosub *et al.*, 1986), Mara Lake, British Columbia (Turner, 1987), and Great Lakes (Creer and Tucholka, 1982) (after Turner, 1987). Inflections in the curves are letter coded to indicate proposed correlations.

1987). This dilemma highlights the critical role of radiocarbon and other age determinations in efforts to establish master curves of secular variation and the problems which can arise when radiocarbon dates are not backed by independent age control. The lake sediment records for the last 10 ky from Minnesota (Lund and Banerjee, 1985) and Fish Lake, Oregon (Verosub *et al.*, 1986) can be convincingly correlated with radiocarbon-dated lava flows and archaeomagnetic records (see Brandsma *et al.*, 1989). Elk Lake in Minnesota has yielded a high-fidelity secular variation record with a varve chronology independent of radiocarbon (Sprowl and Banerjee, 1989).

Marine records of secular variation are confined to high sedimentation rate environments such as parts of the Mediterranean or Black Sea (Opdyke *et al.*, 1972); Creer, 1974) and to coastal hemipelagic environments (Brandsma *et al.*, 1989; Levi and Karlin, 1989) and drift deposits (Lund and Keigwin, 1994) where sedimentation rates are of the order of tens of centimeters per thousand years. Lake sediment accumulation rates are often of the order of 1 m/ky, whereas pelagic marine sedimentation rates are characteristically about 1 cm/ky. The relatively enhanced fidelity of lake sediment records of secular variation is partly due to their higher sedimentation rates and partly due to the general absence of bioturbation. At DSDP Site 480 in the Gulf of California, the estimated sedimentation rate is 1 m/ky and the 50-m core was sampled at 10-cm intervals for secular variation studies (Levi and Karlin, 1989). The record cannot be correlated in detail to the lake records although high-amplitude fluctuations in inclination in the 20–50 ka interval may be related to low paleointensites at the time of the Mono Lake and Laschamp excursions (Levi and Karlin, 1989). In the Santa Catalina Basin (California Borderlands), estimated sedimentation rates range from 13 to 86 cm/ky (Brandsma *et al.*, 1989). A 3-cm sampling interval in 3–4 m-long piston cores has yielded secular variation records that can be correlated among three cores and with North American lake records (Brandsma *et al.*, 1989), thereby improving the chronology of the sediments from the Santa Catalina Basin for the last 10 ky. Lund and Keigwin (1994) noted that secular variation records from the Bermuda Rise are relatively subdued for the Holocene (where sedimentation rates are about 10 cm/ky) compared to the Late Pleistocene where sedimentation rates are 2–3 times higher.

Secular variation records from Australia (Barton and McElhinny, 1979; Constable and McElhinny, 1985) are more difficult to correlate because of the generally lower amplitude of the directional changes. The records from Victoria (Barton and McElhinny, 1979) cannot be matched with those from Queensland on a swing-by-swing basis (Constable and McElhinny, 1985); however, some features of the VGP paths can be correlated.

The original records from Lake Windermere and Lough Neagh were NRM records, and the observed oscillations in declination and inclination were considered to be periodic with a periodicity of about 2800 years. Partial alternating field demagnetization helped to produce the high-fidelity magnetic record from Loch Lomond (Turner and Thompson, 1979), which, together with an improved time scale, showed that the geomagnetic field changes did not follow a simple oscillationary pattern during the past 7000 years. Turner and Thompson (1981) noted that the calibrated master curve for Britain (Fig. 13.1) has similarities with other European lake and archaeomagnetic records but differs from North American and Japanese records. Creer and Tucholka (1982) considered that the similarities between the UK and North American Great Lakes master curves were sufficient to indicate westward drift throughout most of postglacial time at about the historically observed rate. The general lack of similarity of record from Argentina (Creer et al., 1983b), Australia (Barton and McElhinny, 1979), UK (Turner and Thompson, 1981), and the North American Great Lakes (Creer and Tucholka, 1982) confirms that the secular variation is not due to wobble of the main dipole field axis but to nondipole components with different sources at different sites. For this reason, patterns of secular variation cannot be expected to be correlative over large distances. This is also demonstrated by the lack of similarity of the record from the Sea of Galilee with the UK master curve (Thompson et al., 1985). The Western European record does, however, appear to match over distances of up to about 2000 km, for example, from Western Europe to Iceland (Thompson and Oldfield, 1986).

Runcorn (1959) pointed out that westward-drifting nondipole sources will tend to cause the north-seeking geomagnetic vector to precess in a clockwise sense and therefore give a clockwise sense of looping of the VGP paths; however, this interpretation is not unique, and the converse can be true with some source configurations (Dodson, 1979). Presently available European secular variation data indicate a dominantly clockwise sense of looping of the magnetic vectors and of the VGP paths implying that westward drift dominated the past 7000 years or so (with possibly a few hundred years of eastward drift beginning between 1000 and 1500 yr B.P.) preceded by a period of eastward drift (Turner and Thompson, 1981; Thompson and Oldfield, 1986). In the record from British Columbia, clockwise looping from 4.75 ka to 1.5 ka is followed by a period of anticlockwise looping (Turner, 1987). This is broadly consistent with the European record. In the Elk Lake record, clockwise looping predominates with anticlockwise looping in the 500–1500 yr B.P. and 6100–7000 year B.P. intervals (Sprowl and Banerjee, 1989). The Australian record shows predominantly clockwise looping, with periods of anticlockwise looping between 5.7 and 4.0 ka and between 10.5 and 8.8 ka. Within the age constraints, this sense of looping is consistent with the Argentinean record (Constable and McElhinny, 1985). The record from Lac du Bouchet for the 10 ka to 30 ka interval is consistent with predominantly westward-drifting geomagnetic sources (Smith and Creer, 1986). Although there is some consistency between inferred drift direction in the northern hemisphere and in the southern hemisphere, there is no clear global consistency, which has led Creer and Tucholka (1982) to infer that the looping may be largely due to "standing" sources which fluctuate in intensity rather than westward or eastward drift. Cross-correlation of the Bulgarian archaeomagnetic record (Kovacheva, 1983) with the Elk Lake (Minnesota) record yields a peak at an offset of 520 years, implying a westward drift rate of 0.23°/yr; however, there is little consistency in the drift rate or the drift direction when the same exercise is performed for Eurasian and North American records (Sprowl and Banerjee, 1989). In a recent review of North American Holocene secular variation, Lund (1996) found no evidence for westward or eastward drift. According to Lund (1996), repetitive vector loops may indicate a recurring core genera-tion process manifest in the secular variation record.

A global compilation of local changes in declination and inclination for the past 400 years mainly from observatory records (Fig. 13.4) (Thomp-

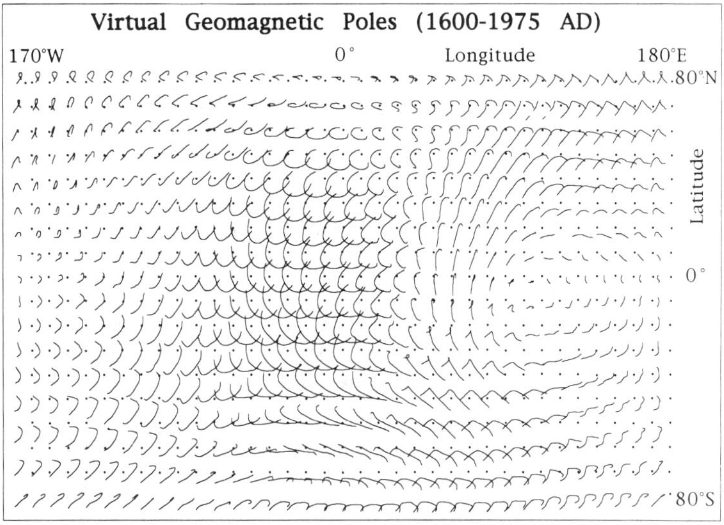

Figure 13.4 Historical virtual geomagnetic poles over the Earth's surface for the 1600–1975 interval. The sense of motion is clockwise except in the region of the Indian Ocean (after Thompson and Barraclough, 1982).

son and Barraclough, 1982) illustrates that the amplitude of the secular variation and the timing of distinctive turning points in the records vary from place to place, with the most rapid high-amplitude secular changes and clearest turning points occurring over Africa and Europe. These are therefore the most promising regions for the use of secular variation as a historical correlation tool.

Although high-quality relative paleointensity records have been obtained from marine sediments (see Ch. 14), paleointensity studies in lake sediments have generally been less successful. Roberts *et al.* (1994b) presented a paleointensity record for the 65–105 ka interval from Lake Chewaucan (Oregon) which indicates an NRM/ARM paleointensity high close to 80 ka, consistent with marine paleointensity records (Meynadier *et al.*, 1992; Tric *et al.*, 1992). The paleointensity records from Lake Baikal (Peck *et al.*, 1996) derived from NRM/ARM in three ~ 9 m cores yield an excellent match to marine paleointensity records for the last 84 ky. In lake sediments, anhysteretic remanence (ARM) and saturation isothermal remanence (SIRM) are often inadequate as a means of normalizing for variable concentration of remanence-contributing grains, possibly because magnetic grains included in a larger nonmagnetic grains contribute to ARM and SIRM but not to remanence (Turner and Thompson, 1979). Tucker (1980, 1981) has pointed out the difference in nature of the natural DRM and of ARM and has suggested the use of stirred remanent magnetization (StRM) as a normalizing parameter. Thouveny (1987) has demonstrated that StRM can have the same stability against magnetic cleaning as the NRM and that this can be the most suitable normalizing parameter. However, laboratory redeposition of natural samples in known magnetizing fields may not adequately mimic the relationship between NRM and magnetizing field because of the role of organic gels, which appear to be important in stabilizing grain orientation in the natural wet sediment (Stober and Thompson, 1979). Clearly, it would be highly desirable to have an adequate means of determining paleointensities, not only as an additional means of correlation but also to have a complete appreciation of the spatial and temporal nature of geomagnetic secular variation.

Within an individual lake, cores can be correlated using tie-points derived from lithological matching and magnetic susceptibility records. A general procedure is to transpose the depth scales for each core to that of a lake master core by linear interpolation between correlation tie-points. In order to correct for variations in tilt of the corer between holes and to find the optimal match between the directional secular variation records from the same lake, several workers (e.g., Constable and McElhinny, 1985; Turner, 1987) have used the cross-correlation technique suggested by Denham (1981). At each step in the iteration, the angle of rotation between two records is calculated to maximize the sum of the scalar products between corresponding unit vectors. In order to use this method, data points of equal depth in the cores must be (linearly) interpolated for each record. The cross-correlation technique applies simple rotations to the whole data set, making no alteration to the depth or time scales. Correction for the tilt of the corer is a critical step as Mackereth-type corers have operated efficiently at tilts up to 35° (Turner, 1987). Matched magnetization vectors from different cores can then be averaged to produce a smoothed master curve for the lake, with estimates of confidence limits using Fisherian statistics.

Smoothing of individual records can be achieved by fitting polynomials or splines. Spline functions are generally preferred to polynomials for approximating smooth empirical functions. The commonly used cubic spline function comprises a piecewise polynomial of degree 3 joined at "knots." For a given set of knots, the parameters of the best-fitting cubic spline can be estimated by least squares. The number of knots determines the degree of smoothing and, if equally spaced knots are used, the optimum knot spacing (bandwidth) can be determined by the cross-validation technique (Clark and Thompson, 1978), which involves the repeated comparison of the smoothed record with randomly chosen data points temporally deleted from the record. The optimum bandwidth is that which minimizes the average sum of squares of the differences between excluded observations and the smoothed curve. This process also permits an estimate of the variance of the error measuring the scatter of observations from the smoothed curve, and this is essential for the estimation of confidence limits (Clark and Thompson, 1978). The method can be applied to the smoothing of scalar quantities such as susceptibility or to directional data.

Smoothed records can then be correlated by visual matching of distinctive features such as turning points in the directional records. Widely available computer programs such as *Analyseries* and *Corepack* are useful for optimizing core correlation by matching distinctive features in the records. A less subjective method of correlation of between-lake records involves determining the stretching function (transformation of depth scales) which results in the optimal fit between records (Clark and Thompson, 1979). The stretching function may be nonlinear, and the method can give valid confidence limits.

13.3 Geomagnetic Excursions in the Brunhes Chron

There is a large number of papers in the literature advocating geomagnetic "excursions" (Opdyke, 1972) within the Brunhes Chron, and the evidence for their existence is controversial. Some have suggested that a full reversal of the Earth's magnetic field has occurred many times within the Brunhes and that the excursions should be classified as subchrons (Champion *et al.,*

1988). These features may be either extreme excursions in secular variation, aborted reversals of the geomagnetic field, or short polarity subchrons with duration $\sim 10^4$ years. In general, the excursions have proved to be too short in duration and difficult to detect to be particularly useful for stratigraphic correlation purposes, but they are potentially very important for our understanding the nature of the geomagnetic field.

It is clear that some of the proposed geomagnetic excursions are based on spurious data. For example, the Gothenberg excursion originated from studies of Swedith varved sediments (Morner et al., 1971). Attempts to replicate the results in sediments from nearby lakes were unsuccessful (Thompson and Berglund, 1976). Both the Gothenberg excursion and the Erieu excursion (Creer et al., 1976) can be attributed to the effects of bottom currents and abrupt facies changes (Thompson and Berglund, 1976; Banerjee et al., 1979). Physical disturbance of sediments, inversion of core segments, and self-reversal mechanisms in igneous rocks may all yield results that could be erroneously interpreted as geomagnetic field behavior.

There is a major problem associated with the recognition of geomagnetic events which may have very short duration, local manifestation, and differing character from one place to another. Inability to duplicate the record of such geomagnetic events may not mean that the events do not exist; however, the lack of duplication casts doubt upon the validity of the original record. Nonetheless, four Brunhes Chron excursions have been documented at a number of locations and are now generally considered to represent perturbations of the geomagnetic field (Table 13.1).

Mono Lake Excursion: The Mono Lake excursion, first documented by Denham and Cox (1971) and subsequently investigated by Liddicoat and Coe (1979), is arguably the best documented excursion of the geomagnetic field during the Brunhes Chron. The excursion (dated at 27–28 ka) is followed by four successive recurrences of the excursion waveform (Fig.

Table 13.1
Confirmed Excursions

Excursion	Age	References
Mono Lake	27–28,000 years B.P.	Denham and Cox (1971); Liddicoat and Coe (1979)
Laschamp	42,000 years B.P.	Bonhomet and Babkine (1967); Levi et al. (1990)
Blake	108–112,000 years B.P.	Smith and Foster (1969; Tric et al. (1991b); Zhu et al. (1994)
Pringle Falls	218,000 ± 10,000 years B.P.	Herrero-Bervera et al. (1989, 1994)
Big Lost	565,000 years B.P.	Negrini et al. (1987); Champion et al. (1988)

13.2) (Lund *et al.*, 1988). The directional changes have been closely duplicated at four sites on the shores of Mono Lake, although some sites at this locality with identical tephrostratigraphy failed to record the excursion, exemplifying the problem associated with the recognition of such events. The excursion is recorded as a declination swing to the west by 60° followed by a swing to the east by about 40° and a swing in inclination to −30° (Liddicoat and Coe, 1979). The duration of the excursion has been estimated to be about 10^3 years. Lack of recognition of the Mono Lake excursion at Clear Lake (Verosub, 1977b) or Pyramid Lake (Verosub *et al.*, 1980), located about 300 and 200 km from Mono Lake, respectively, has been used as evidence that the Mono Lake record is spurious (Verosub *et al.*, 1980). These authors considered contemporaneous nondeposition or erosion at Clear Lake and Mono Lake to be highly unlikely. This argument depends critically on the duration of the event; clearly, if the duration was short, recording of the excursion may be due to fortuitously rapid accumulation rates at Mono Lake at this precise time. Two lake records appear to show geomagnetic excursions of comparable age to that recorded at Mono Lake. These records are from Lake Tahoe (Palmer *et al.*, 1979) and from Summer Lake (Negrini *et al.*, 1984), which are 1000 and 600 km from Mono Lake, respectively. Although there are no good age constraints or demagnetization data on the Lake Tahoe record, the Summer Lake excursion is precisely dated by tephrostratigraphy and the magnetic record is well documented. The character of the excursion at Summer Lake is similar to *part* of that recorded at Mono Lake. The fact that the entire excursion is apparently not recorded at Summer Lake and that the record of the excursion has proved so elusive in lakes with a similar sedimentation history indicates that the excursion has very short duration, probably less than 10^3 years. This in turn leads us to the conclusion that this excursion will be of little use as a means of correlation, although it is well dated both at Mono Lake and Summer Lake by tephrostratigraphy as being in the 24–29 ka range (Liddicoat, 1992; Negrini *et al.*, 1984). The fact that this age correlates closely with the 27–30 ka age for the Lake Mungo excursion (Barbetti and McElhinny, 1976) is an interesting coincidence, but most probably fortuitous.

Laschamp Excursion: This excursion was originally detected by Bonhommet and Babkine (1967) in recent lavas of the Puy de Dome region of the Massif Central, France (see also Bonhommet and Zahringer, 1969). The best estimate for its age seems to be 42 ka (Condomes *et al.*, 1982). Detailed studies on these lavas by Heller and Petersen (1982) indicate that they possess self-reversing properties; therefore, the existence of the excursion may be in doubt. The observed directions of magnetization make large angles with the present geomagnetic field direction but are not fully reverse polarity. Coring in nearby lake sediments which span the same time interval failed to detect the excursion (Thouveny and Creer, 1992). Recent studies of the Lonchadiere volcanic flow from central France yielded deviating magnetization directions and the lava is the same age as the basalts at Laschamp and Olby. Icelandic lava flows which yield ages indistinguishable from those for the Laschamp excursion give paleomagnetic results which are in accord with the Laschamp excursion (Kristjansson and Gundmundsson, 1980; Levi *et al.*, 1990). Based on these results, the geomagnetic excursion appears to be confirmed.

Blake Excursion: Another geomagnetic excursion within the Brunhes Chron which has been well documented is the so-called Blake Event. This excursion was first recorded in the Atlantic (Smith and Foster, 1969) and subsequently both in the Atlantic and in the Caribbean (Denham, 1976). This excursion appears to be a complete, albeit short-lived, reversal of the geomagnetic field and therefore may qualify as a subchron. The appearance of the Blake Event in the sedimentary record is variable, depending, presumably, on the completeness of the record rather than vagaries in the remanence acquisition process. It appears as a single reversal (Tucholka *et al.*, 1987) and two reverse intervals (Creer *et al.*, 1980b; Tric *et al.*, 1991b) in Mediterranean cores and as three more complete reversals in Atlantic cores (Denham, 1976) and in Chinese loess (Fig. 13.5) (Zhu *et al.*, 1994). The Blake Event has now been recorded in China and in the Atlantic and Mediterranean and therefore appears to be a global event. Age estimates for the Blake Event from Atlantic cores (100 ka, Denham, 1976) compare closely to age estimates from Chinese loess (111–117 ka, Zhu *et al.*, 1994). In Mediterranean piston cores, negative inclinations correlate to oxygen isotopic substages 5e and 5d (Tucholka *et al.*, 1987; Tric *et al.*, 1991b). Tephrochronology and $\delta^{18}O$ stratigraphy in these cores indicate fully reverse magnetization directions over a 40-cm interval which the authors estimate to represent ~4000 years. The event is centered at about 110 ka, close to the age originally suggested by Smith and Foster (1969), and best estimates of the duration of the event lie in the 4–6 ky range.

Pringle Falls Excursion: The three partially reverse polarity intervals recorded in lake sediments from near Pringle Falls, Oregon (Herrero, Bervera *et al.*, 1989; Herrero-Bervera and Helsley, 1993) were first thought to be a record of the Blake Event. The excursion at Pringle Falls has now been correlated to similar directional records of field excursions at Summer Lake (Oregon) and Long Valley (California), and $^{40}Ar/^{39}Ar$ dating and tephrochronology indicate an age of 218 ± 10 ka (Fig. 13.6) (Herrero-Bervera *et al.*, 1994). Anomalous magnetization directions from the Mamaku ignimbrite in New Zealand (Shane *et al.*, 1994) yield virtual geomagnetic poles which lie close to the VGP path from Oregon/California records

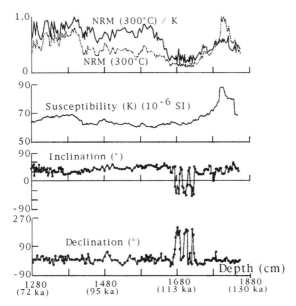

Figure 13.5 Record of the Blake Event from Chinese loess (after Zhu *et al.*, 1994).

of the Pringle Falls excursion. Isothermal plateau fission track age on glass (230 ± 12 ka) (Shane *et al.*, 1994) and the $^{40}Ar/^{39}Ar$ age on plagioclase (223 ± 3 ka) (McWilliams, personal communication, 1995) from the Mamaku ignimbrite are consistent with the estimated age of the Pringle Falls excursion in Oregon/California.

Big Lost Excursion: Champion *et al.* (1988) have identified reverse directions in bore holes and outcrops from Idaho dated at 565 ± 14 ka. They called this event the Big Lost subchron. The duration of this possible excursion (or subchron) is not known. A reverse interval recorded in Humbolt Canyon (Negrini *et al.*, 1987; Champion *et al.*, 1988) and low-latitude poles from volcanic rocks in southwest Germany dated at 510 ± 30 ka have been correlated to this excursion (Schnepp and Hradetzky, 1994).

Apart from those listed above, Champion *et al.* (1988) have postulated the existence of five additional subchrons within the Brunhes based on data from the Lake Biwa cores (Kawai, 1984) and from old piston cores (e.g., Goodell and Watkins, 1968; Stearwald *et al.*, 1968; Ryan and Flood, 1972; Watkins, 1968), most of which should not be used for this purpose. Early piston core data containing anomalous magnetization directions (e.g., Ninkovich *et al.*, 1966) were reinterpreted by Wollin *et al.* (1977) as recording field-related behavior. The cores in question had sedimentation rates of 1 cm/ky or less and it is, therefore, very unlikely that short-term geomagnetic field behavior would have been recorded.

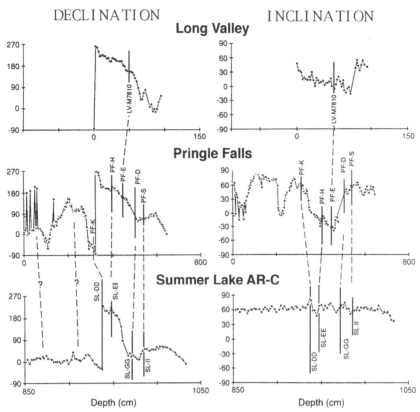

Figure 13.6 Declination and inclination of the magnetization in sediments from California and Oregon which record the Pringle Falls geomagnetic excursion, with the correlation of tephra among the sections (after Herrero-Bervera *et al.,* 1994). Tephras are letter coded to aid correlation from the inclination to the declination records.

In recent years, abundant data have become available from Brunhes age sediments with rates of sedimentation above 1 cm/ky from both conventional piston cores and HPC/APC cores of the Ocean Drilling Program. High-quality data are also available from loess sections of China. In Figure 13.7, these are plotted as straight-line segments assigning the top of the cores to zero age and the Brunhes/Matuyama boundary to 780 ka. In at least one study of loess in China (Heller and Liu, 1984), dense sampling was carried out in an attempt to detect the Blake and Laschamp subchrons; however, no deviating directions were observed. The core with the highest rate of sedimentation is from ODP Leg 107 (Site 650) in the Tyrrhenian Sea, which has a sedimentation rate of 56 cm/ky (Channell and Torii, 1990). Two zones of negative (upward) magnetic inclination are seen in this core;

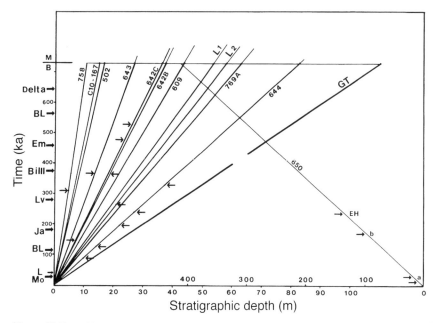

Figure 13.7 Positions of reverse polarity intervals in Brunhes age sediment cores and loess sections (L1 and L2). Brunhes sections range in length from 8 to 400 m. Site 650 from the Tyrrhenian Sea is plotted from right to left with an independent depth scale increased by a factor of 4. The positions of the reverse polarity samples or intervals are indicated by the arrows or in the case of GT (Italy) by a gap in the straight line. It can be seen that a correlation of reverse polarity intervals is not compelling. The positions in time of the Mono Lake (Mo), Laschamp (L), Blake (BL), Jamaica (Ja), Levantine (Lv), Biwa III (BiIII), Emperor (Em), Big Lost (BL), and Delta excursions are shown.

the first (A) occurs at a depth of 10.39 m to 12.31 m and the second occurs at a depth of 100.16 to 100.45 m. When plotted on a time versus depth plot, event A would seem to correspond to the Mono Lake excursion. Event B occurs at an extrapolated age of about 175-ky, an apparent age older than expected for the Blake excursion. Hole 642B (Fig. 13.7) has zones of negative inclination which appear to correlate with the Mono Lake and Laschamp excursions. The Blake Event, on the other hand, does not appear to correlate with negative inclinations in any of the cores and sections examined. This is also true of the Big Lost excursion.

Studies of reversal transitions indicate that the length of time taken for a transition to occur is ~5–10 ky. The shortest subchron duration you could expect in which the field goes completely reverse would be from 10 to 20 ky. This is, in fact, the length of time estimated for the Cobb Mountain and Réunion subchrons. If full reversals of the magnetic field can take as little as 1 ky, as seems to be indicated from the results by Tric *et al.* (1991b), then very short polarity subchrons may occur. For correlative purposes, however, such short subchrons would seem to be of marginal value except in regions of very high sedimentation rates. Excursion records and records of short polarity events such as the Cobb Mountain subchron (see Clement and Martinson, 1992) imply that the geomagnetic field exhibits a complete spectrum of behavior from high-amplitude secular variation to full polarity reversals. Brief polarity reversals which never stabilize, where the dipole field immediately reverts back to its initial polarity, have also been recorded. Detailed high-fidelity records of the process of polarity reversal, recorded in sediments with high accumulation rates, have occasionally indicated that the dipole configuration of the field is maintained during the reversal process for certain polarity transitions (see Clement, 1992; Clement *et al.*, 1995) and that the paleomagnetic pole follows preferred longitudinal paths during the reversal process (Clement, 1991; Laj *et al.*, 1991) (Ch. 2).

14

Rock Magnetic Stratigraphy and Paleointensities

14.1 Introduction

Rock magnetic stratigraphy involves the utilization of magnetic properties of sediments and sedimentary rocks as a means of (1) stratigraphic correlation, (2) identification of sediment sources and transport mechanisms, (3) characterization/detection of paleoenvironmental change.

The first rock magnetic parameter to be widely used for correlation was initial (low field) magnetic susceptibility (k). Its usefulness has long been recognized particularly for correlating volcanoclastic-rich horizons in marine sediments (e.g., Radhakrishnamurty *et al.*, 1968), and it is now a standard correlation technique in deep-sea sediments (e.g., Robinson *et al.*, 1995), lake sediments (e.g., Thompson *et al.*, 1975; Peck *et al.*, 1994), and in loess (e.g., Kukla *et al.*, 1988). Magnetic susceptibility essentially monitors the concentration of ferrimagnetic minerals such as magnetite and maghemite, and therefore the correlation method is sensitive in sediments which show variations in concentration of either of these magnetic minerals. Magnetic susceptibility measurements are part of the standard core scanning procedure on R/V *Joides Resolution* and other oceanographic research vessels. High-resolution paleoceanographic studies require not only multiple cores at each site (ensuring complete recovery of the sediment sequence) but also the ability to make high-resolution correlation among cores. An important step forward in shipboard core correlation was accomplished during ODP Leg 138 (Lyle *et al.*, 1992; Hagelberg *et al.*, 1992). Specialized computer software linked to core scanning systems recording saturated bulk (GRAPE) density, magnetic susceptibility, and digital color reflectance spectroscopy produced optimal correlation of physical parameters between cores. This was the basis for "real-time" core correlation within site, the construction of a composite section for the site, and correlation to downhole logs. Real time within site correlation can be used to determine the number of holes necessary to ensure complete recovery at a site and to compensate for the poorly understood phenomenon of core expansion (often up to 10–20%) during recovery.

The field of rock magnetic stratigraphy has broadened considerably in the last 15 years with the use of magnetic parameters which are sensitive not only to magnetic mineral concentration but also to the grain size and mineralogy. These techniques were largely developed to study environmental changes in lake catchments by monitoring changes in lake sediment magnetic mineralogy which are sensitive to changes in land use, erosion, industrialization, and climate (see Thompson and Oldfield, 1986). Similar methods have been applied to marine sediments (Robinson, 1986; Bloemendal *et al.*, 1988; Doh *et al.*, 1988) and loess deposits (Maher and Thompson, 1991, 1992) and have been shown to provide not only a sensitive means of correlation but also information on changing detrital and biogenic fluxes. The magnetic parameters useful for correlation and/or environmental modeling depend on the nature of the variations in magnetic mineralogy within the sediment.

14.2 Magnetic Parameters Sensitive to Concentration, Grain Size, and Mineralogy

The magnetic parameters listed below and in Table 14.1 are the principal parameters used in studies of marine and lake sediments (see Thompson *et al.*, 1975; King *et al.*, 1982; Thompson and Oldfield, 1986; Robinson, 1986; Bloemendal *et al.*, 1988; Doh *et al.*, 1988; King and Channell, 1991; Stoner *et al.*, 1996).

a. Magnetic Susceptibility (k)

Volume (initial) magnetic susceptibility, defined as volume magnetization divided by applied field, is often considered to be a measure of ferrimagnetic mineral concentration; however, susceptibility (in magnetite) shows a grain size dependence (see Maher, 1988) increasing through the SD-PSD-MD range. As the susceptibility of the ferrimagnetic minerals magnetite and maghemite is 3 or 4 orders of magnitude greater than that of common antiferromagnetic minerals, such as hematite and goethite, the susceptibility is usually dominated by the ferrimagnetic grains and therefore gives information on the presence or absence of this group of magnetic minerals. For magnetite- and maghemite-bearing sediments, magnetic susceptibility is a reasonable measure of the *concentration* of the magnetic mineral, and is not in itself a sensitive measure of grain size.

b. Anhysteretic Remanent Magnetization (ARM) or Anhysteretic Susceptibility (k_{arm}) and Saturation Remanent Magnetization (SIRM)

These parameters are also concentration dependent and, as for susceptibility, are dominated by the ferrimagnetic grains present in the sample. These parameters are particularly useful because their magnitude tends to increase with decreasing *grain size* of the ferrimagnetic grains (see Maher, 1988). Normalizing these parameters by initial susceptibility partially compensates for variations in concentration and hence the SIRM/k ratio (Thompson, 1986) and ARM/k or k_{arm}/k ratio (King *et al.*, 1982) vary inversely with magnetic particle size and are therefore useful granulometric parameters. The former is likely to be more sensitive to the multidomain grain size range and the latter to the pseudo-single-domain range. The granulometric method utilizes the fact that ARM and SIRM decrease with increasing grain size whereas susceptibility is relatively insensitive, showing slight increase with increasing grain size. Interpretation is complicated by the non-linear response of these parameters to changes in grain size. An additional drawback of all three granulometric ratios is that ultrafine superparamagnetic (SP) grains will have negligible remanence (SIRM, ARM, k_{arm}) but relatively high k, resulting in low values of the ratios (SIRM/k ARM/k, or k_{arm}/k), comparable to those seen in coarse multidomain grains.

c. *S*-Ratio and "Hard" IRM (HIRM)

The *S*-ratio (Stober and Thompson, 1979; Bloemendal, 1983) is found by measuring the SIRM and then placing the sample in a back field (typically 0.1 T) and remeasuring the IRM.

$$S_{-0.1} = (IRM_{-0.1}/SIRM)$$

$S_{-0.1}$ is a convenient measure of the proportion of (coarse) low-coercivity ferrimagnetic grains to higher coercivity grains. As the proportion of low-coercivity grains saturating in fields less than 0.1 T increases, this *S*-ratio approaches unity. For discriminating ferrimagnetic grains (such as magnetite) from high-coercivity antiferromagnetic grains such as geothite or hematite, $S_{-0.3}$ (determined using a back field of 0.3 T) would be more appropriate.

The HIRM ("hard" IRM) is the difference between the $IRM_{-0.3}$ (IRM in a 0.3-T field) and SIRM divided by 2 (Robinson, 1986; Bloemendal *et al.*, 1988).

$$HIRM_{-0.3} = (SIRM + IRM_{-0.3})/2$$

This parameter is also a measure of the concentration of antiferromagnetic minerals with high coercivity. The two parameters (S-ratio and HIRM) complement each other and might be expected to show an inverse relationship in cores exhibiting a wide range of magnetic mineralogy and grain size. The combination of these two parameters provides a means of monitoring changes in *magnetic mineralogy* and magnetic mineral concentrations. The coercivity range in which the S-ratio and HIRM are sensitive will depend not only on the value of the back field but also the magnetizing field used to acquire the SIRM. Note that, in practice, the "SIRM" is often acquired in a magnetizing field of about 1 T, which is close to the maximum field attainable by most small electromagnets and pulse magnetizers. This field may not fully saturate the sample, particularly in the presence of high-coercivity antiferromagnetic minerals such as goethite or hematite.

d. Frequency-Dependent Susceptibility

The susceptibility of fine-grained magnetite, with grain sizes close to the superparamagnetic/single domain boundary, is frequency dependent, decreasing with increasing frequency of the applied field (Stephenson, 1971). The susceptibility of single-domain and multidomain grains is relatively unaffected by changes in frequency. The frequency-dependent susceptibility (k_{fd}) is often expressed as a percentage, where $k_{fd} = (k_{hf} - k_{lf})/k_{lf}$ and k_{hf} and k_{lf} are the susceptibility in the high-frequency (typically about 5–10 kHz) and low-frequency field (typically about 1 kHz), respectively. The frequency-dependent susceptibility is a convenient measure of the *concentration of very fine grained (SP) magnetite* in the sample.

14.3 Rock Magnetic Stratigraphy in Marine Sediments

Over 15 years ago, the recognition of orbital frequencies in NRM sediment records led to the proposal that geomagnetic field intensity was orbitally modulated (e.g., Wollin *et al.*, 1977). It has since been demonstrated that these observations are best explained by orbitally controlled climatic varia-

Table 14.1
Generalized Table of Downcore Magnetic Parameters and Their Interpretations

Parameter	Interpretation
Bulk Magnetic Measurements	
Natural Remanent Magnetization (NRM): The fossil (remanent) magnetization preserved within the sediment. NRM recorded as declination, inclination, and intensity.	Dependent on mineralogy, concentration, and grain size of the magnetic material as well as mode of acquisition of remanence and intensity and direction of the geomagnetic field.
Volumetric Magnetic Susceptibility (k): A measure of the concentration of magnetizable material. Defined as the ratio of induced magnetization intensity (M) per volume to the strength of the applied weak field (H): $k = M/H$.	k is a first-order measure of the amount of ferrimagnetic material (e.g., magnetite): k is particularly enhanced by superparamagnetic (SP) magnetite (<0.03 μm) and by large magnetite grains (>10 μm). When the concentration of ferrimagnetic material is low, k responds to antiferromagnetic (e.g., hematite), paramagnetic (e.g., Fe, Mg silicates), and diamagnetic material (e.g., calcium carbonate, silica) which may complicate the interpretation.
Isothermal Remanent Magnetization (IRM): Magnetic remanence acquired under the influence of a strong DC field. Commonly expressed as a saturation IRM or SIRM when a field greater than 1 T is used. A back field IRM (BIRM) is that acquired in a reversed DC field after SIRM acquisition.	SIRM primarily depends on the concentration of magnetic, principally ferrimagnetic, material. It is grain size dependent, being particularly sensitive to magnetite grains smaller than a few tens of microns.
Anhysteretic Remanent Magnetization (ARM): Magnetization acquired in a biasing DC field within a decreasing alternating field. Commonly expressed as anhysteretic susceptibility (k_{ARM}) when normalized by the biasing field used.	k_{ARM} is primarily a measure of the concentration of ferrimagnetic material: however, it is also strongly grain size dependent. k_{ARM} preferentially responds to smaller magnetite grain sizes (<10 μm) and is useful in the development of grain size dependent ratios.
Constructed Magnetic Parameters	
The "hard" IRM (HIRM): This is derived by imparting a back field, typically 0.1 or 3.0 T, on a sample previously given a SIRM. The resulting BIRM, which has a negative sign, is used to derive the HIRM by the formula: HIRM = (SIRM + BIRM)/2.	HIRM is a measure of the concentration of magnetic material with higher coercivity than the back field. This commonly gives information on the concentration of antiferromagnetic (e.g., hematite) or very fine grained ferrimagnetic (e.g., magnetite) grains depending on the back field used.

Table 14.1 *continued*

Parameter	Interpretation
S ratios: These are derived by imparting a back field, typically 0.1 or 0.3 T, on a sample previously given a SIRM. The resulting BIRM, which has a negative sign, is normalized by the SIRM; S = BIRM/SIRM. This provides a measure of the proportion of saturation at the back field applied.	The S ratios can be used to estimate the magnetic mineralogy (e.g., magnetite or hematite). Downcore variations may be associated with changing mineralogy. Values close to -1 indicate lower coercivity and a ferrimagnetic mineralogy (e.g., magnetite): values closer to zero indicate a higher coercivity, possibly an antiferromagnetic (e.g., hematite) mineralogy. S ratio with a back field of 0.1 T may be sensitive to mineralogic and grain-size changes, whereas the S ratio with the 0.3 T back field is more sensitive to mineralogical changes (e.g., proportion of magnetite to hematite).
k_{ARM}/k: Indicates changes in magnetic grain size, if the magnetic mineralogy is dominantly magnetite.	If the magnetic mineralogy is dominantly magnetite, k_{ARM}/k varies inversely with magnetic grain size, particularly in the 1–10 μm grain-size range. However, the interpretation of this ratio may be complicated by significant amounts of superparamagnetic (SP) or paramagnetic material.
SIRM/k: Indicates changes in magnetic grain size, if the magnetic mineralogy is dominantly magnetite.	If the magnetic mineralogy is dominantly magnetite, SIRM/k varies inversely with magnetic particle size. SIRM/k is more sensitive than k_{ARM}/k to changes in the proportion of large (>10 μm) grains. SIRM/k may also be compromised by SP or paramagnetic material.
SIRM/k_{ARM}: Indicates changes in magnetic grain size, if the magnetic mineralogy is dominantly magnetite.	SIRM/k_{ARM} increases with increasing magnetic grain size but is less sensitive and can be more difficult to interpret than the two ratios above. A major advantage of SIRM/k_{ARM} is that it only responds to remanence carrying magnetic material and is therefore not affected by SP or paramagnetic material.
Frequency Dependent Magnetic Susceptibility (k_f): The ratio of low-frequency k (0.47 kHz) to high frequency k_{hf} (4.7 kHz) calculated by $k_f = 100 \times (k - k_{hf})/k$.	k_f is used to indicate the presence of SP material. SP material in high concentrations can compromise the grain-size interpretation made using k_{ARM}/k and SIRM/k.
Hysteresis Measurements[a]	
Saturation Magnetization (M_s): M_s is the magnetization within a saturating field; Saturation Remanence (M_{rs}): M_{rs} is the remanence remaining after removal of the saturating field.	M_{rs}/M_s decreases with increasing magnetite grain size in the submicron to few tens of microns grain size range.

(*continues*)

Table 14.1 *continued*

Parameter	Interpretation
Coercive Force (H_c): The back field required to rotate saturation magnetization to zero within an applied field; Coercivity of Remanence (H_{cr}): The back field required to rotate saturation magnetization to zero remanence.	For magnetite, the ratio H_{cr}/H_c increases with increasing grain size (in the submicron to several hundred micron grain-size range) due to the strong grain-size dependence of both parameters, particularly H_c. H_{cr} is a useful guide to magnetic mineralogy.

Note: After Stoner *et al.*, 1996.

[a] Hysteresis parameters provide a means of monitoring grain-size variations in magnetite. Sediments should be homogeneous because of the small sample size (<0.05 g) typically used for hysteresis measurements. Mixed magnetic mineral assemblages greatly complicate the interpretations. The generalized interpretations listed above are based on a ferrimagnetic (e.g., magnetite) mineral assemblage.

tions affecting magnetic mineral concentrations, which in turn affect NRM intensities (Amerigian, 1974; Chave and Denham, 1979; Kent, 1982). In deep-sea sediments, the principal climatically controlled variable which influences NRM intensity is carbonate content, which dilutes the magnetic mineral flux (Kent, 1982). Detailed studies of the magnetic mineralogy of marine sediment cores (Robinson, 1986; Bloemendal *et al.*, 1988; Doh *et al.*, 1988) have shown that the climatically induced carbonate dilution of magnetic mineral concentrations is not the only paleoclimatic process influencing the magnetic properties and that changes in eolian, biogenic, and ice-rafted fluxes of magnetic minerals can be deduced from changes in magnetic properties.

Susceptibility records at ODP sites off west Africa and in the Arabian Sea are related to terrigeneous (eolian) flux (Bloemendal and deMenocal, 1989). Spectral analysis of these Late Pliocene/Pleistocene susceptibility records reveal 100 ky, 41 ky, 23 ky, and 19 ky periodicities (Fig. 14.1). There is an apparent shift in power at about 2.4 Ma, coincident with the onset of Northern Hemisphere Glaciation. Prior to this time, the 23 ky and 19 ky (precession) periodicities dominate, and after this time the 41 ky (obliquity) periodicity dominates. The presence of the 41 ky periodicity during the Matuyama Chron is also a feature of the North Atlantic $\delta^{18}O$ record (Raymo *et al.*, 1989). The shift in power in the off-Africa susceptibility records implies a change in monsoon intensity and/or eolian source area climate related to the onset of Northern Hemisphere Glaciation (Bloemendal and deMenocal, 1989). A similar change in power at this time (~2.4 Ma) has been noted in foraminiferal abundance data from the Mediterranean (Sprovieri, 1992).

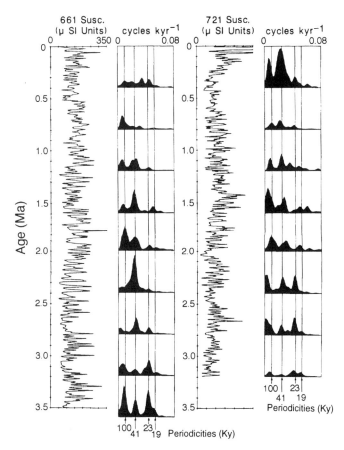

Figure 14.1 Susceptibility records from the central Atlantic (Site 661) and Arabian Sea (Site 721) coasts of Africa. Spectral analysis over 0.4 My intervals indicates power in the Milankovitch periodicities (after Bloemendal and deMenocal, 1989).

In the King's Trough Flank area of the North Atlantic (41–43°N), peaks in susceptibility (k) tend to coincide with glacial intervals (Fig. 14.2) as indicated by the oxygen isotope record, cold water foraminiferal assemblages, and lows in calcium carbonate concentrations (Robinson, 1986; Robinson *et al.*, 1995). In core BOFS-5K, located further north at 50.6°N (Fig. 14.2), the susceptibility record is dominated by the ice-rafted detritus (IRD) associated with Heinrich events, as observed in other piston cores from this region (Grousset *et al.*, 1993). Robinson *et al.* (1995) have used the ratio of susceptibility at the Last Glacial Maximum (LGM) (~18–19 ka) to susceptibility in the Holocene as a means of mapping the distribu-

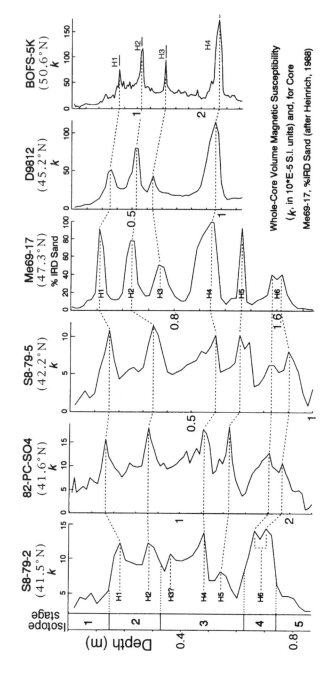

Figure 14.2 Susceptibility records from North Atlantic piston cores illustrating the change in the record from the carbonate content–controlled record in the King's Trough area (~41–43°N) to the region close to 50°N where Heinrich (IRD) events dominate the record (after Robinson *et al.*, 1995).

tion of IRD during the LGM (Fig. 14.3). The ice sheet instability which leads to high-susceptibility Heinrich (IRD) events in the North Atlantic (Broecker *et al.*, 1992) has a different manifestation in the Labrador Sea. In this region, detrital carbonate (DC) and low detrital carbonate (LDC) layers, some of which correlate to Heinrich events and other IRD layers in the North Atlantic, have been associated with turbiditic activity in the North Atlantic Mid-Ocean Channel (NAMOC) (Stoner *et al.*, 1996). Some

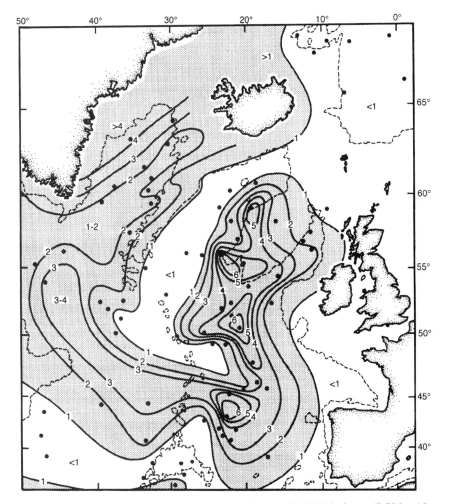

Figure 14.3 Ratios of magnetic susceptibility at the Last Glacial Maximum (LGM, ~18–19 ka) to susceptibility in the Holocene, used to map terrigeneous ice-rafted detritus during the LGM (after Robinson *et al.*, 1995).

DC and LDC events are characterized by high susceptibility (due to their association with IRD) although the detrital events are more easily recognized by ferrimagnetic grain size–sensitive ratios such as k_{arm}/k and SIRM/k (Fig. 14.4). Such records provide a link between (Laurentide) ice sheet variability and the marine record which are critical to the understanding of climate change during the last glacial cycle.

Parameters which are not strictly concentration dependent and therefore independent of calcium carbonate content, such as the S ratio and the

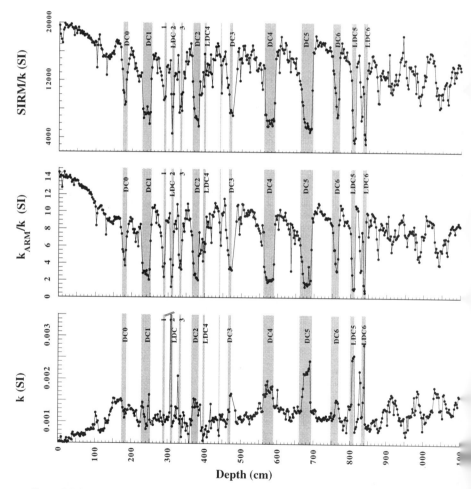

Figure 14.4 In Labrador Sea piston core P-094, the grain size–sensitive parameters (k_{arm}/k and SIRM/k) are more sensitive than susceptibility (k) for recognition of detrital carbonate (DC) and low detrital carbonate (LDC) layers (after Stoner *et al.*, 1995a).

ARM/k and ARM/SIRM ratios, show variations which reflect not only the principal glacial/interglacial variations but also additional fine-scale variations, particularly within interglacials (Fig. 14.5). Robinson (1986) interpreted these data in terms of three magnetic mineral fluxes: (1) antiferromagnetic mineral flux mainly of eolian origin associated with vigorous atmospheric circulation and aridity in glacial periods; (2) coarse ferrimagnetic mineral flux associated with ice rafting in glacial periods; and (3) a constant flux of fine-grained (single-domain) magnetite of eolian or biogenic origin, the effect of which is noticeable when fluxes (1) and (2) are low in interglacial periods.

Changes in k and k_{arm}/k at Termination I and II for a core located off the southern tip of Greenland are interpreted in terms of progressive deglaciation as the continental ice sheet melted back to the shoreline and then retreated into the continental interior (Stoner *et al.*, 1995a). At Termination I in this core, k_{arm}/k perturbations record the Younger Dryas (YD), Heinrich Layer 1 (H_1), and an additional melt event (ME_1) prior to a broad

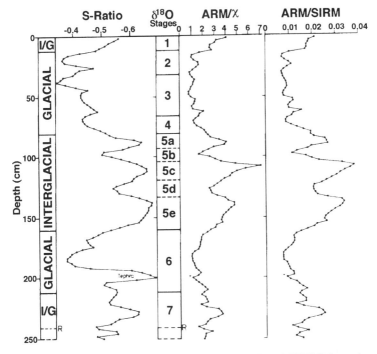

Figure 14.5 Magnetic parameters (S-ratio, ARM/k, and ARM/SIRM) for a piston core (S8-79-4) from the King's Trough region of the North Atlantic (after Robinson, 1986).

low in the k_{arm}/k record (Fig. 14.6). ME_1 is interpreted to signify detrital influx at the time of initial detachment of grounded sea ice from the shelf. The subsequent broad low in k_{arm}/k is interpreted as the detrital influx during continental deglaciation on Greenland (Stoner *et al.*, 1995a). This

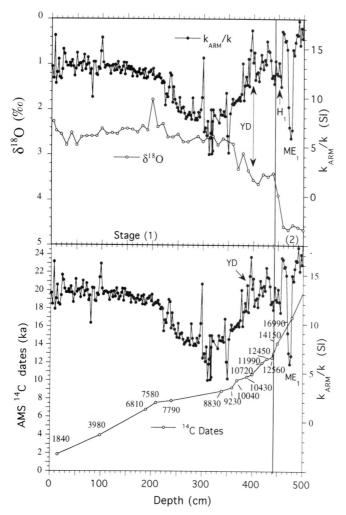

Figure 14.6 The grain size–sensitive parameter (k_{arm}/k) as a monitor of continental derived meltwater detritus during the last deglaciation in piston core P-013 from off SW Greenland (after Stoner *et al.*, 1995a). The timing of the decrease in k_{arm}/k (increase in ferrimagnetic grain size) coincides with the retreat of continental ice from the coastline into the continental interior of Greenland.

interpretation is supported by radiocarbon age control tying the low in k_{arm}/k to geomorphological evidence for the timing of continental ice retreat on Greenland (Funder, 1989).

Bloemendal *et al.* (1988) used seven rock magnetic parameters as a means of detailed correlation in Quaternary cores from the eastern central Atlantic. Rock magnetic parameters correlate to oxygen isotope records, with HIRM and $k_{arm}/SIRM$ showing the strongest correlation with the oxygen isotope curve. Magnetic parameters are computed both as magnetic accumulation rates (computed by dividing the parameter by the estimated time taken for each individual sample to accumulate; Bloemendal, 1983) and as accumulation rates of the noncarbonate content. This allows the variations in magnetic properties independent of total sedimentation rate and carbonate sedimentation rate to be assessed. An antiferromagnetic and coarse ferrimagnetic detrital flux is dominant in glacial intervals, and a fine-grained ferrimagnetic flux is largely confined to warm interglacial events. The principal indicator for this latter component (k_{arm}) shows a strong 21–23 My precessional variance in one of the cores.

Bivariate scatter plots of k_{arm} against k and $S_{-0.3}$ against k_{arm}/k have been used as a means of discriminating the magnetic mineralogy of late Neogene and Pleistocene sediments from different sedimentological environments (Bloemendal *et al.*, 1992). The k_{arm} against k plot is largely controlled by the grain size spectrum of ferrimagnetic grains (magnetite and maghemite), and the $S_{-0.3}$ against k_{arm}/k plot by the ratio of antiferromagnetic (hematite) to ferrimagnetic grains. Differing sedimentary environments tend to reveal characteristic distributions in these bivariate plots. For example, eolian detritus is rich in antiferromagnetic grains, ice-rafted detritus is poorly sorted and rich in coarse ferrimagnetic material, and current-controlled deposition tends to be relatively well sorted.

Doh *et al.* (1988) studied a 25-m core (LL44-GPC3) from the central North Pacific which covers the entire Cenozoic. The lithology is the unfossiliferous pelagic clay which is characteristic of the Pacific midlatitudes (Davies and Gorsline, 1976). The stratigraphic control in this facies is particularly poor due not only to the almost total lack of microfossils but also to the unstable nature of the remanent magnetization. The magnetic parameters show coherent fluctuations downcore which are potentially useful for correlation. In addition, fluctuations in concentration-dependent parameters (such as HIRM) plotted as accumulation rates correlate well with the sedimentologically determined total eolian accumulation rate (Fig. 14.7), indicating that these parameters can be used as a proxy indicator for eolian

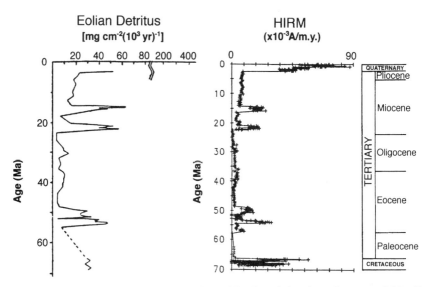

Figure 14.7 HIRM as a proxy for eolian detrital flux in red clays from the equatorial Pacific (after Doh *et al.*, 1988).

flux (Doh *et al.*, 1988). Changes in eolian flux are related to changes in paleogeography and paleoclimate, as the Pacific plate moved northward, and can provide the basis for *local* stratigraphic correlation; however, the precision of the correlation technique depends critically on whether the magnetic properties are due to changes in detritus or to later diagnetic growth/alteration of magnetic minerals. In core LL44-GPC3, polarity intervals (stable magnetizations) are recorded above 4.33 m (late Pliocene/ Pleistocene) and below 19 m (Paleocene/early Eocene) (Prince *et al.*, 1980). The intervening interval is characterized by unstable, highly viscous remanence. The intervals of stable remanent magnetization coincide with low values of ARM/*k*, indicating that increases in grain size of magnetic minerals may be the cause of remanence instability (Doh *et al.*, 1988). In these intervals, magnetic grain size (from ARM/*k*) and eolian grain size (from grain observation) are high and vary consistently, suggesting that most of the magnetic minerals in these intervals are of eolian origin. This is not the case for the interval of unstable remanence, where eolian grain size fluctuations are not recorded in the ARM/*k* profile, suggesting diagenetic alteration of magnetic minerals in this interval. Indeed, the instability of the remanent magnetization in this particular facies has been attributed to oxidation of primary magnetite to maghemite downcore (Kent and Lowrie,

1974; H. P. Johnson *et al.*, 1975) or alternatively the growth of authigenic ferromanganese phases during diagenesis (Henshaw and Merrill, 1980). The critical observation is whether the boundary between the stable and the unstable remanence record, which is controlled by the magnetic mineralogy, occurs at a consistent time horizon or stratigraphic level. Available data indicate that it tends to occur very close to the Gauss/Matuyama boundary (Opdyke and Foster, 1970; Prince *et al.*, 1980; Yamazaki and Katsura, 1990) although synchroneity has not been well established. The change in magnetic mineralogy in the late Pliocene may be either directly or indirectly due to change in eolian flux, which in turn may be related to the onset of Northern Hemisphere Glaciation. If so, these mineralogical changes represent time lines suitable for regional stratigraphic correlation.

14.4 Rock Magnetic Stratigraphy in Loess Deposits

Heller and Liu (1982) pointed out that the susceptibility variations in the Chinese loess match the paleosol/loess lithologic variability, and Heller and Liu (1984) matched the record to the Brunhes oxygen isotope record from the Pacific of Shackleton and Opdyke (1976). Kukla *et al.* (1988) showed that the susceptibility record closely parallels the orbitally tuned SPECMAP oxygen isotope record derived from deep-sea sediments (Fig. 14.8), indicating a close linkage between eolian flux, global ice volume, and climate. Rock magnetic data have played an important role in understanding the origin of the climatic record in Chinese loess. The present consensus is that the higher susceptibility of the paleosols in the Chinese loess is the result of "magnetic enhancement" associated with authigenic production of SP and SD magnetite during pedogenesis (Zhou *et al.*, 1990; Heller *et al.*, 1991; Maher and Thompson, 1991, 1992; Banerjee *et al.*, 1993; Evans and Heller, 1994). Beer *et al.* (1993) showed that the pedogenic and detrital susceptibility components in the loess can be quantitatively distinguished by comparing the susceptibility variations with [10]Be concentrations. According to these calculations, the susceptibility in the paleosols due to magnetite formed *in situ* is often ~50% greater than the detrital susceptibility. Taking present-day precipitation/susceptibility data (Liu *et al.*, 1992) as the means of calibration, Heller *et al.* (1993) used estimates of the pedogenic susceptibility contribution to estimate paleoprecipitation. Maher *et al.* (1994) constructed a paleoprecipitation record based on bulk susceptibility differences between paleosols and intervening unweathered loess and a logarithmic relationship between rainfall and susceptibility in nine modern paleosols across the Chinese loess plateau.

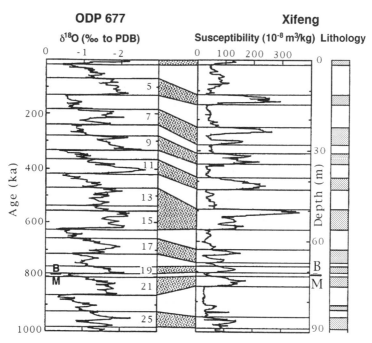

Figure 14.8 Correlation of susceptibility records from Chinese loess with the benthic oxygen isotope record from ODP Site 677 (after Heller and Evans, 1995). Shading in lithology indicates paleosols separated by loess.

The pedogenic model which explains the magnetic variability in Chinese loess cannot be applied to all loess deposits. The susceptibility record in Alaskan loess has orbital variability, but the intervals of high susceptibility correlate to cold climate intervals, opposite to the relationship in the Chinese loess (Begét and Hawkins, 1989). A close relationship between mean loess grain size and susceptibility values implied that the high-susceptibility intervals can be attributed to greater wind velocities during glacial intervals, which transported larger magnetite grains which dominate the susceptibility record (Begét et al., 1990).

14.5 Rock Magnetic Stratigraphy in Lake Sediments

Magnetic parameters provide a basis for correlation and for determining changes of magnetic mineral fluxes in both marine and lake sediments. However, because lake sedimentation rates are typically 2 or 3 orders of magnitude greater than those of deep-sea sediments, the time scale of the observations is very different. Whereas magnetic mineral flux changes in the marine setting can give us information on long-term climate change, the same information from lake sediments is generally giving us information on Holocene and recent anthropogenic environmental change. There are, however, lake sediment cores which extend back through the last glacial cycle such as those from Lac du Bouchet (Thouveny *et al.*, 1994) and Lake Baikal (Peck *et al.*, 1994). In these studies the susceptibility record is an important means of monitoring lithologic changes associated with the glacial-interglacial cycling.

Volume susceptibility measurements are a very important, rapid, and nondestructive means of core correlation in lake sediments (Thompson *et al.*, 1975; Bloemendal *et al.*, 1979; Peck *et al.*, 1994). Distinctive features in the records, which are controlled by variations in concentrations of ferrimagnetic minerals, can be matched to monitor variations in sediment accumulation. The method can be useful even in lakes which show wide variations in sedimentation rate (Thompson and Morton, 1979). Other magnetic parameters, such as SIRM, can also be used for core correlation and may be useful when susceptibility values are uniformly low, but they may require subsampling of the whole-core and therefore measurements may be considerably slower than whole-core susceptibility measurements.

The use of magnetic susceptibility profiles for correlation in lake sediments led to the observation, at Loch Neagh, that progressively increasing susceptibility values parallel pollen variations signifying progressive deforestation (Fig. 14.9) (O'Sullivan *et al.*, 1972). Thompson *et al.* (1975) interpreted these data to indicate increased erosion and a resulting flux of ferrimagnetic grains from the basaltic bedrock. It has since been demonstrated that the susceptibility of topsoil can be greater than that of the deeper soil horizons and of the bedrock. This "magnetic enhancement" has been attributed to the formation of magnetite and maghemite in the surface soil (Özdemir and Banerjee, 1982; Mullins, 1977; Maher and Taylor, 1988; Evans and Heller, 1994; Eyre and Shaw, 1994). Increases in susceptibility associated with pollen evidence for deforestation would now be interpreted to reflect increased erosion of topsoil as a result of changes in land use (e.g., Dearing *et al.*, 1986; Thompson and Oldfield, 1986). It has been demonstrated that susceptibility of lake sediments can be a very sensitive indicator not only of changes in erosion style in the catchment but also of changes in climate (Thompson and Oldfield, 1986), changes in stream discharge (Dearing and Flower, 1982), natural burning or burning for land clearance (Rummery *et al.*, 1979; Rummery, 1983), and industrial burning of fossil fuels (Oldfield *et al.*, 1983). Burning contributes to the lake sediment susceptibility record because magnetite is an important product of both soil burning (Longworth *et al.*, 1979) and industrial burning (Chaddha and

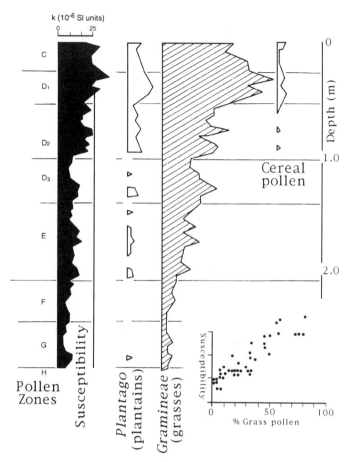

Figure 14.9 6000-year record from Lough Neagh (Northern Ireland). Forest clearance and farming are indicated by increases in plantains, grasses, and cereal pollen. Increases in susceptibility are attributed to increased soil erosion in the lake catchment (after Thompson and Oldfield, 1986).

Seehra, 1983). As many of these anthropogenic processes produce changes in ferrimagnetic influx, concentration-dependent parameters such as susceptibility and SIRM are most applicable (rather than mineralogy-dependent magnetic parameters). However, for detailed studies of erosion history and land use changes, a more complete characterization of sediment source material and of the sediment itself is necessary. Magnetic parameters can be used as tracers to determine changes in sediment source and thereby provide markers of environmental and land use change. In this type of application, there is a need to establish a more formal, *quantitative* approach to the study of magnetic mixtures, in order to delineate the proportion of individual magnetic constituents. Thompson (1986) has used the Simplex search method to determine optimal mixing models on the basis of particular magnetic characteristics (such as IRM acquisition curves) of individual sediment samples and of possible source materials. Alternatively, the magnetic characteristics of the sediment can be matched to combinations of standard curves based on synthetic samples.

14.6 Future Prospects for Rock Magnetic Stratigraphy

Until recently, rock magnetic stratigraphy was confined to the use of susceptibility as a convenient method of core correlation in volcanogenic marine sediments. During the past ten years, rock magnetic studies of lake sediments have demonstrated that rock magnetic parameters are not only useful for core correlation but can provide detailed information on changes in magnetic mineral flux that can be related to change in land use (Thompson and Oldfield, 1986). The techniques used in this developing field of environmental magnetism are now being applied to the marine and terrestrial sedimentary record. Susceptibility data from Chinese loess deposits (Kukla *et al.*, 1988) and from the eolian-dominated equatorial sites off Africa (Bloemendal and deMenocal, 1989) have variance in which the dominant "heartbeat" is very similar to that seen in marine oxygen isotope records, providing a fascinating link between marine and terrestrial climate records.

Rock magnetic methods are generally conducted on subsamples from sediment cores often collected as 8-cm^3 plastic cubes. The widely used 2G-755R cryogenic magnetometer (with 4.2-cm-diameter access) is designed for discrete sample measurement with homogeneous response of SQUID sensing coils over a volume large compared to the discrete sample volume. Weeks *et al.* (1993) describe a modified 2G-755R magnetometer in which the SQUID sensing coils are arranged for optimal spatial resolution of magnetization from continuous "u-channels" tracked through the sensing region. Deconvolution of the signal in the time domain (Constable and Parker, 1991; Oda and Shibuya, 1996) gives resolution comparable to that achieved by back-to-back discrete sampling. The u-channels (see Tauxe *et al.*, 1983b; Nagy and Valet, 1993) are typically 1.5 m in length with a 2 × 2 cm cross-section. Stepwise demagnetization of NRM, acquisition and stepwise demagnetization of IRM and ARM can be carried out on a single automated measurement track (Weeks *et al.*, 1993). Rock magnetic measurement tracks with this capability will greatly increase the speed at which such data can be acquired and will result in increased use of these data for stratigraphic and paleoenvironmental studies.

14.7 Paleointensity Determinations

Over the 40-year history of modern paleomagnetic research, the vast majority of the studies have dealt with directional variability of the ancient geomagnetic field with very little attention given to paleointensity studies. In recent years, there has been a renewed interest in paleointensity studies for several reasons: (1) Paleointensity information is an important component of understanding the field generation process, the reversal process, and factors controlling reversal frequency. (2) Paleointensity data from sediments appear to record the time-average dipole field intensity, the variations of which may be a useful means of high-resolution global correlation. (3) The atmospheric production of radionuclides such as ^{14}C and ^{10}Be is partly controlled by geomagnetic field intensity, due to the role of the field in shielding the low-latitude atmosphere from incoming cosmic rays. Paleointensity records provide a potential means of monitoring variations in ^{14}C and ^{10}Be production, improving the potential precision of radiocarbon ages and possibly providing calibration for variations in ^{10}Be concentrations in recent sediments and ice cores (see Mazaud *et al.*, 1994).

The method developed by Thellier and Thellier (1959) has been widely used as a means of determining paleointensities from igneous rocks. The basis of the method is the assumption that the ratio of the TRM produced in the laboratory to the original TRM (NRM) is equal to the ratio of the laboratory field to the ancient field in which the rock cooled. The method as modified by Coe *et al.* (1978) and Prévot *et al.* (1985) involves heating the sample to progressively higher temperatures at 10–20°C increments (Fig. 14.10). At each temperature step, the sample is first cooled in field-free space to record the NRM lost; the sample is then reheated at the same temperature and cooled in a known field (typically 40 μT) to record the p-TRM gain. The slope of the line on the so-called Arai plot (Nagata *et al.*, 1963) yields the paleointensity estimate (Fig. 14.10). The bane of the Thellier method is the likelihood of alteration of the magnetic mineralogy during the incremental heating/ cooling cycles. Following Coe *et al.* (1978), some p-TRM acquisition steps are repeated after heating at higher temperatures (Fig. 14.10), and if the intensity of the p-TRM is reproducible, then it is less likely that alteration has occurred. Other criteria for a reliable paleointensity determination (see Pick and Tauxe, 1993a) include: (1) the plot of p-TRM against NRM loss should be linear (Fig. 14.10); (2) susceptibility

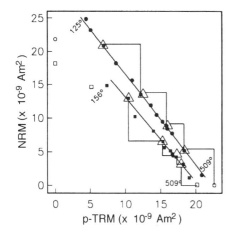

Figure 14.10 Arai paleointensity plot from the glassy rind of a basaltic pillow basalt (after Pick and Tauxe, 1993a).

should remain constant during the incremental heat treatment; (3) thermal demagnetization of the NRM, measured during the first of the two heating steps, should show a single well-defined component; (4) remanence should be carried by single-domain magnetite.

A compilation by Prévot *et al.* (1990) of 12 Mesozoic and Cenozoic Thellier paleointensity determinations (6 of them from papers by Bol'shalov, Solodovnikov, and co-workers) indicates a relative paleointensity low for Early Cretaceous time. Four paleointensity determinations from ODP pillow lavas by Pick and Tauxe (1993b) imply that this paleointensity low continues through most of the Cretaceous with a rather abrupt doubling of the virtual axial dipole moments (VADMs) at the end of the Cretaceous. These results debunk the notion that extended periods of constant geomagnetic polarity (such as the Late Cretaceous long normal) are associated with relatively high paleointensities (e.g., Larson and Olson, 1991). The present paucity of paleointensity data for Cenozoic and Mesozoic time will probably improve considerably in the near future if the recently derived data from volcanic glass (see Pick and Tauxe, 1993a,b) lives up to its early promise.

In contrast to the paucity of Thellier paleointensity determinations for pre-Holocene time, there is an abundance of determinations for the last 5 ky with a few determinations dating back to 30 ka (Fig. 14.11). There is considerable scatter in the data due partly to the influence of secular variation, but probably more importantly to alteration of magnetic minerals during the experiment and use of unsuitable archaeological and volcanic materials. Nonetheless, the volcanic data show similarities with the sedimentary relative paleointensity data (Fig. 14.11).

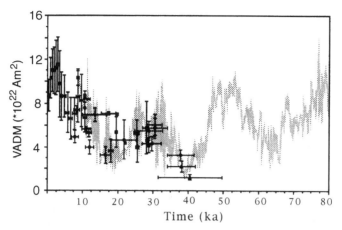

Figure 14.11 Comparison of sedimentary relative paleointensity records from Mediterranean Sea (continuous shaded record) with paleointensity data from volcanic rocks and archeomagnetic artefacts (points with error bars) (after Tric *et al.*, 1992).

Deep-sea sediment cores hold considerable promise for giving long continuous records of relative geomagnetic paleointensities. The method relies on the proposition that the intensity of DRM is a linear function of ambient field over the range of geomagnetic field intensity (Kent, 1973; Barton *et al.*, 1980; Tucker, 1980) and that variations in concentrations of magnetic minerals in the sediment can be compensated by a suitable normalization factor (see review by Tauxe, 1993). The ideal sediment for paleointensity determinations is one in which the remanence is a DRM carried by PSD magnetite in the 1–15 μm grain size range and in which the concentration of magnetite does not vary by more than a factor of 20 or 30 (King *et al.*, 1983). In most studies, the paleointensity record is derived from the NRM intensity (after partial AF demagnetization to eliminate VRM) normalized for variations in magnetite concentration by dividing by ARM, k, SIRM, low-field IRM, or StRM (stirred remanence from redeposition experiments) (Kent and Opdyke, 1977; Tauxe, 1993). More than one normalization factor can be used, strengthening the record in the case of concordance. ARM is often the normalization factor of choice as it is particularly sensitive to SD/PSD magnetite grain sizes, whereas SIRM and k are more sensitive to MD magnetite grains. Meynadier *et al.* (1992) demonstrated a similarity in coercivity of the NRM and ARM in three piston cores from the Somali Basin, thereby showing that the grain size fraction of magnetite carrying the ARM is similar to that carrying the NRM. For these 5 cm/ky sedimentation rate cores, the NRM_{20mT}/ARM

ratio was used as the measure of relative paleointensity in the 20–140 ka interval. The Somali Basin records are consistent among the three cores and compare well with the 80 ky record of Tric *et al.* (1992) from the Mediterranean Sea (Fig. 14.11) based on stacked NRM_{20mT}/k data from four cores with sedimentation rates of about 10 cm/ky. These records can be compared with variation seen in Thellier archaeomagnetic and volcanic records for the last 40 ky (e.g., McElhinny and Senanayake, 1982), with relative paleointensities for the last 20 ky from box cores (Constable and Tauxe, 1987), and with longer records from Ontong–Java (Tauxe and Wu, 1990; Tauxe and Shackleton, 1994), Sulu Sea (Schneider, 1993), western equatorial Pacific (Yamazaki and Ioka, 1994), central North Atlantic (Weeks *et al.*, 1995; Lehman *et al.*, 1996), and Labrador Sea (Stoner *et al.*, 1995b). The correlation of the marine paleointensity records to a 84 ky record from Lake Baikal (Peck *et al.*, 1996) is quite straight forward, but correlation to Lau du Bouchet (Thouveny *et al.*, 1990) is less clear. Comparison of the Labrador Sea record with a composite record compiled from the Mediterranean and Somali Basin records (Fig. 14.12) leads to two conclusions. (1) Relative paleointensity records for the last 100 ky can be correlated over large distances, implying that marine sediments can record paleointensities of the geomagnetic dipole field. (2) The AMS ^{14}C age control for the Labrador Sea record is inconsistent with the oxygen isotope age control from the Mediterranean/Somali basins.

Paleointensity studies in marine sediments, which hitherto had been restricted to the last ~200 ky, have been greatly expanded by an important study of ODP Leg 138 cores from the equatorial Pacific (Valet and Meynadier, 1993). The record covers the last 3.9 My in cores with mean sedimentation rates not exceeding 2.5 cm/ky (Fig. 14.13). Relative paleointensities were converted into virtual axial dipole moments (VADMs) by calibration to Thellier determinations of paleointensities for the last 0.5 My. Several important features are apparent in the records: (1) paleointensity lows in the Brunhes Chron appear to correlate in age to geomagnetic directional excursions; (2) a sawtooth pattern of paleointensity fluctuations appears to be related to reversal boundaries; reversals tend to coincide with lows in the VADMs with abrupt increases immediately post reversal. The longer the duration of the polarity chron, the greater the abrupt increase in VADM at its onset; and each polarity chron appears to be characterized by decay of VADM leading to the next reversal. This has led to the suggestion that the reversal frequency, and the occurrence of polarity excursions, is ultimately controlled by the intensity of the dipole field, which also controls the stability of a particular polarity state.

The correlation of widely spaced sedimentary paleointensity records for the last 140 ky supports the contention that they are reliable records

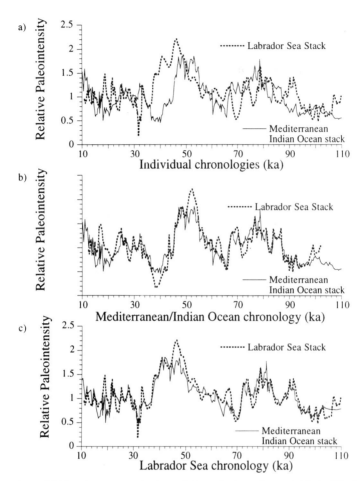

Figure 14.12 (a) The Labrador Sea relative paleointensity stack (dashed line) with age model derived mainly from AMS ^{14}C dates and a planktic δ^{18}O record. Mediterranean/Indian Ocean (MIO) paleointensity stack (solid line) (Meynadier *et al.*, 1992; Tric *et al.*, 1992) with ages based on δ^{18}O and tephrochronology. (b) Labrador Sea paleointensity stack correlated to MIO stack using visual ties and plotted on the MIO time scale. (c) MIO paleointensity stack correlated to Labrador Sea stack using visual ties and plotted on the Labrador Sea time scale (after Stoner *et al.*, 1995b).

Figure 14.13 Composite paleointensity (virtual axial dipole moment, VADM) record from ODP Leg 138 (Site 848 and Site 851) in the equatorial Pacific Ocean (after Valet and Meynadier, 1993).

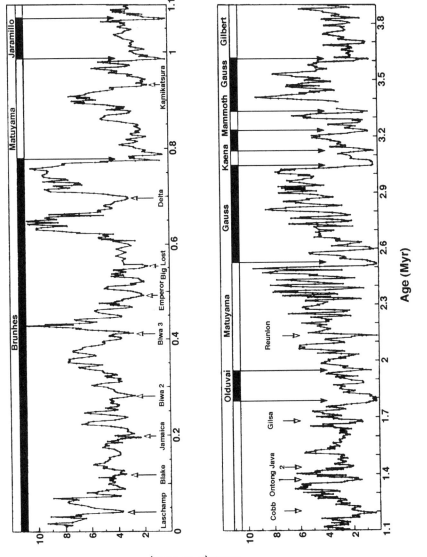

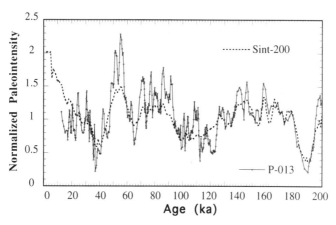

Figure 14.14 Comparison of the relative paleointensity stack (Sint-200, Guyodo and Valet, 1996) with the P-013 paleointensity record from the Labrador Sea (Stoner *et al.*, 1995b). The correlation allows the oxygen isotope chronology from Sint-200 to be imported into the Labrador Sea, where obtaining such chronologies directly is problematic.

of the VADM. The presence in the paleointensity record of orbital periodicities (see Meynadier *et al.*, 1992) suggests, however, that climate-controlled lithologic variability may be influencing the records. Tauxe and Wu (1990) advocate use of a coherence function to test whether paleointensity records are coherent (at the 95% confidence level) with the normalizing factor used, and with other magnetic parameters (such as ARM or *k*) which may be affected by climate-controlled lithologic cycles. Recent paleointensity studies from the Atlantic and Indian Oceans for the Jaramillo to early Brunhes interval (Valet *et al.*, 1994) appear to confirm the general structure of the Leg 138 record (Valet and Meynadier, 1993), indicating that paleointensity records will not only provide us with significant insights into the nature of the geomagnetic field and the mechanism for polarity reversals and polarity excursions but also provide us with a new means of high-resolution stratigraphic correlation. An example of such a correlation is depicted in Figure 14.14. The relative paleointensity stack compiled by Guyodo and Valet (1996), based on paleointensity records that can be directly correlated to the oxygen isotope chronology, is correlated to the paleointensity record from the Labrador Sea (Stoner *et al.*, 1995b) where oxygen isotope data are compromised by lack of benthos and meltwater influence on the planktic record. The correlation in Fig. 14.14 allows the oxygen isotope chronology to be imported into the Labrador Sea, heralding a new era of high resolution magnetic stratigraphy based on geomagnetic paleointensity.

Bibliography

Aissaoui, D. M., McNeill, D. F., and Kirschvink, J. L. (1990). Magnetostratigraphic dating of shallow-water carbonates from Mururoa Atoll, French Polynesia: Implications for global eustasy. *Earth Planet. Sci. Lett.* **97**, 102–112.

Algeo, T. J. (1996). Geomagnetic polarity bias patterns through the Phanerozoic. *J. Geophys. Res.* **101**, 2785–2814.

Alvarez, L. W., Alvarez, W., Asaro, P., and Michel, H. V. (1980). Extraterrestrial cause for the Cretaceous-Tertiary extinction. *Science* **208**, 1095–1108.

Alvarez, W., and Lowrie, W. (1978). Upper Cretaceous paleomagnetic stratigraphy at Moria (Umbrian Apennines, Italy), verification of the Gubbio section. *Geophys. J. R. Astron. Soc.* **55**, 1–17.

Alvarez, W., and Lowrie, W. (1984). Magnetic stratigraphy applied to synsedimentary slumps, turbidities and basin analysis: The Scaglia limestone at Furlo (Italy). *Geol. Soc. Am. Bull.* **95**, 324–336.

Alvarez, W., Arthur, M. A., Fischer, A. G., Lowrie, W., Napoleone, G., Premoli-Silva, I., and Roggenthen, W. M. (1977). Upper Cretaceous–Paleocene magnetic stratigraphy at Gubbio, Italy. V. Type section for the Late Cretaceous–Paleocene geomagnetic reversal time scale. *Geol. Soc. Am. Bull.* **88**, 383–389.

Amerigian, C. (1974). Sea-floor dynamic processes as the possible cause of correlations between paleoclimatic and paleomagnetic indices in deep-sea sedimentary cores. *Earth Planet. Sci. Lett.* **21**, 321–326.

Amerigian, C. (1977). Measurement of the effect of particle size variation on the DRM to ARM ratio in some abyssal sediments. *Earth Planet. Sci. Lett.* **36**, 434–442.

An, Z., and Ho, C. K. (1989). New magnetostratigraphic dates of Lantian Homo erectus. *Q. Res.* **32**, 213–221.

An, Z., Bowler, J. M., Opdyke, N. D., Macumber, P. G., and Firman, J. B. (1986). Paleomagnetic stratigraphy of Lake Bungunnia: Plio-Pleistocene precursor of arridity in the Murry Basin, southeastern Australia. *Palaeogeogr., Palaeoclimatol., Palaeoecol.* **54**, 219–239.

An, Z., Liu, T. S., Kan, X., Sun, J., Wang, J., Kaowanyi, Z. Y., and Wei, M. (1987). Loess-paleosol sequences and chronology at Lautian Mau localities. *In* "Aspects of Loess Research" (T. S. Liu, ed.), pp. 192–203. China Ocean Press, Beijing.

Andreieff, P., Bellon, H., and Westercamp, D. (1976). Chronometrie et stratigraphie comparée des edifices volcaniques et formations sedimentaires de la Martinique (Antilles françaises). *Bull. Bur. Rech. Geol. Min.* (*Fr.*) **4**, 335–346.

Anonymous (1979). Magnetostratigraphic polarity units. A supplementary chapter of the International Subcommission on Stratigraphic Classification, International Stratigraphic Guide. *Geology* **7**, 578–583.

Appel, E., Rösler, W., and Corvinus, G. (1991). Magnetostratigraphy of the Miocene-Pleistocene Surai Khola Siwaliks in West Nepal. *Geophys. J. Int.* **105**, 191–198.

Archibald, J. D., Butler, R. F., Lindsay, E. H., Clemens, W. A., and Dingus, L. (1982). Upper Cretaceous–Paleocene biostratigraphy and magnetostratigraphy, Hell Creek and Tullook Formations, northeastern Montana. *Geology* **10**, 153–159.

Armstrong, R. L. (1982). Late Triassic–Early Jurassic time-scale calibration in British Columbia, Canada. *In* "Numerical Dating in Stratigraphy" (G. S. Odin, ed.), pp. 509–514. Wiley, Chichester.

Aubry, M. P., Berggren, W. A., Kent, D. V., Flynn, J. J., Klitgord, K. D., Obradovich, J. D., and Prothero, D. R. (1988). Paleogene geochronology: An integrated approach. *Paleoceanography* **3**, 707–742.

Azzarolli, A., and Napoleone, G. (1982). Magnetostratigraphic investigation of the Upper Siwaliks near Pinjor India. *Riv. Ital. Paleontol.* **87**(4), 739–762.

Backman, J., Schneider, D. A., Rio, D., and Okada, H. (1990). Neogene low-latitude magnetostratigraphy from site 710 and revised age estimates of Miocene nannofossil datum events. *In* "Proceedings of the Ocean Drilling Program, Scientific Results" (R. A. Duncan, J. Backman, L. C. Peterson *et al.*, eds.), Vol. 115, pp. 271–276. ODP, College Station, TX.

Badgley, C., and Tauxe, L. (1990). Paleomagnetic stratigraphy and time in sediment studies in alluvial Siwalik rocks of Pakistan. *J. Geol.* **98**, 451–477.

Baksi, A. K. (1993). A geomagnetic polarity time scale for the period 0–17 Ma, based on ^{40}Ar/^{39}Ar plateau ages for selected field reversals. *Geophys. Res. Lett.* **17**, 1117–1120.

Baksi, A. K. (1994). Concordant sea-floor spreading rates obtained from geochronology, astrochronology and space geodesy. *Geophys. Res. Lett.* **21**, 133–136.

Baksi, A. K., Hsu, A. K. V., McWilliams, M. O., and Farrar, E. (1992). ^{40}Ar/^{39}Ar dating of the Brunhes-Matuyama geomagnetic field reversal. *Science* **256**, 356–357.

Baksi, A. K., Hoffman, K. A., and Farrar, E. (1993a). A new calibration point for the late Miocene section of the GPTS: ^{40}Ar/^{39}Ar plateau ages from Akaroa volcano, New Zealand. *Geophys. Res. Lett.* **20**, 667–670.

Baksi, A. K., Hoffman, K. A., and McWilliams, M. (1993b). Testing the accuracy of the Geomagnetic Polarity Time-Scale (GPTS) at 0–5 Ma utilizing ^{40}Ar/^{39}Ar incremental heating data on whole-rock basalts. *Earth Planet. Sci. Lett.* **118**, 135–144.

Baldwin, R. T., and Langel, R. (1993). Tables and maps of the DGRF 1985 and IGRF 1990. *IAGA Bull.* **54**, 117.

Banerjee, S. K., Lund, S. P., and Levi, S. (1979). Geomagnetic record in Minnesota lake sediments—absence of the Gothenburg and Erieu excursions. *Geology* **7**, 588–591.

Banerjee, S. K., Hunt, C. P., and Liu, X. M. (1993). Separation of local signals from the regional paleomonsoon record of the Chinese loess plateau: A rock magnetic approach. *Geophys. Res. Lett.* **20**, 843–846.

Bannon, J. L., Bottjer, D. J., Lund, S. P., and Saul, L. R. (1989). Campanian/Maastrichtian stage boundary in southern California: Resolution and implications for large-scale depositional patterns. *Geology* **17**, 80–83.

Barbera, X., Pares, J. M., Cabrera, L., and Anadón, P. (1994). High-resolution magnetic stratigraphy across the Oligocene-Miocene boundary in an alluvial-lacustrine succession (Ebro Basin, northeast Spain). *Phys. Earth Planet. Inter.* **85**, 181–193.

Barbetti, M. F., and McElhinny, M. W. (1976). The Lake Mungo geomagnetic excursion. *Philos. Trans. R. Soc. London Ser. A* **281**, 515–542.

Barendregt, R. W., Thomas, F. F., Irving, E., Baker, J., Macs, A., Stalker, and Churcher, L. S. (1991). Stratigraphy and paleomagnetism of the Jaw Face section, Wellsch Valley site Saskatchewan. *Can. J. Earth Sci.* **28**, 1353–1364.

Barghorn, S. (1981). Magnetic polarity stratigraphy of the Miocene type Tesuque Formation, Santa Fe Group, in the Espanola Valley, New Mexico. *Geol. Soc. Am. Bull.* **92**, 1027–1041.

Barndt, J., Johnson, N. M., Johnson, G. D., Opdyke, N. D., Lindsay, E. H., Pilbearn, D., and Tahirkheli, R. A. K. (1978). The magnetic polarity stratigraphy and age of the Siwalik Group near Dhok Patham Village, Potwar Plateau, Pakistan. *Earth Planet. Sci. Lett.* **41**, 355–364.

Barry, J. C., Lindsay, E. H., and Jacobs, L. L. (1982). A biostratigraphic zonation of the middle and upper Siwaliks of the Potwar Plateau of Northern Pakistan. *Palaeogeogr., Palaeoclimatol., Palaeoecol.* **37**, 95–139.

Barry, J. C., Johnson, N. M., Raza, S. M., and Jacobs, L. L. (1985). Neogene mammalian faunal change in southern Asia: Correlations with climatic tectonic and eustatic events. *Geology* **13**, 637–640.

Barton, C. E. (1989). Geomagnetic secular variation: Direction and intensity. In "The Encyclopedia of Solid Earth Geophysics" (D. E. James, ed.), pp. 560–577. Van Nostrand-Reinhold, New York.

Barton, C. E., and Bloemendal, J. (1986). Paleomagnetism of sediments collected during Leg 90, southern west Pacific. In "Initial Report of the Deep Sea Drilling Project" (J. P. Kennett, C. C. von der Borch *et al.*, eds.), Vol. 90, Part 2, pp. 1273–1316. U.S. Govt. Printing Office, Washington, DC.

Barton, C. E., and McElhinny, M. W. (1979). Detrital remanent magnetization in five slowly redeposited long cores of sediment. *Geophys. Res. Lett.* **6**, 229–232.

Barton, C. E., and McElhinny, M. W. (1980). A 10,000 yr geomagnetic secular variation record from three Australian maars. *Geophys. J. R. Astron. Soc.* **67**, 465–85.

Barton, C. E., McElhinny, M. W., and Edwards, D. J. (1980). Laboratory studies of depositional DRM. *Geophys. J. R. Astron. Soc.* **61**, 355–377.

Baumgartner, P. O. (1987). Age and genesis of Tethyan Jurassic radiolarites. *Eclogae Geol. Helv.* **80**, 831–8879.

Bazylinski, D. A., Frankel, R. B., and Jannasch, H. W. (1988). Anaerobic magnetite production by a marine, magnetotactic bacterium. *Nature (London)* **334**, 518–519.

Beer, J. A. (1990). Steady sedimentation and lithologic completeness, Bermejo Basin Argentina. *J. Geol.* **98**, 501–518.

Beer, J. A., Shen, C. D., Heller, F., Liu, T. S., Bonani, G., Dittrich, B., Suter, M., and Kubik, P. W. (1993). ^{10}Be and magnetic susceptibility in Chinese loess. *Geophys. Res. Lett.* **20**, 57–60.

Begét, J. E., and Hawkins, D. B. (1989). Influence of orbital parameters on Pleistocene loess deposition in central Alaska. *Nature (London)* **337**, 151–153.

Begét, J. E., Stone, D. B., and Hawkins, D. B. (1990). Paleoclimatic forcing of magnetic susceptibility variations in Alaskan loess during the late Quaternary. *Geology* **18**, 40–43.

Behrensmeyer, A. K., and Tauxe, L. (1982). Isochronous fluvial systems in miocene deposits of Northern Pakistan. *Sedimentology* **29**, 331–352.

Benson, R. H., and Hodell, D. A. (1994). Comment on "A critical re-evaluation of the Miocene/Plocene boundary as defined in the Mediterranean" by F. J. Hilgen and C. G. Langereis. *Earth Planet. Sci. Lett.* **124**, 245–250.

Berger, A. (1988). Milankovitch theory and climate. *Rev. Geophys.* **26**, 624–657.

Berger, A., and Loutre, M. F. (1991). Insolation values for the climate of the last 10 my. *Quat. Sci. Rev.* **10**, 297–317.

Berger, W. H. (1982). Deep-sea stratigraphy Cenozoic climatic steps and the search for chemoclimatic feedback. In "Cyclic Event Stratification" (G. Einsele and A. Seilacher, eds.), pp. 121–157. Springer-Verlag, New York.

Berger, W. H., Vincent, E., and Thierstein, H. R. (1981). The deep-sea record: Major steps. *In* "Cenozoic Ocean Evolution," SEPM Society for Sedimentary Geology Spec. Publ., No. 37, pp. 489–504.

Berggren, W. A., Burckle, L. H., Cita, M. B., Cooke, H. B. S., Funnell, B. M., Gartner, S., Hays, J. D., Kennett, J. P., Opdyke, N. D., Pastouret, L., Shackleton, N. J., and Takayanagi, Y. (1980). Towards a Quaternary time scale. *Quat. Res.* **13,** 277–302.

Berggren, W. A., Aubry, M. P., and Hamilton, N. (1983a). Neogene magnetobiostratigraphy of deep sea drilling project site 516 (Rio Grande Rise, South Atlantic). *In* "Initial Reports of the Deep Sea Drilling Project" (P. F. Barker, R. L. Carlson, D. A. Johnson *et al.,* eds.), Vol. 72, pp. 675–705. U.S. Govt. Printing Office, Washington, DC.

Berggren, W. A., Hamilton, N., Johnson, D. A., Pujol, C., Weiss, W., Cepek, P., and Gombos, A. M., Jr. (1983b). Magnetobiostratigraphy of deep sea drilling project leg 172, sites 515–518, Rio Grande rise (South Atlantic). *In* "Initial Reports of the Deep Sea Drilling Project" (P. F. Barker, R. L. Carlson, D. A. Johnson *et al.,* eds.), Vol. 72, pp. 939–948. U.S. Govt. Printing Office, Washington, DC.

Berggren, W. A., Kent, D. V., and Flynn, J. J. (1985a). Jurassic to Paleogene. Part 2. Paleogene geochronology and chronostratigraphy. *In* "The Chronology of the Geological Record" (N. J. Snelling, ed.), pp. 141–195. Blackwell, Oxford.

Berggren, W. A., Kent, D. V., and van Couvering, J. A. (1985b). The Neogene. Part 2: Neogene geochronology and chronostratigraphy. *In* "The Chronology of the Geological Record" (N. J. Snelling, ed.), pp. 211–260. Blackwell, Oxford.

Berggren, W. A., Kent, D. V., Obradovich, J. D., and Swisher, C. C. (1992). Toward a revised Paleogene geochronology. *In* "Eocene-Oligocene Climatic and Biotic Evolution" (D. R. Prothero and W. A. Berggren, eds.), pp. 29–45. Princeton Univ. Press, Princeton, NJ.

Berggren, W. A., Kent, D. V., Swisher, C. C., and Aubry, M. P. (1995). A revised Cenozoic geochronology and chronostratigraphy in time scales and global stratigraphic correlations: A unified temporal framework for an historical geology. *In* "Geochronology, Timescales, and Stratigraphic Correlation," (W. A. Berggren, D. V. Kent, M. Aubry, and J. Hardenbol, eds.), SEPM Spec. Publ. No. 54, pp. 129–212.

Beske-Diehl, S., and Shive, P. N. (1978). The rock magnetism of the Mississippian Madison Limestone north-central Wyoming. *Geophys. J. R. Astron. Soc.* **55,** 351–362.

Bice, D. A., and Montanari, A. S. (1988). Magnetic stratigraphy of the Massignano section across the Eocene-Oligocene boundary. *Int. Subcomm. Paleogene Stratigr., E/O Meet.,* Ancona, Spec. Publ. II, Vol. 4, pp. 111–117.

Bingham, C. (1974). An antipodally symmetric distribution on a sphere. *Ann. Stat.* **2,** 1201–1225.

Blackett, P. M. S. (1952). A negative experiment relating magnetism and the Earth's rotation. *Philos. Trans. R. Soc. London, Ser. A* **245,** 309–370.

Blackett, P. M. S. (1956). "Lectures on Rock Magnetism." Weizman Science Press, Jerusalem.

Blakely, R. L. (1974). Geomagnetic reversals and crustal spreading rates during the Miocene. *J. Geophys. Res.* **79,** 2979–2985.

Blakemore, R. P. (1975). Magnetic bacteria. *Science* **190,** 377–379.

Blakemore, R. P., Short, K. A., Bazylinski, D. A., Rosenblatt, C., and Frankel, R. B. (1985). Microaerobic conditions are required for magnetite formation within *Aquaspirillum magnetotacticum. Geomicrobiol. J.* **4,** 53–71.

Bleil, U. (1985). The magnetostratigraphy of northwest Pacific sediments, Deep Sea Drilling Project Leg 86. *In* "Initial Reports of the Deep Sea Drilling Project" (G. R. Heath, L. H. Burkle *et al.,* eds.), Vol. 86, pp. 441–458. U.S. Govt. Printing Office, Washington, DC.

Bleil, U. (1989). Magnetostratigraphy of Neogene and Quaternary sediment series from the Norwegian Sea: Ocean Drilling Program Leg 104, 1989. *In* "Proceedings of the Ocean Drilling Program, Scientific Results" (O. Eldholm, J. Theide, E. Taylor *et al.*, eds.), Vol. 104, pp. 829–850. ODP, College Station, TX.

Bloemendal, J. (1983). Paleoenvironmental implications of the magnetic characteristics of sediments from DSDP Site 514, Southeast Argentine Basin. *In* "Initial Reports of the Deep Sea Drilling Project" (W. J. Ludwig, V. A. Krasheninnikov *et al.*, eds.), Vol. 71, pp. 1097–1108. U.S. Govt. Printing Office, Washington, DC.

Bloemendal, J., and deMenocal, P. B. (1989). Evidence for a change in the periodicity of tropical climate cycles at 2.4 Myr from whole-core magnetic susceptibility measurements. *Nature* (*London*) **342,** 897–900.

Bloemendal, J., Oldfield, F., and Thompson, R. (1979). Magnetic measurements used to assess sediment influx at Llyn Goddionduon. *Nature* (*London*) **280,** 50–53.

Bloemendal, J., Lamb, B., and King, J. W. (1988). Paleoenvironmental implications of rock-magnetic properties of Late Quaternary sediment cores from the eastern equatorial Atlantic. *Paleoceanography* **3,** 61–87.

Bloemendal, J., King, J. W., Tauxe, L., and Valet, J. P. (1989). Rock magnetic stratigraphy of Leg 108 (Eastern tropical Atlantic) sites 658, 659, 661, and 665. *In* "Proceedings of the Ocean Drilling Program, Scientific Results" (W. Ruddiman, M. Sarthein *et al.*, eds.), Vol. 108, pp. 415–428. ODP, College Station, TX.

Bloemendal, J., King, J. W., Hall, F. R., and Doh, S. J. (1992). Rock magnetism of Late Neogene and Pleistocene deep-sea sediments: Relationship to sediment source, diagenetic processes, and sediment lithology. *J. Geophys. Res.* **97,** 4361–4375.

Bloemendal, J., King, J. W., Hunt, A., deMenocal, P. B., and Hayahida, A. (1993). Origin of the sedimentary magnetic record at ocean drilling program sites on the Owen Ridge, Western Arabian Sea. *J. Geophys. Res.* **98,** 4199–4219.

Bloxham, J. and Gubbins, D. (1985). The secular variation of the Earth's magnetic field. *Nature* (*London*) **317,** 777–781.

Bloxham, J., and Gubbins, D. (1987). Thermal core-mantle interactions. *Nature* (*London*) **325,** 511–513.

Bonhommet, N. and Babkine, J. (1967). Sur la présence d'aimentation inverse dans la Chaine des Puys. *C.R. Heb. Seances Acad. Sci., Ser. B* **264,** 92–94.

Bonhommet, N., and Zahringer, J. (1969). Paleomagnetism and potassium argon determinations of the Laschamp geomagnetic polarity event. *Earth Planet. Sci. Lett.* **6,** 43–46.

Bralower, T. J. (1987). Valanginian to Aptian calcareous nannofossil stratigraphy and correlation with the upper M-sequence magnetic anomalies. *Mar. Micropaleontol.* **11,** 293–310.

Bralower, T. J., Monechi, S., and Thierstein, H. R. (1989). Calcareous nannofossil zonation of the Jurassic-Cretaceous boundary interval and correlation with the geomagnetic polarity timescale. *Mar. Micropaleontol.* **14,** 153–235.

Bralower, T. J., Ludwig, K. R., Obradovich, J. D., and Jones, D. L. (1990). Berriasian (Early Cretaceous) radiometric ages from the Grindstone Creek section, Sacramento Valley, California. Earth Planet. Sci. Lett. **98,** 62–73.

Brandsma, D., Lund, S. P., and Henyey, T. L. (1989). Paleomagnetism of Late Quaternary marine sediments from Santa Catalina Basin, California Continental Borderland. *J. Geophys. Res.* **94,** 547–564.

Bressler, S. L., and Butler, R. F. (1978). Magnetostratigraphy of the later Tertiary Verde Formation, central Arizona. *Earth Planet. Sci. Lett.* **38,** 319–330.

Broecker, W., Bond, G., Klas, M., Clark, E., and McManus, J. (1992). Origin of the northern Atlantic's Heinrich events. *Clim. Dyn.* **6,** 265–273.

Brown, F. H., Shuey, R. T., and Croes, M. K. (1978). Magnetostratigraphy of the Shungura and Usno formations, southwestern Ethiopia: New data and comprehensive reanalysis. *Geophys. J. R. Astron. Soc.* **54,** 519–538.

Brunhes, B. (1906). Recherches sur le direction d'aimentation des roches volcaniques. *J. Phys.* **5,** 705–724.

Bryan, N. B., and Duncan, R. A. (1983). Age and provenance of clastic horizons from Hole 516F. *In* "Initial Reports of the Deep Sea Drilling Project" (P. F. Barker, R. L. Carlson, D. A. Johnson *et al.,* eds.), Vol. 72, pp. 75–477. U.S. Govt. Printing Office, Washington, DC.

Bullard, E. C. (1949). The magnetic field within the Earth. *Proc. R. Soc. London, Ser. A* **197,** 433–453.

Burbank, D. W. (1983). The chronology of intermontane basin development in the northwestern Himalaya and the evolution of the northwest syntaxis. *Earth Planet. Sci. Lett.* **64,** 77–92.

Burbank, D. W., and Barnosky, A. D. (1990). The magnetochronology of Barstovian mammals in southwestern Montana and implications for the initiation of Neogene crustal extension in the northern Rocky Mountains. *Geol. Soc. Am. Bull.* **102,** 1093–1104.

Burbank, D. W., and Li, J. (1985). Age and palaeoclimatic significance of the loess of Lanzhou, North China. *Nature (London)* **316,** 429–431.

Burbank, D. W., and Johnson, G. D. (1983). The late Cenozoic chronologic and stratigraphic development of the Kashmire intermontane basin, northwestern Himalaya. *Palaeogeogr., Palaeoclimatol., Palaeoecol.* **43,** 205–235.

Burbank, D. W., and Raynolds, R. G. H. (1984). Sequential late Cenozoic structural disruption of the northern Himalayan foredeep. *Nature (London)* **311,** 114–118.

Burbank, D. W., and Raynolds, R. G. H. (1988). Stratigraphic keys to the timing of thrusting in Terrestrial Forland basins: Applications in the Northwestern Himalaya. *In* "New Perspectives in Basin Analysis" (K. L. Kleinspehn and C. Paola, eds.), pp. 331–351. Springer-Verlag, New York.

Burbank, D. W., and Tahirkheli, R. A. K. (1985). The magnetostratigraphy, fission-track dating, and stratigraphic evolution of the Peshawar intermontane basin, northern Pakistan. *Geol. Soc. Am. Bull.* **96,** 539–552.

Burbank, D. W., and Whistler, D. P. (1987). Temporally contained tectonic rotation derived from magnetostratigraphic data: Implications for the initiation of the Garlock Fault, California. *Geology* **15,** 1172–1175.

Burbank, D. W., Engesser, B., Matter, A., and Weidman, M. (1992a). Magnetostratigraphic chronology, mammalian faunas, and stratigraphic evolution of the Lower Freshwater Molasse, Haut-Savoie France. *Eclogae Geol. Helv.* **85,** 399–431.

Burbank, D. W., Verges, J., Muñoz, J. A., and Bentham, P. (1992b). Coeval hindward and forward-imbricating thrusting in the south-central Pyrenees, Spain: Timing and rates of shortening and deposition. *Geol. Soc. Am. Bull.* **104,** 3–17.

Burbank, D. W., Puig de Fabregas, C., and Muñoz, J. A. (1992c). The chronology of the Eocene tectonic and stratigraphic development of the eastern Pyrenean foreland basin, northeast Spain. *Geol. Soc. Am. Bull.* **104,** 1101–1120.

Burek, P. J. (1964). Korrelation revers magnatisierter Gesteinfolgen in Oberen Bandsandstein S. W. Deutschlands. *Geol. Jahrb.* **84,** 591–616.

Burek, P. J. (1970). Magnetic reversals: Their applications to stratigraphic problems. *Am. Assoc. Pet. Geol. Bull.* **54,** 1120–1139.

Butler, R. F. (1992). "Paleomagnetism." Blackwell, Oxford.

Butler, R. F., and Lindsay, E. H. (1985). Mineralogy of magnetic minerals and revised magnetic polarity stratigraphy of continental sediments, San Juan Basin, New Mexico. *J. Geol.* 535–554.

Butler, R. F., Lindsay, E. H., Jacobs, L. L., and Johnson, N. M. (1977). Magnetostratigraphy of the Cretaceous-Tertiary boundary in the San Juan Basin, New Mexico. *Nature (London)* **267**, 318–323.

Butler, R. F., Gingerich, P. D., and Lindsay, E. H. (1981a). Magnetic polarity stratigraphy and biostratigraphy of Paleocene and lower Eocene continental deposits, Clark's Fork Basin, Wyoming. *J. Geol.* **89**, 299–316.

Butler, R. F., Lindsay, E. H., and Johnson, N. M. (1981b). Paleomagnetic polarity stratigraphy of the Cretaceous/Tertiary boundary. San Juan Basin, New Mexico. In Papers presented to the conference on large body impacts and terrestrial evolution: Geological, climatological, and biological implications. *LPI Contrib.* **449**, 7.

Butler, R. F., Marshall, L. G., Drake, R. E., and Curtis, G. A. (1984). Magnetic polarity stratigraphy and ^{40}K-^{40}Ar dating of late Miocene and early Pliocene continental deposits, Catamarca Province, N.W. Argentina. *J. Geol.* **92**, 623–636.

Butler, R. F., Krause, D. W., and Gingerich, P. D. (1987). Magnetic polarity stratigraphy and biostratigraphy of middle-late Paleocene continental deposits of South Central Montana. *J. Geol.* **95**, 647–658.

Cande, S. C., and Kent, D. V. (1992a). A new geomagnetic polarity timescale for the late Cretaceous and Cenozoic. *J. Geophys. Res.* **97**, 13917–13951.

Cande, S. C., and Kent, D. V. (1992b). Ultrahigh resolution marine magnetic anomaly profiles: A record of continuous paleointensity variations? *J. Geophys. Res.* **97**, 15,075–15,083.

Cande, S. C., and Kent, D. V. (1995). Revised calibration of the geomagnetic polarity timescale for the Late Cretaceous and Cenozoic. *J. Geophys. Res.* **100**, 6093–6095.

Cande, S. C., and Kristoffersen, Y. (1977). Late Cretaceous magnetic anomalies in the North Atlantic. *Earth Planet. Sci. Lett.* **35**, 215–224.

Cande, S. C., and LaBrecque, J. L. (1974). Behavior of the Earth's paleomagnetic field from small scale marine magnetic anomalies. *Nature (London)* **247**, 26–28.

Cande, S. C., Larson, R. L., and LaBrecque, J. L. (1978). Magnetic lineations in the Pacific Jurassic Quiet Zone. *Earth Planet. Sci. Lett.* **41**, 434–440.

Cande, S. C., LaBrecque, J. L., and Haxby, W. F. (1988). Plate kinematics of the South Atlantic: Chron C34 to present. *J. Geophys. Res.* **93**, 13479–13492.

Cao, Z., Xing, L., and Yu, Q. (1985). The age and boundary of magnetic strata of the Yushe Formation. *Bull. Inst. Geomech. C.A.G.S.* **6**, 143–153.

Cassie, R. A. (1978). Palaeomagnetic studies of the Suva Marl. B. Sc. Honors Thesis, University of Sydney, Sydney.

Castillo, J., Gose, W. A., and Perarnan, A. (1991). Paleomagnetic results from Mesozoic strata in the Merida Andes, Venezuela. *J. Geophys. Res.* **96**, 6011–6022.

Catalano, R., DiStefano, P., and Kozur, H. (1991). Permian circumpacific deep water faunas from the western Tethys (Sicily, Italy)—new evidence for the position of the Permian Tethys. *Palaeogeogr., Palaeoclimatol., Palaeoecol.* **87**, 75–108.

Cecca, F., Pallini, G., Erba, E., Premoli-Silva, I., and Coccioni, R. (1994). Hauterivian-Barremian chronostratigraphy based on ammonites, nannofossils, planktonic foraminifera, and magnetic chrons from the Mediterranean domain. *Cretaceous Res.* **15**, 457–467.

Cerveny, P. R., Naeser, N. D., Zeitler, P. K., Naeser, C. W., and Johnson, N. M. (1988). History of uplift and relief of the Himalaya during the past 18 million years: Evidence from fission-track ages of detrital group. *In* "New Perspectives in Basin Analysis" (K. L. Kleinspehn and C. Paola, eds.), pp. 43–62. Springer-Verlag, New York.

Chaddha, G., and Seehra, M. S. (1983). Magnetic components and particle size distribution of coal fly ash. *J. Phys. D.* **16**, 1767–1776.

Chamalaun, F. H., and McDougall, I. (1966). Dating geomagnetic polarity episodes in Reunion. *Nature (London)* **210**, 1212–1214.

Champion, D. E., Lanphere, M. A., and Kuntz, M. A. (1988). Evidence for a new geomagnetic reversal from lava flows in Idaho: Discussion of short polarity reversals in the Brunhes and late Matuyama polarity chrons. *J. Geophys. Res.* **93**, 11667–11680.

Chang, S. R., and Kirschvink, J. L. (1989). Magnetofossils, the magnetization of sediments, and the evolution of magnetite biomineralization. *Annu. Rev. Earth Planet. Sci.* **17**, 169–195.

Chang, S. R., Allen, C. R., and Kirchvinck, J. L. (1987). Magnetic stratigraphy and a test for block rotation of sedimentary rocks within the San Andreas fault zone, Mecca Hills southeastern California. *Quat. Res.* **27**, 30–40.

Channell, J. E. T., and Dobson, J. P. (1989). Magnetic stratigraphy and magnetic mineralogy at the Cretaceous-Tertiary boundary section, Braggs, Alabama. *Palaeogeogr. Palaeoclimatol., Palaeocol.* **69**, 267–277.

Channell, J. E. T., and Erba, E. (1992). Early Cretaceous polarity chrons CM0 to CM11 recorded in northern Italian land sections near Brescia. *Earth Planet. Sci. Lett.* **108**, 161–179.

Channell, J. E. T., and Grandesso, P. (1987). A revised correlation of Mesozoic polarity chrons and calpionellid zones. *Earth Planet. Sci. Lett.* **85**, 222–240.

Channell, J. E. T., and Hawthorne, T. (1990). Progressive dissolution of titanomagnetites at ODP Site 653 (Tyrrhenian Sea). *Earth Planet. Sci. Lett.* **96**, 469–480.

Channell, J. E. T., and McCabe, C. (1994). Comparison of magnetic hysteresis parameters of unremagnetized and remagnetized limestones. *J. Geophys. Res.* **99**, 4613–4623.

Channell, J. E. T., and Medizza, F. (1981). Upper Cretaceous and Paleogene magnetic stratigraphy and biostratigraphy from the Venetian (Southern) Alps. *Earth Planet. Sci. Lett.* **55**, 419–432.

Channell, J. E. T., and Torii, M. (1990). Two events recorded in the Brunhes Chron at Hole 650A (ODP Leg 107, Tyrrhenian Sea): Geomagnetic phenomena? *In* Proceedings of the Ocean Drilling Program, Scientific Results" (K. A. Kastens, J. Muscle *et al.*, eds.), Vol. 107, pp. 347–359. College Station, TX.

Channell, J. E. T., Lowrie, W., and Medizza, F. (1979). Middle and Early Cretaceous magnetic stratigraphy from the Cismon section, northern Italy. *Earth Planet. Sci. Lett.* **42**, 133–166.

Channell, J. E. T., Freeman, R., Heller, F., and Lowrie, W. (1982a). Timing of diagenetic hematite growth in red pelagic limestones from Gubbio (Italy). *Earth Planet. Sci. Lett.* **58**, 189–201.

Channell, J. E. T., Ogg, J. G., and Lowrie, W. (1982b). Geomagnetic polarity in the early Cretaceous and Jurassic. *Philos. Trans. R. Soc. London, Ser. A* **306**, 137–146.

Channell, J. E. T., Lowrie, W., Pialli, P., and Venturi, F. (1984). Jurassic magnetic stratigraphy from Umbrian (Italian) land section. *Earth Planet. Sci. Lett.* **68**, 309–325.

Channell, J. E. T., Bralower, T. J., and Grandesso, P. (1987). Biostratigraphic correlation of Mesozoic polarity chrons (CM1 to CM23) at Capriolo and Xausa (Southern Alps, Italy). *Earth Planet. Sci. Lett.* **85**, 203–321.

Channell, J. E. T., Rio, D., and Thunell, R. C. (1988). Miocene/Pliocene boundary magneto-stratigraphy at Capo Spartivento, Calabria, Italy. *Geology* **16**, 1096–1099.

Channell, J. E. T., Massari, F., Benetti, A., and Pezzoni, N. (1990a). Magnetostratigraphy and biostratigraphy of Callovian-Oxfordian limestones from the Trento plateau, (Monti Lessini, northern Italy). *Palaeogeogr., Palaeoclimatol., Palaeoecol.* **79**, 289–303.

Channell, J. E. T., Torii, M., and Hawthorne, T. (1990b). Magnetostratigraphy of sediments recovered at sites 650, 651, 652, and 654, (Leg 107 in the Tyrrhenian sea). *In* "Proceedings of the Ocean Drilling Program, Scientific Results" (K. A. Kastens, J. Muscle *et al.*, eds.), Vol. 107, pp. 335–346. ODP, College Station, TX.

Channell, J. E. T., Erba, E., and Lini, A. (1993). Magnetostratigraphic calibration of the Late Valanginan carbon isotope event in pelagic limestones from northern Italy and Switzerland. *Earth Planet. Sci. Lett.* **118**, 145–166.

Channell, J. E. T., Poli, M. S., Rio, D., Sprovieri, R., and Villa, G. (1994). Magnetic stratigraphy and biostratigraphy of Pliocene "argille azzurre" (Northern Apennines, Italy). *Palaeogeogr. Palaeoclimatol., Palaeoecol.* **110**, 83–102.

Channell, J. E. T., Erba, E., Nakanishi, M., and Tamaki, K. (1995a). A Late Jurassic–Early Cretaceous timescale and oceanic magnetic anomaly block models. *In* "Geochronology, Timescales, and Stratigraphic Correlation," (W. A. Berggren, D. V. Kent, M. Aubry, and J. Hardenbol, eds.), SEPM Spec. Publ. No. 54, pp. 51–64.

Channell, J. E. T., Cecca, F., and Erba, E. (1995b). Correlations of Hauterivian and Barremian (Early Cretaceous) stage boundaries to polarity chrons. *Earth Planet. Sci. Lett.* **134**, 125–140.

Chave, A. D. (1984). Lower Paleocene–Upper Cretaceous magnetostratigraphy, Site 525, 527, 528 and 529, Deep Sea Drilling Project Leg 74. *In* "Initial Reports of the Deep Sea Drilling Project" (T. C. Moore, Jr., P. D. Rabinowitz *et al.*, eds.), Vol. 74, pp. 525–531. U.S. Govt. Printing Office, Washington, DC.

Chave, A. D., and Denham, C. R. (1979). Climatic changes, magnetic intensity variations and fluctuations of the eccentricity of the Earth's orbit during the past 2,000,000 years and a mechanism which may be responsible for the relationship—a discussion. *Earth Planet. Sci. Lett.* **44**, 150–152.

Chevallier, R. (1925). L'aimentation des lavas de l'Etna et l'orientation du champ terrestre en Sicile du XII au XVII Sicile. *Ann. Phys. (Paris)* [10] **4**, 5–162.

Christie-Blick, N., Mountain, G. S., and Miller, K. G. (1990). Stratigraphic and seismic stratigraphy record of sea level change. *In* "Sea-level Change, National Research Council Studies in Geophysics" (R. Revelle, ed.), pp. 116–140. National Research Council, Washington, DC.

Chugaeva, M. N. (1976). Ordovician in the North-Eastern U.S.S.R. *In* "The Ordovician System" (M. G. Bassett, ed.), Proc. Palaeontilogical Assoc. Symp., Birmingham, 1974, pp. 283–292. Univ. of Washington Press, Seattle.

Cirilli, S., Marton, P., and Vigli, L. (1984). Implications of a combined biostratigraphic and paleomagnetic study of the Umbrian Maiolica formation. *Earth Planet. Sci. Lett.* **69**, 203–214.

Cisowski, S. (1981). Interacting vs. non-interacting single domain behavior in natural and synthetic samples. *Phys. Earth Planet. Inter.* **26**, 56–62.

Claoué-Long, J. C., Zichao, Z., Guogan, M., and Schaohua, D. (1991). The age of the Permian-Triassic boundary. *Earth Planet. Sci. Lett.* **105**, 182–190.

Claoué-Long, J. C., Compston, W., Roberts, J., and Fanning, C. M. (1995). Two carboniferous ages: A comparison of SHRIMP zircon dating with conventional zircon ages and ^{40}Ar/^{39}Ar analysis. *In* "Geochronology, Timescales, and Stratigraphic Correlation," SEPM Spec. Publ. No. 54.

Clark, D. A. (1984). Hysteresis properties of sized dispersed monoclinic pyrrhotite grains. *Geophys. Res. Lett.* **11**, 173–176.

Clark, R. M., and Thompson, R. (1978). An objective method for smoothing paleomagnetic data. *Geophys. J. R. Astron. Soc.* **52**, 205–213.

Clark, R. M., and Thompson, R. (1979). A new approach to the alignment of time series. *Geophys. J. R. Astron. Soc.* **58**, 593–607.

Clement, B. M. (1991). Geographical distribution of transitional V.G.P.'s: Evidence for non-zonal equatorial symmetry during the Matuyama-Brunhes geomagnetic reversal. *Earth Planet. Sci. Lett.* **104**, 48–58.

Clement, B. M. (1992). Evidence for dipolar fields during the Cobb Mountain geomagnetic polarity reversals. *Nature* (*London*) **358**, 405–407.

Clement, B. M., and Hailwood, E. A. (1991). Magnetostratigraphy of sediments from sites 701 and 702. *In* "Proceedings of the Ocean Drilling Program. Scientific Results" (P. F. Ciesielski, Y. Kristofferson *et al.*, eds.), Vol. 114, pp. 359–365. ODP, College Station, TX.

Clement, B. M., and Kent, D. V. (1984). A detailed record of the Lower Jaramillo polarity transition from a southern hemisphere deep sea sediment core. *J. Geophys. Res.* **89**, 1049–1058.

Clement, B. M., and Kent, D. V. (1987). Short polarity intervals within the Matuyama: Transition field records from hydraulic piston cored sediments from the North Atlantic. *Earth Planet. Sci. Lett.* **91**, 253–264.

Clement, B. M., and Martinson, D. G. (1992). A quantitative comparison of two paleomagnetic records of the Cobb Mountain subchron from North Atlantic deep-sea sediments. *J. Geophys. Res.* **97**, 1735–1752.

Clement, B. M., and Robinson, R. (1986). The magnetostratigraphy of Leg 94 sediments. *In* "Initial Reports of the Deep Sea Drilling Project" (W. Ruddiman *et al.*, eds.), Vol. 94, pp. 635–650. U.S. Govt. Printing Office, Wasington, DC.

Clement, B. M., and Stixrude, L. (1995). Inner core anisotropy, anomalies in the time-averaged paleomagnetic field, and polarity transition paths. *Earth Planet. Sci. Lett.* **130**, 75–85.

Clement, B. M., Kent, D. V., and Opdyke, N. D. (1982). Brunhes-Matuyama transition in three deep sea sediment cores. *Philos. Trans. R. Soc. London, Ser. A* **306**, 113–119.

Clement, B. M., Hall, F. J., and Jarrard, R. D. (1989). The magnetostratigraphy of Ocean Drilling Program Leg 105 sediments. *In* "Proceedings of The Ocean Drilling Program, Scientific Results" (S. P. Srivastava, M. Arthur, B. Clement, eds.), Vol. 105, pp. 583–595. College Station, TX.

Clement, B. M., Rhodda, P., Smith, E., and Sierra, L. (1995). Recurring transitional geomagnetic field geometries: evidence from sediments and lavas. *Geophys. Res. Lett.* **22**, 3171–3174.

Clyde, W. C., Stamatakos, J., and Gingerich, P. D. (1994). Chronology of the Wasatchian land-mammal age (early Eocene): Magnetostratigraphic results from the McCullough Peaks Section, Northern Bighorn Basin, Wyoming. *J. Geol.* **102**, 367–377.

Coccioni, R., Erba, E., and Premoli-Silva, I. (1992). Barremian-Aptian calcareous plankton biostratigraphy from the Gorgo Cerbara section (Marche, central Italy) and implications for plankton evolution. *Cretaceous Res.* **13**, 517–537.

Coe, R. S., and Liddicoat, J. C. (1994). Overprinting of a natural magnetic remanence in lake sediments by a subsequent high-intensity field. *Nature* (*London*) **367**, 57–59.

Coe, R. S., Grommé, S., and Mankinen, E. A. (1978). Geomagnetic paleointensities from excursion sequences in lavas on Oahu, Hawaii. *J. Geophys. Res.* **89**, 1059–1069.

Coleman, D. S., and Bralower, T. J. (1993). New U-Pb zircon age constraints on the Early Cretaceous time scale. *EOS, Trans. Am. Geophys. Union* **73**, 556.

Collinson, D. W. (1965). The remanent magnetism and magnetic properties of red sediments. *Geophys. R. Astron. Soc.* **10**, 105–126.

Collinson, D. W. (1974) The role of pigment and specularite in the remanent magnetism of red sandstones. *Geophys. J. R. Astron. Soc.* **38**, 253–264.

Collinson, D. W. (1983). "Methods in Rock-magnetism and Paleomagnetism: Techniques and Instrumentation." Chapman & Hall, London.

Condomes, M., Moraud, P., Camus, G., and Duthon, L. (1982). Chronological and geochemical study of lavas from the Chaine des Puys, Massif Central, France: Evidence for crustal contamination. *Contrib. Mineral. Petrol.* **81**, 296–303.

Constable, C. G. (1985). Eastern Australian geomagnetic field intensity over the past 14,000 yr. *Geophys. J. R. Astron. Soc.* **81,** 121–130.

Constable, C. G. (1992). Link between geomagnetic reversal paths and secular variation of the field over the past 5 Myr. *Nature (London)* **358,** 230–233.

Constable, C. G., and McElhinny, M. W. (1985). Holocene geomagnetic secular variation records from north-eastern Australian lake sediments. *Geophys. J. R. Astron. Soc.* **81,** 103–120.

Constable, C. G., and Parker, R. (1991). Deconvolution of long-core paleomagnetic measurements, spline therapy for the linear problem. *Geophys. J. Int.* **104,** 453–468.

Constable, C. G., and Tauxe, L. (1987). Palaeointensity in the pelagic realm: Marine sediment data compared with archaeomagnetic and lake sediment records. *Geophys. J. R. Astron. Soc.* **90,** 43–59.

Cope, J. C. W., Getty, T. A., Howarth, M. K., Morton, N., and Torrens, H. S. (1980a). A correlation of Jurassic rocks in the British Isles. Part I: Introduction and Lower Jurassic. *Geol. Soc. London, Spec. Rep.* **14,** 73.

Cope, J. C. W., Duff, K. L., Parsons, C. F., Torrens, H. S., Wimbledon, W. A., and Wright, J. K. (1980b). A correlation of Jurassic rocks in the British Isles. Part 2: Middle and Upper Jurassic. *Geol. Soc. London, Spec. Rep.* **15,** 109.

Courtillot, V., and Besse, J. (1987). Magnetic field reversals, polar wander, and core mantle coupling. *Science* **237,** 1140–1147.

Cox, A. (1968). Length of geomagnetic polarity intervals. *J. Geophys. Res.* **73,** 3247–3260.

Cox, A. (1969). Geomagnetic reversals. *Science* **163,** 237–245.

Cox, A., and Dalrymple, G. B. (1967). Statistical analysis of geomagnetic reversal data and the precision of potassium-argon dating. *J. Geophys. Res.* **72,** 2603–2614.

Cox, A., Doell, R. R., and Dalrymple, G. B. (1963a). Geomagnetic polarity epochs and Pleistocene geochonometry. *Nature (London)* **198,** 1049–1051.

Cox, A., Doell, R. R., and Dalrymple, G. B. (1963b). Geomagnetic polarity epoch: Sierra Nevada II. *Science* **142,** 382–385.

Cox, A., Doell, R. R., and Dalrymple, G. B. (1964). Reversals of the Earth's magnetic field. *Science* **144,** 1537–1543.

Cox, A., Doell, R. R., and Dalrymple, G. B. (1965). Quaternary paleomagnetic stratigraphy *In* "The Quaternary of the United States" (H. E. Wright, Jr., and D. G. Frey eds.), pp. 817–830. University Press, Princeton, N.J.

Cox, A., Hopkins, D. M., and Dalrymple, G. B. (1966). Geomagnetic polarity epochs: Pribilot Islands, Alaska. *Geol. Soc. Am. Bull.* **77,** 883–910.

Creer, K. M. (1974). Geomagnetic variations for the interval 7000–25,000 yr BP as recorded in a core of sediment for station 1474 of the Black Sea cruise of "Atlantic II." *Earth Planet. Sci. Lett.* **23,** 34–42.

Creer, K. M., and Tucholka, P. (1982). Construction of type curves of geomagnetic secular variation for dating lake sediments from east central North America. *Can. J. Earth Sci.* **19,** 1106–1115.

Creer, K. M., Irving, E., and Runcorn, S. K. (1954). The direction of the geomagnetic field in remote epochs in Great Britain. *J. Geomagn. Geoelectr.* **6,** 164–168.

Creer, K. M., Mitchell, J. G., and Valencio, D. A. (1971). Evidence for a normal geomagnetic field polarity event at 263 ± 5 My B. P. within the Late Paleozoic reversal interval. *Nature (London)* **233,** 87–89.

Creer, K. M., Thompson, R., Molyneaux, L., and Mackereth, F. J. H. (1972). Geomagnetic secular variation recorded in the stable magnetic remanence of recent sediments. *Earth Planet. Sci. Lett.* **14,** 115–127.

Creer, K. M., Anderson, T. W., and Lewis, C. F. M. (1976). Late Quaternary geomagnetic stratigraphy recorded in Lake Erie sediments. *Earth Planet. Sci. Lett.* **31,** 37–47.

Creer, K. M., Hogg, T. E., Readman, P. W., and Reynaud, C. (1980a). Paleomagnetic secular variation curves extending back to 13,400 years BP recorded by sediments deposited in Lac de Joux, Switzerland. *J. Geophys.* **48,** 139–147.

Creer, K. M., Readman, P. W., and Jacobs, A. M. (1980b). Paleomagnetic and paleontological dating of a section at Gioa Tauro, Italy: Identification of the Blake event. *Earth Planet. Sci. Lett.* **50,** 289–300.

Creer, K. M., Readman, P. W., and Papamarinopoulos, S. (1981). Geomagnetic secular variation in Greece through the last 6000 years obtained from lake sediment studies. *Geophys. J. R. Astron. Soc.* **66,** 193–219.

Creer, K. M., Tucholka, P., and Barton, C. E., eds. (1983a). "Geomagnetism of Baked Clays and Recent Sediments." Elsevier, Amsterdam.

Creer, K. M., Valencio, D. A., Sinito, A. M., Tucholka, P., and Vilas, J. F. A. (1983b). Geomagnetic secular variations 0–14000 yr BP as recorded by lake sediments from Argentina. *Geophys. J. R. Astron. Soc.* **74,** 199–221.

Cunningham, K. J., Farr, M. R., and Rakic-El Beid, K. (1994). Magnetostratigraphic dating of an Upper Miocene shallow-marine and continental sedimentary succession in northeastern Morocco. *Earth Planet. Sci. Lett.* **127,** 77–93.

Dalrymple, G. B., Cox, A., Doell, R. R., and Grommé, G. S. (1967). Pliocene geomagnetic polarity epochs. *Earth Planet. Sci. Lett.* **2,** 163–173.

Davies, T. A., and Gorsline, D. S. (1976). Oceanic sediments and sedimentary processes. *In* "Chemical Oceanography" (J. P. Riley and R. Chester, eds.), pp. 1–80. Academic Press, New York.

Day, R., Fuller, M., and Schmidt, V. A. (1977). Hysteresis properties of titanomagnetites: Grain-size and compositional dependence. *Phys. Earth Planet. Inter.* **13,** 260–267.

Dearing, J. A., and Flower, R. A. (1982). The magnetic susceptibility of sedimenting material trapped in Lough Neagh, Northern Ireland and its erosional significance. *Limnol. Oceanogr.* **17,** 969–975.

Dearing, J. A., Morton, R. I., Price, T. W., and Foster, I. D. L. (1986). Tracing movements of topsoil by magnetic measurements: Two case studies. *Phys. Earth Planet. Inter.* **42,** 93–104.

De Boer, P. L. (1982). Cyclicity and storage of organic matter in middle Cretaceous pelagic sediments. *In* "Cyclic and Event Stratification" (G. Einsele and A. Seilacher, eds.), pp. 456–475. Springer-Verlag, Berlin.

Deino, A., Tauxe, L., Monaghan, M., and Drake, R. (1990). ^{40}Ar/^{39}Ar age calibration of the litho and paleomagnetic stratigraphies of the Ngorora Fm., Kenya. *J. Geol.* **98,** 567–587.

Dekkers, M. J. (1988). Magnetic properties of natural pyrrhotite. Part I: Behaviour of initial susceptibility and saturation-magnetization related rock-magnetic parameters in a grain-size dependent framework. *Phys. Earth Planet. Inter.* **52,** 376–393.

Dekkers, M. J. (1989). Magnetic properties of natural pyrrhotite. II. High- and low-temperature behaviour of Jrs and TRM as a function of grain size. *Phys. Earth Planet. Inter.* **57,** 266–283.

Dekkers, M. J., Mattei, J. L., Fillion, G., and Rochette, P. (1989). Grain size dependence of the magnetic behavior of pyrrhotite during its low-temperature transition at 34 K. *Geophys. Res. Lett.* **16,** 855–858.

de Menocal, P. B., Ruddiman, W. F., and Kent, D. V. (1990). Depth of post-depositional remanence acquisition in deep-sea sediments: A case study of the Brunhes-Matuyama reversal and oxygen isotopic stage 19.1. *Earth Planet. Sci. Lett.* **99,** 1–13.

Denham, C. R. (1974). Counter-clockwise motion of paleomagnetic directions 24,000 years ago at Mono Lake, California. *J. Geomagn. Geoelectr.* **26,** 487–498.

Denham, C. R. (1976). Blake polarity episode in two cores from the Greater Antilles outer ridge. *Earth Planet Sci. Lett.* **29**, 422–443.

Denham, C. R. (1981). Numerical correlation of recent paleomagnetic records in two Lake Tahoe cores. *Earth Planet Sci. Lett.* **54**, 48–52.

Denham, C. R., and Cox, A. (1971). Evidence that the Laschamp Polarity Event did not occur 13300–30400 years ago. *Earth Planet. Sci. Lett.* **13**, 181–190.

DePaolo, J. J., and Ingram, B. L. (1985). High-resolution stratigraphy with strontium isotopes. *Science* **227**, 938–940.

Diehl, J. F., and Shive, P. N. (1979). Paleomagnetic studies of the early Permian Ingelside formation of northern Colorado. *Geophys. J. R. Astron. Soc.* **56**, 278–282.

Diehl, J. F., and Shive, P. N. (1981). Paleomagnetic results from the Late Carboniferous/ Early Permian Casper formation: Implications for northern Appalachian tectonics. *Earth Planet. Sci. Lett.* **54**, 281–291.

DiVenere, V. J., and Opdyke, N. D. (1990). Paleomagnetism of the Maringouin and Shepody formations, New Brunswick: A Namurian magnetic stratigraphy. *Can. J. Earth Sci.* **27**, 803–810.

DiVenere, V. J., and Opdyke, N. D. (1991a). Magnetic polarity stratigraphy in the uppermost Mississippian Mauch Chunk formation, Pottsville, Pennsylvania. *Geology* **19**, 127–130.

DiVenere, V. J., and Opdyke, N. D. (1991b). Magnetic polarity stratigraphy and carboniferous paleopole positions from the Joggins Section, Cumberland Basin, Nova Scotia. *J. Geophys. Res.* **96**, 4051–4064.

Dodson, M. A., and McClelland-Brown, E. (1980). Magnetic blocking temperatures of single domain grains during cooling. *J. Geophys. Res.* **85**, 2625–2637.

Dodson, R. E. (1979). Counterclockwise precession of the geomagnetic field vector and westward drift of the nondipole field. *J. Geophys. Res.* **84**, 637–644.

Doell, R. R., and Dalrymple, G. B. (1966). Geomagnetic polarity epochs: A new polarity event and the age of the Brunhes-Matuyama boundary. *Science* **152**, 1060–1061.

Doell, R. R., Dalrymple, G. B., and Cox, A. (1966). Geomagnetic polarity epochs: Sierra Nevada Data, 3. *J. Geophys. Res.* **71**, 531–541.

Doh, S. J., King, J. W., and Leinen, M. (1988). A rock-magnetic study of Giant Piston Core LL44-GPC from the central North Pacific and its paleocenagraphic significance. *Paleoceanography* **3**, 89–111.

Douglas, D. M. (1988). Paleomagnetics of Ringerike Old Red Sandstone and related rocks, southern Norway: Implications for pre-Carboniferous separation of Baltica and British terranes. *Tectonophysics* **148**, 11–27.

Dunlop, D. J. (1972). Magnetic mineralogy of unheated and heated red sediments by coercivity spectrum analysis. *Geophys. J. R. Astron. Soc.* **27**, 37–55.

Dunlop, D. J. (1979). On the use of Zijderveld vector diagrams in multicomponent paleomagnetic studies. *Phys. Earth Planet. Inter.* **20**, 12–24.

Dunlop, D. J. (1983). Viscous magnetization of 0.04–100 μm magnetites. *Geophys. J. R. Astron. Soc.* **27**, 37–55.

Dunlop, D. J. (1986). Hysteresis properties of magnetite and their dependence on particle size: A test of the pseudo-single-domain remanence models. *J. Geophys. Res.* **91**, 9569–9584.

Dunlop, D. J. (1990). Developments in rock magnetism. *Rep. Prog. Phys.* **53**, 707–792.

Dunning, G. R., and Hodych, J. P. (1990). U/Pb zircon and baddeleyite ages for the Palisades and Gettysburg sills of the northeastern United States: Implications for the age of the Triassic/Jurassic boundary. *Geology* **18**, 795–798.

Dziewonski, A. M. (1984). Mapping the lower mantle: Determination of lateral heterogeneity in P velocity up to degree and order 6. *J. Geophys. Res.* **39**, 5929–5952.

Dziewonski, A. M., and Woodhouse, J. H. (1987). Global images of the earth's interior. *Science* **236**, 37–48.

Eardley, A. J., Shuey, R. T., Gvosdetsky, V., Nash, W. P., Picard, M. D., Grey, D. C., and Kukla, G. J. (1973). Lake cycles in the Bonneville Basin, Utah. *Geol. Soc. Am. Bull.* **84**, 211–216.

Eddy, J. A. (1976). The Maunder minimum. *Science* **192**, 1189–1202.

Egbert, G. (1992). Sampling bias in VGP longitudes. *Geophys. Res. Lett.* **19**, 2353–2356.

Einarsson, T., and Sigurgeirsson, T. (1955). Rock magnetism in Iceland. *Nature (London)* **175**, 892.

Ellis, B. J., Levi, S., and Yeats, R. S. (1993). Magnetic stratigraphy of the Morales formation: Late Neogene clockwise rotation and compression in the Cuyama Basin, California Coast Ranges. *Tectonics* **12**, 1170–1179.

Elsasser, W. M. (1946). Induction effects in terrestrial magnetism. I. Theory. *Phys. Rev.* **69**, 106–116.

Elston, D. P., and Perucker, M. (1979). Detrital magnetization in red beds of the Moenkopi formation. *J. Geophys. Res.* **84**, 1653–1665.

Embleton, B. J. J., and McDonnell, K. L. (1981). Magnetostratigraphy in the Sydney Basin, southeastern Australia. *In* "Global Reconstructions and the Geomagnetic Field During the Paleozoic" (M. W. McElhinny *et al.*, eds.), Adv. Earth Planet. Sci., Vol. 10, pp. 1–10 Center for Academic Publ., Tokyo.

Emiliani, C. (1955). Pleistocene temperatures. *J. Geol.* **63**, 538–578.

Ernesto, M., Pacca, I. G., Hiodo, F. Y., and Nardy, A. J. R. (1990). Palaeomagnetism of the Mesozoic Sierra Geral formation southern Brazil. *Phys. Earth Planet. Inter.* **64**, 153–175.

Eusley, R. A., and Verosub, K. L. (1982). A magnetostratigraphic study of the sediments of the Ridge Basin, southern California and its tectonic and sedimentologic implications. *Earth Planet. Sci. Lett.* **59**, 192–207.

Evanoff, E., Prothero, D. R., and Lander, R. H. (1992). Eocene-Oligocene climatic change in North America: The White River Formation near Douglas, east central Wyoming. *In* "Eocene-Oligocene Climatic and Biotic Evolution" (D. R. Prothero and W. A. Berggren, eds.), pp. 116–130. Princeton Univ. Press, Princeton, NJ.

Evans, M. E., and Heller, F. (1994). Magnetic enhancement and paleoclimate: Study of a loess/paleosol couplet across the Loess Plateau of China. *Geophys. J. Int.* **117**, 257–264.

Evans, M. E., Wang, Y., Rutter, N., and Ding, Z. (1991). Preliminary magnetic stratigraphy of the red clay underlying the loess sequence at Baoji, China. *Geophys. Res. Lett.* **18**(8), 1409–1412.

Evernden, J. F., Savage, D. E., Curtis, G. H., and James, G. T. (1964). Potassium–argon dates and the Cenozoic mammalian chronology of North America. *Am. J. Sci.* **262**, 145–198.

Eyre, J. K., and Shaw, J. (1994). Magnetic enhancement of Chinese loess—the role of γFe_2O_3. *Geophys. J. Int.* **117**, 265–271.

Fang, W., Van der Voo, R., and Liang, Q. Z. (1989). Reconnaissance magnetostratigraphy of the Precambrian-Cambrian boundary section at Meichueun, Southwest China. *Cuad. Geol. Iber.* **12**, 205–222.

Farr, M. R., Sprowl, D. R., and Johnson, J. (1993). Identification and initial correlation of magnetic reversal in the Lower to Middle Ordovician of northern Arkansas. *In* "Applications of Paleomagnetism to Sedimentary Geology" (D. M. Aissaoui, D. F. McNeill, and N. F. Hurley, eds.), SEPM Society for Sedimentary Geology Spec. Publ. No. 49, pp. 83–93.

Fisher, R. A. (1953). Dispersion on a sphere. *Proc. R. Astron. Soc.* **A217**, 295–305.

Flynn, J. J. (1986). Correlation and geochronology of middle Eocene strata from the western U.S. *Palaeogeogr., Palaeoclimatol., Palaeoecol.* **55**, 335–406.

Flynn, J. J., MacFadden, B. J., and McKenna, M. C. (1984). Land mammal ages, faunal heterochroneity, and temporal resolution in Cenozoic terrestrial sequences. *J. Geol.* **92,** 687–705.

Foland, K. A., Gilbert, L. A., Sebring, C. A., and Jiang-Feng, C. (1986). $^{40}Ar/^{39}Ar$ ages for plutons of the Monteregian Hills, Quebec: Evidence for a single episode of Cretaceous magmatism. *Geol. Soc. Am. Bull.* **97,** 966–974.

Forster T. H., and Heller, F. (1994). Loess deposits from the Tujik depression (central Asia): Magnetic properties and paleoclimate. *Earth Planet. Sci. Lett.* **128,** 501–512.

Foster, J. H., and Symons, D. T. A. (1979). Defining a paleomagnetic polarity pattern in the Monteregian intrusives. *Can. J. Earth Sci.* **16,** 1716–1725.

Foster, J. H. (1966). A paleomagnetic spinner magnetometer using a flux gate gradiometer. *Earth Planet. Sci. Lett.* **1,** 463–466.

Foster, J. H., and Opdyke, N. D. (1970). Upper Miocene to Recent magnetic stratigraphy in deep-sea sediments. *J. Geophys. Res.* **75,** 4465–4473.

Frankel, R. B. (1987). Anaerobes pumping iron. *Nature (London)* **330,** 208.

Freeman, R. (1986). Magnetic mineralogy of pelagic limestones. *Geophys. J. R. Astron. Soc.* **85,** 455–452.

Friedman, R., Gee, J., Tauxe, L., Downing, K., and Lindsay, E. (1992). The magnetostratigraphy of the Chitarwata and lower Vihowa formations of the Dera Ghazi Khan area, Pakistan. *Sediment. Geol.* **81,** 253–268.

Friend, P. F., Johnson, N. M., and McRae, L. E. (1989). Time-level plots and accumulation rates of sediment sequences. *Geol. Mag.* **126,** 491–498.

Fry, G. J., Bottjer, D. J., and Lund, S. P. (1985). Magnetostratigraphy of displace Upper Cretaceous strata in southern California. *Geology* **13,** 648–651.

Fuller, M. D., and Kobayashi, K. (1967). Identification of magnetic phases in certain rocks by low-temperature analysis. *In* "Methods in Paleomagnetism" (D. W. Collinson, K. M. Creer, and S. K. Runcorn, eds.), pp. 529–534. Elsevier, New York.

Fuller, M. D., Williams, I., and Hoffman, K. A. (1979). Paleomagnetic records of geomagnetic field reversals and the morphology of the transition fields. *Rev. Geophys. Space Phys.* **17,** 179–203.

Fulton, R. J., Irving, E., and Wheadon, P. M. (1992). Stratigraphy and paleomagnetism of Brunhes and Matuyama, Quaternary deposits at Merritt, British Columbia. *Can. J. Earth Sci.* **2,** 76–92.

Funder, S. (1989). Quaternary geology of the ice-free areas and adjacent shelves of Greenland. *In* "Quaternary Geology of Canada and Greenland" (R. J. Fulton, ed.), Vol. 1, pp. 743–792. Geol. Surv. Can.

Gaiber-Puertas, C. (1953). Varacion secular del campo geomagnetico. *Observ. Elso Memo.* **11,** 4–16.

Galbrun, B. (1985). Magnetostratigraphy of the Berriasian Stratotype section (Berrias, France). *Earth Planet Sci. Lett.* **74,** 130–136.

Galbrun, B. (1989). Résultats magnetostratigraphiques à la limite Rognacien-Vitrollien: Précisions de la limite Cretace-Tertiar dans le Bassin D'Aix-en-Provenence. *Cah. Reserve Geol. Haute-Provence, Digne* **1,** 34–37.

Galbrun, B., Gabilly, J., and Raspins, L. (1988a). Magnetostratigraphy of the Toarcian stratotype sections at Thouars and Airvault (Deux-Sevres, France). *Earth Planet. Sci. Lett.* **87,** 453–462.

Galbrun, B., Baudin, F., Comas-Rengito, M. J., Faucault, A., Fourcade, E., Goy, A., Moutercade, R., and Ruget, C. (1988b). Resultats magnetostratigraphiques préliminaires sur le Toarcian de la Sierra Palomera (Cahine ibérique, Espagne). *Bull. Soc. Geol. Fr.* **8,** 193–198.

Galbrun, B., Baudin, F., Faurcade, E., and Rivas, P. (1990). Magnetostratigraphy of the Toarcian Ammonitico Rosso limestone at Iznallez, Spain. *Geophys. Res. Lett.* **17**, 2441–2444.

Galbrun, B., Crasquin-Soleau, S., and Jauquey, J. (1992). Magnétostratigraphic des sédiments triasiques des Sites, 279 et 760, ODP Leg 122, Plateau de Wombat, Nord Ouest de l'Australie. *Mar. Geol.* **107**, 293–198.

Galbrun, B., Feist, M., Colombo, F., Rocchia, R., and Tambarean, Y. (1993). Magnetostratigraphy and biostratigraphy of Cretaceous-Tertiary continental deposits, Ager Basin province of Levida, Spain. *Palaeogeogr., Palaeoclimatol., Palaeoecol.* **102**, 41–52.

Gallet, Y., Besse, J., Krystyn, J., Marcoux, J., and Thereniant, H. (1992). Magnetostratigraphy of the Late Triassic Bolücektasi Tepe section (southwestern Turkey): Implications for changes in magnetic reversal frequency. *Phys. Earth Planet. Inter.* **73**, 85–108.

Gallet, Y., Besse, J., Krystyn, L., Theveniant, H., and Marcoux, J. (1993). Magnetostratigraphy of the Kavur Tepe section (southwestern Turkey): A magnetic polarity time scale for the Norian. *Earth Planet. Sci. Lett.* **117**, 443–456.

Gallet, Y., Besse, J., Krystyn, L., Théveniant, H., and Marcoux, J. (1994). Magnetostratigraphy of the Mayerling section (Austria) and Erenkolu Mezarlik Turkey sections: Improvement of the Carnian (Late Triassic) magnetic polarity time scale. *Earth Planet. Sci. Lett.* **125**, 173–191.

Galusha, T., Johnson, N. M., Lindsay, E. H., Opdyke, N. D., and Tedford, R. H. (1984). Biostratigraphy and magnetostratigraphy, late Plicoene rocks, 111 Ranch Arizona. *Geol. Soc. Am. Bull.* **95**, 714–722.

Gauss, C. F. (1838). Allgemeine Theorie des Erdmagnetismus. Leipzig.

Gee, J., Kootwijk, C. T., and Smith, G. M. (1991). Magnetostratigraphy of Paleocene and Upper Cretaceous sediments from Broken Ridge, eastern Indian Ocean. *In* "Proceedings of the Ocean Drilling Program, Scientific Results" (J. Weissel, J. Pierce, E. Taylor, J. Alt *et al.*, eds.), Vol. 121, pp. 359–375. ODP, College Station, TX.

Gilbert, W. (1600). "De Magnete," (P. F. Mottely, trans.) 1893; Dover, New York, 1958.

Giovanoli, F. (1979). A comparison of the magnetization of detrital and chemical sediments from Lake Zurich. *Geophys. Res. Lett.* **6**, 233–235.

Glass, B. P., Kent, D. V., Schneider, D. V., and Tauxe, L. (1991). Ivory Coast microtektite strewn field: Description and relation to the Jaramillo geomagnetic event. *Earth Planet. Sci. Lett.* **107**, 182–196.

Goodell, H. G., and Watkins, N. D. (1968). The paleomagnetic stratigraphy of the Southern Ocean: 20° West to 160° East longitude. *Deep-Sea Res.* **15**, 89–112.

Goree, W. S., and Fuller, M. (1976). Magnetometers using RF-driven squids and their applications in rock magnetism and paleomagnetism. *Rev. Geophys. Space Phys.* **14**, 591–608.

Gose, W. A., and Helsley, C. E. (1972). Paleomagnetism and rock magnetism of the Permian Cutler and Elephant Canyon Formations in Utah. *J. Geophys. Res.* **77**, 1534–1548.

Gough, D. I. and Opdyke, N. D. (1963). The paleomagnetism of the Lupata Volcanics of Mozambique. *Geophys. J. R. Astron. Soc.* **7**, 457–468.

Gradstein, F. M., Agterberg, F. P., Ogg, J. G., Hardenbol, J., Van Veen, P., Thierry, J., and Huang, Z. (1994). A Mesozoic time scale. *J. Geophys. Res.* **99**, 24,051–24,074.

Gradstein, F. M., Agterberg, F. P., Ogg, J. G., Hardenbol, J., van Veen, P., Thierry, J., and Huang, Z. (1995). A Triassic, Jurassic, and Cretaceous timescale. *In* "Geochronology, Time Scales and Stratigraphic Correlation" (W. A. Berggen, D. V. Kent, M. Aubry, and J. Hardenbol, eds.), SEPM Spec. Publ. No. 54, pp. 95–128.

Graham, J. W. (1949). The stability and significance of magnetism in sedimentary rocks. *J. Geophys. Res.* **54**, 131–167.

Grommé, C. S., and Hay, R. L. (1963). Magnetization of basalt of bed I, Olduvai Gorge. *Nature (London)* **200**, 560–561.

Grommé, C. S., and Hay, R. L. (1971). Geomagnetic polarity epochs: Age and duration of the Olduvai normal polarity event. *Earth Planet. Sci. Lett.* **18**, 179–185.

Groot, J. J., de Jonge, R. G. C., Langereis, C. G., ten Kate, W. G. H. Z., and Smit, J. (1989). Magnetostratigraphy of the Cretaceous-Tertiary boundary at Agost, Spain. *Earth Planet. Sci. Lett.* **94**, 385–397.

Grousset, F. E., Labeyrie, L., Sinko, J. A., Cremer, M., Bond, G., Duprat. J., Cortijo, E., and Huon, S. (1993). Patterns of ice-rafted detritus in the glacial North Atlantic (40–55°N). *Paleoceanography* **8**, 175–192.

Gubbins, D. (1987a). Thermal core-mantle interactions and time averaged paleomagnetic field. *J. Geophys. Res.* **93**, 3413–3420.

Gubbins, D. (1987b). Mechanism for geomagnetic reversals. *Nature (London)* **326**, 167–169.

Gubbins, D. (1994). Geomagnetic polarity reversals: A connection with secular variation and core-mantle interaction? *Rev. Geophys.* **32**, 61–83.

Gubbins, D., and Coe, R. S. (1993). Longitudinally confined geomagnetic reversed paths from non-dipolar transition fields. *Nature (London)* **362**, 51–53.

Gubbins, D., and Kelly, P. (1993). Persistent patterns in the geomagnetic field during the last 2.5 M yr. *Nature (London)* **365**, 829–832.

Gurevich, Y. L., and Slautsitays, I. P. (1985). A paleomagnetic section in the Upper Permian and Triassic deposits on Novaya Zemlya. *Int. Geol. Rev.* **27**, 168–177.

Guyodo, Y., and Valet, J. P. (1996). Relative variation in geomagnetic intensity from sedimentary records: The past 200 thousand years. *Earth Planet. Sci. Letters,* in press.

Haag, M., and Heller, F. (1991). Late Permian to Early Triassic magnetostratigraphy. *Earth Planet. Sci. Lett.* **107**, 42–54.

Hagelberg, T., Shackleton, N. J., Pisias, N., and Shipboard Scientific Party (1992). Development of composite depth sections for Sites 844 through 854. *In* "Proceedings of the Ocean Drilling Program, Initial Reports" (L. A. Mayer, N. G. Pisias, T. R. Janecek *et al.,* eds.), Vol. 138, pp. 79–85. ODP, College Station, TX.

Haggerty, S. E. (1976a). Oxidation of opaque minerals in basalts. *In* "Oxide Minerals" (D. Rumble, III, ed.), pp. 1–100. Mineralogical Society of America, Washington, DC.

Haggerty, S. E. (1976b). Oxidation of opaque minerals in terrestrial igneous rocks. *In* "Oxide Minerals." (D. Rumble, III, ed.), pp. 101–300. Mineralogical Society of America, Washington, DC.

Hailwood, E. A. (1979). Paleomagnetism of Late Mesozoic to Holocene sediments from the Bay of Biscay and Rockall Plateau, drilled on IPOD Leg 48. *In* Initial Reports of the Deep Sea Drilling Project" (L. Montadent, D. Roberts *et al.,* eds.), Vol. 48, pp. 305–333. U.S. Govt. Printing Office, Washington, DC.

Hailwood, E. A., and Clement, B. M. (1991). Magnetostratigraphy of Sites 703 and 704, Meteor Rise, southeastern South Atlantic. *In* "Proceedings of The Ocean Drilling Project, Scientific Results" (P. F. Ciesielski, Y. Kristoffersen *et al.,* eds.), Vol. 114, pp. 367–386. ODP, College Station, TX.

Hall, C. M., and Farrell, J. W. (1993). Laser $^{40}Ar/^{39}Ar$ age from ash D of ODP Site 758: Dating the Brunhes-Matuyama reversal and oxygen isotope stage 19.1. *EOS, Trans. Am. Geophys. Union,* **74**, 110.

Hall, F. R., Bloemendal, J., King, J. W., Arthur, M. A., and Aksu, A. E. (1989a). Middle to Late Quaternary sediment fluxes in the Labrador Sea, ODP Leg 105, Site 646: A synthesis of rock-magnetic, oxygen isotopic, carbonate and planktonic foraminiferal data. *In* "Proceedings of the Ocean Drilling Program, Scientific Results" (S. P. Srivastava, M. Arthur, B. Clement *et al.,* eds.), Vol. 105, pp. 653–688. ODP, College Station, TX.

Hall, F. R., Busch, W. H., and King, J. W. (1989b). The relationship between variations in rock-magnetic properties and grain size of sediments from ODP hole 645C. *In* "Proceedings of the Ocean Drilling Program, Scientific Results" (S. P. Srivastava, M. Arthur, B. Clement *et al.* eds.), Vol. 105, pp. 837–841. ODP, College Station, TX.

Hallam, A. (1975). "Jurassic Environments." Cambridge Univ. Press, Cambridge, UK.

Halls, H. C. (1976). A least squares method to find a remanence direction from converging remagnetization circles. *Geophys. J. R. Astron. Soc.* **45**, 297–304.

Halls, H. C. (1978). The use of converging remagnetization circles in paleomagnetism. *Phys. Earth Planet. Inter.* **16**, 1–11.

Halvorson, E., Lowandowski, M., and Jelenska, M. (1989). Paleomagnetism of the Upper Carboniferous Stzegom and Karkonose granites and the Kudowa Granitoid from the Sudet Mountains Poland. *Phys. Earth Planet. Inter.* **55**, 54–64.

Hamilton, N. (1990). Mesozoic magnetostratigraphy of Maud Rise, Antarctic, 1990. *In* "Proceedings of the Ocean Drilling Project, Scientific Results" P. F. Barker, J. P. Kennett *et al.*, eds.), Vol. 113, pp. 525–560. ODP, College Station, TX.

Hamilton, N., Suzyumov, A. E., and Shirshiv, P. P. (1983). Late Cretaceous magnetostratigraphy of site 516, Rio Grande Rise, southwestern Atlantic Ocean, Deep Sea Drilling Project Leg 72. *In* "Initial Reports of the Deep Sea Drilling Project" (P. F. Barker, R. L. Carlson, D. A. Johnson *et al.*, eds.), Vol. 72, pp. 723–730. U.S. Govt. Printing Office, Washington, DC.

Handschumacher, D. W., Sager, W. W., Hilde, T. W. C., and Bracey, D. R. (1988). Pre-Cretaceous tectonic evolution of the Pacific plate and extension of the geomagnetic polarity reversal timescale with implications for the origin of the Jurassic "Quiet Zone." *Tectonophysics* **155**, 365–380.

Hanna, R., and Verosub, K. L. (1989). A review of lacustrine paleomagnetic records from western North America: 0–40,000 years BP. *Phys. Earth Planet. Inter.* **56**, 76–95.

Haq, B. U., Worsley, T. R., Burckle, L. H., Douglas, R. G., Keigwin, L. D., Jr., Opdyke, N. D., Savin, S. M., Sommer, M. A., II, Vincent, E., and Woodruff, F. (1980). The late Miocene marine carbon isotope shift and the synchroneity of some phytoplanktonic biostratigraphic datums. *Geology* **8**, 427–432.

Haq, B. U., Hardenbol, J., and Vail, P. R. (1987). Chronology of the fluctuation of sea levels since the Triassic (250 million years ago to present). *Science* **235**, 1156–1167.

Harland, W. B., Cox, A. V., Llewellyn, P. G., Pickton, C. A. G., Smith, A. G., and Walters, R. (1982). "A Geologic Time Scale." Cambridge Univ. Press, Cambridge, UK.

Harland, W. B., Armstrong, R. L., Cox, A. V., Craig, L. E., Smith, A. G., and Smith, D. G. (1990). "A Geologic Time Scale 1989." Cambridge Univ. Press, Cambridge, UK.

Harrison, C. G. A., and Funnell, B. M. (1964). Relationship of paleomagnetic reversals and micropaleontology in two late Cenozoic cores from the Pacific Ocean. *Nature (London)* **204**, 566.

Hartl, P., Tauxe, L., and Constable, C. (1993). Early Oligocene geomagnetic field behavior from Deep Sea Drilling Project Site 522. *J. Geophys. Res.* **98**, 19,649–19,665.

Hartstra, R. L. (1982). A comparative study of the ARM and I_{sr} of some natural magnetites of MD and PSD grain size. *Geophys. J. R. Astron. Soc.* **71**, 497–518.

Hayashida, A., and Bloemendal, J. (1991). Magnetostratigraphy of ODP Leg 117 sediments from the Owen Ridge and Oman Margin, Western Arabian Sea. *In* "Proceedings of the Ocean Drilling Project, Scientific Results" (W. L. Prell, N. Nittsuma *et al.*, eds.), Vol. 117, p. 161. ODP, College Station, TX.

Hays, J. D., and Opdyke, N. D. (1967). Antarctic radiolaria, magnetic reversals, and climatic change. *Science* **158**, 1001–1011.

Hays, J. D., Saito, T., Opdyke, N. D., and Burckle, L. H. (1969). Pliocene and Pleistocene sediments of the equatorial Pacific: Their paleomagnetic, biostratigraphic and climatic record. *Geol. Soc. Am. Bull.* **80,** 1481–1514.

Hays, J. D., Imbrie, J., and Shackleton, N. J. (1976). Variations in the earth's orbit: Pacemaker of the ice ages. *Science* **194,** 1121–1132.

Heath, G. R., Rea, D. H., and Levi, S. (1985). Paleomagnetism and accumulation rates of sediments at sites 576 and 578, Deep Sea Drilling Project Leg 86, western North Atlantic. *In* "Initial Reports of the Deep Sea Drilling Project" (H. R. Heath, L. H. Burckle *et al.,* eds.), Vol. 86, pp. 459–502. U.S. Govt. Printing Office, Washington, DC.

Hedley, I. G. (1971). The weak ferromagnetism of geothite (α FeOOH). *J. Geophys.* **37,** 409–420.

Heider, F., Dunlop, D. J., and Sugiura, N. (1987). Magnetic properties of hydrothermally recrystallized magnetite crystals. *Science* **236,** 1287–1290.

Heiniger, C., and Heller, F. (1976). A high temperature vector magnetometer. *Geophys. J. R. Astron. Soc.* **44,** 281–287.

Heirtzler, J. R., Dickson, G. O., Herron, E. M., Pittman, W. C., III, and LePichon, X. (1968). Marine magnetic anomalies, geomagnetic field reversal and motions of the ocean floor and continents. *J. Geophys. Res.* **73,** 2119–2136.

Heller, F. (1978). Rock magnetic studies of Upper Jurassic limestones from southern Germany. *J. Geophys.* **44,** 525–543.

Heller, F. (1980). Self reversal of natural remanent magnetization in the Olby-Laschamp lavas. *Nature (London)* **284,** 334–335.

Heller, F., and Channell, J. E. T. (1979). Paleomagnetism of Upper Cretaceous limestones from the Munster Basin, Germany. *J. Geophys.* **46,** 413–427.

Heller, F. and Evans, M. E. (1995). Loess magnetism. *Rev. Geophys.* **33,** 211–240.

Heller, F., and Liu, T. S. (1982). Magnetostratigraphical dating of loess deposits in China. *Nature (London)* **300,** 161–163.

Heller, F., and Liu, T. (1984). Magnetism of Chinese loess deposits. *Geophys. J. R. Astron. Soc.* **77,** 125–141.

Heller, F., and Petersen, N. (1982). Self-reversal explanation for the Laschamp/Olby geomagnetic field excursion. *Phys. Earth Planet. Int.* **30,** 358–372.

Heller, F., Lowrie, W., and Channell, J. E. T. (1984). Late Miocene magnetic stratigraphy at Deep Sea Drilling Project hole 521A. *In* "Initial Reports of the Deep Sea Drilling Project" (K. J. Hsu, J. L. LaBrecque *et al.,* eds.), Vol. 73, pp. 637–644. U.S. Govt. Printing Office, Washington, DC.

Heller, F., Carracedo, J. C., and Soler, V. (1986). Reversed magnetization in pyroclastics from the 1985 eruption of Nevado del Ruiz, Columbia. *Nature (London)* **324,** 241–242.

Heller, F., Lowrie, W., Li, H., and Wang, J. (1988). Magnetostratigraphy of the Permi-Triassic boundary section at Shangsi. *Earth Planet. Sci. Lett.* **88,** 348–356.

Heller, F., Liu, X. M., Liu, T. S., and Xu, T. C. (1991). Magnetic susceptibility of loess in China. *Earth Planet. Sci. Lett.* **103,** 301–310.

Heller, F., Shen, C. D., Beer, J., Liu, X. M., Liu, T. S., Bronger, A., Suter, M., and Bonani, G. (1993). Quantitative estimates and palaeoclimatic implications of pedogenic ferromagnetic mineral formation in Chinese loess. *Earth Planet. Sci. Lett.* **114,** 385–390.

Helsley, C. E. (1965). Paleomagnetic results from the Lower Permian Dunkard series of West Virginia. *J. Geophys. Res.* **70,** 413–424.

Helsley, C. E. (1969). Magnetic reversal stratigraphy of the Lower Triassic Moenkope formation of western Colorado. *Geol. Soc. Am. Bull.* **80,** 2431–2450.

Helsley, C. E., and Steiner, M. B. (1969). Evidence for long intervals of normal polarity during the Cretaceous period. *Earth Planet. Sci. Lett.* **5,** 325–332.

Helsley, C. E., and Steiner, M. B. (1974). Paleomagnetism of the lower Triassic Moenkopi formation. *Geol. Soc. Am. Bull.* **85**, 457–464.

Henshaw, P. C., and Merrill, R. T. (1980). Magnetic and chemical changes in marine sediments. *Rev. Geophys. Space Phys.* **18**, 483–504.

Henthorn, D. L. (1981). The magnetostratigraphy of the Lebombo Group along the Olifents River, Krugar National Park. *Ann. Geol. Surv. S. Afr.* **15**(2), 1–10.

Herbert, T. D. (1992). Paleomagnetic calibration of Milankovitch cyclicity in lower Cretaceous sediments. *Earth Planet. Sci. Lett.* **112**, 15–28.

Herbert, T. D., and D'Hondt, S. L. (1990). Precessional climate cyclicity in Late Cretaceous–Early Tertiary marine sediment: A high resolution chronometer of Cretaceous-Tertiary bondary events. *Earth Planet. Sci. Lett.* **99**, 263–275.

Herbert, T. D., and Fischer, A. G. (1986). Milankovitch climatic origin of mid-Cretaceous black shale rhythms from central Italy. *Nature (London)* **321**, 739–743.

Herbert, T. D., D'Hondt, S. L., Premoli-Silva, I., Erba, E., and Fischer, A. G. (1995). Orbital chronology of Cretaceous–Early Paleocene marine sediments. *In* "Geochronology, Time Scales and Stratigraphic Correlation" (W. A. Berggren, D. V. Kent, M. Aubry, and J. Hardenbol, eds.) SEPM Spec. Publ. No. 54, pp. 81–94.

Herrero-Bervera, E., and Helsley, C. E. (1993). Global paleomagnetic correlation of the Blake geomagnetic polarity episode. *In* "Applications of Paleomagnetism to Sedimentary Geology" (D. M. Aissaoui, D. F. McNeill, and N. F. Hurley, eds.), SEPM Spec. Publ. No. 49, pp. 71–82.

Herrero-Bervera, E., Helsley, C. E., Hammond, S. R., and Chitwood, L. A. (1989). A possible lacustrine paleomagnetic record of the Blake episode from Pringle Falls, Oregon, USA. *Phys. Earth Planet. Inter.* **56**, 112–123.

Herrero-Bervera, E., Helsley, C. E., Sarna-Wojcicki, A. M., Lajoie, K. R., Meyer, C. E., McWilliams, M. O., Negrini, R. M., Turrin, B. D., Donnelly-Nolan, J. M., and Liddicoat, J. C. (1994). Age and correlation of a paleomagnetic episode in the western United States by ^{40}Ar/^{39}Ar dating and tephrochronology: The Jamaica, Blake, or a new polarity episode? *J. Geophys. Res.* **99**, 24,091–24,103.

Hess, H. H. (1962). History of ocean basins. *In* "Petrologic Studies, A volume in honor of A. F. Buddington" (A. E. J. Engel *et al.*, eds.), pp. 599–620. Geol. Soc. Am., Boulder, CO.

Hess, J. C., and Lippolt, H. J. (1986). ^{40}Ar/^{39}Ar ages of Toustein and tuff sanidimes: New calibration points for the improvement of the Upper Carboniferous time scale. *Chem. Geol., Isot. Geosci. Sect.* **59**, 143–154.

Hickey, L., West, R. M., Dawson, M. R., and Chai, D. K. (1983). Arctic terrestrial biota: Paleomagnetic evidence of age disparity with mid-northern latitudes during the late Cretaceous and early Tertiary. *Science* **224**, 1153–1156.

Hicks, J. F., Obradovich, J. D., and Tauxe, L. (1995). A new calibration point for the Late Cretaceous time scale: The ^{40}Ar/^{39}Ar isotopic age of the C33r/C33n geomagnetic reversal from the Judith River Formation (Upper Cretaceous), Elk Basin, Wyoming, USA. *J. Geol.* **103**, 243–256.

Hilde, T. W. C., Isezaki, N., and Wageman, J. M. (1976). Mesozoic seafloor spreading in the North Pacific. *Geophys. Monogr., Am. Geophys. Union* **19**, 205–226.

Hilgen, F. J. (1991a). Astronomical calibration of Gauss to Matuyama sapropels in the Mediterranean and implication for the geomagnetic polarity time scale. *Earth Planet. Sci. Lett.* **104**, 226–244.

Hilgen, F. J. (1991b). Extension of the astronomically calibrated (polarity) time scale to the Miocene/Pliocene boundary. *Earth Planet. Sci. Lett.* **107**, 349–368.

Hilgen, F. J., and Langereis, G. G. (1989). Periodicities of CaCO$_3$ cycles in the Mediterranean Pliocene: Discrepancies with the quasi-periods of the Earth's orbital cycles. *Terra Nova* **1**, 409–415.

Hilgen, F. J., Krijgsman, W., Langereis, C. G., Lourens, L. J., Santarelli, A., and Zachariasse, W. J. (1995). Extending the astronomical (polarity) time scale into the Miocene. *Earth Planet. Sci. Lett.* **136,** 495–510.

Hillhouse, J., and Cox, A. (1976). Brunhes-Matuyama polarity transition. *Earth Planet. Sci. Lett.* **29,** 51–64.

Hillhouse, J. W., Cerling, T. E., and Brown, F. H. (1986). Magnetostratigraphy of the Koobi Fora Formation, Lake Turkana, Kenya. *J. Geophys. Res.* **91,** 11581–11595.

Hoare, J. M., Condon, W. H., Cox, A., and Dalrymple, G. B. (1968). Geology, paleomagnetism, and potassium-argon ages of basalts from Nunivak Island, Alaska. *Mem. Geol. Soc. Am.* **116,** 377–413.

Hodell, D. A., and Woodruff, F. (1994). Variations in the strontium isotope ratio of seawater during the Miocene: Stratigraphic and geochemical implications. *Paleoceanography* **9,** 405–426.

Hodell, D. A., Mueller, P. A., and Garrido, J. R. (1991). Variations in the strontium composition of sea water during the Neogene. *Geology* **19,** 24–27.

Hodell, D. A., Benson, R. H., Kent, D. V., Boersma, A., and Rakic-El Bied. (1994). Magneto-stratigraphic, biostratigraphic, and stable isotope stratigraphy of an Upper Miocene drill core from the Sale Brigueterie (north western Morroco): A high-resolution chronology for the Messinian stage. *Paleoceanography* **9,** 835–855.

Hoffman, K. A. (1977). Polarity transition records and the geomagnetic dynamo. *Science* **196,** 1329–1332.

Hoffman, K. A. (1989). Geomagnetic polarity reversals: Theory and models. *In* "The Encyclo-pedia of Solid Earth Geophysics" (D. E. James, ed.), pp. 547–555. Van Nostrand-Reinhold, New York.

Hoffman, K. A. (1991). Long-lived transitional states of the geomagnetic field and the two dynamo families. *Nature* (*London*) **354,** 273–277.

Hoffman, K. A. (1992). Dipolar reversal states of the geomagnetic field and core-mantle dynamics. *Nature* (*London*) **359,** 789–794.

Hoffman, K. A., and Day, R. (1978). Separation of multi-component NRM: A general method. *Earth Planet. Sci. Lett.* **40,** 433–438.

Horner, F., and Heller, F. (1983). Lower Jurassic magnetostratigraphy at the Breggia Gorge (Ticino, Switzerland) and Alpe Turat (Como, Italy). *Geophys. J. R. Astron. Soc.* **73,** 705–718.

Hospers, J. (1951). Remanent magnetization of rocks and the history of the geomagnetic field. *Nature* (*London*) **168,** 1111–1112.

Hospers, J. (1953). Reversals of the main geomagnetic field. I, II, and III. *Proc. K. Ned. Akad. Wet., Ser. B: Phys. Ser.* **56,** 467–491.

Hospers, J. (1954). *Proc. K. Ned. Akad. Wet., Ser. B: Phys. Sci.* **57,** 112–121.

Hsu, V., Shibuya, H., and Merrill, D. L. (1991). Paleomagnetism study of deep sea sediments from the Cagayan Ridge in the Sulu Sea: Results of Leg 124. *In* "Proceedings of the Ocean Drilling Program, Scientific Results" (E. A. Silver, C. Rangin, M. T. von Breymann *et al.,* eds.), Vol. 124. ODP, College Station, TX.

Hunkins, K., Be, A. W. H., Opdyke, N. D., and Mathieu, G. (1971). "The Late Cenozoic History of the Arctic Ocean, the Late Cenozoic Glacial Ages," Turekian, ed., pp. 215–237. Yale Univ. Press, New Haven, CT.

Hurley, N. F., and Van der Voo, R. (1990). Magnetostratigraphy, Late Devonian iridium anomaly, and impact hypothesis. *Geology* **18,** 291–294.

Hyodo, M. (1984). Possibility of reconstruction of the past geomagnetic field from homoge-neous sediments. *J. Geomagn. Geoelectr.* **36,** 45–62.

Ikebe, N., Chiji, M., Tsuchi, R., Morozumi, Y., and Kawata, T. (1981). *In* "Proceedings of the IGCP-114 International Workshop in Pacific Neogene Biostratigraphy," Nos. i-iv, pp. 1–150.

Imbrie, J., Hays, J. D., Martinson, D. G., McIntyre, A., Mix, A. C., Morley, J. J., Pisias, N. G., Prell, W. L., and Shackleton, N. J. (1984). The orbital theory of Pleistocene climate: Support from a revised chronology of the marine $\partial^{18}O$ Record. *NATO ASI Ser., Ser. C* **126**, 269–305.

Imlay, R. W. (1961). Late Jurassic Ammonites from the Western Sierra Nevada California. *Geol. Surv. Prof. Pap. (U.S.)* **374D**, D1-D30.

Irving, E. (1964). "Paleomagnetism and Its Application to Geological and Geophysical Problems." Wiley, New York and London.

Irving, E. (1966). Paleomagnetism of some Carboniferous rocks from New South Wales and its relation to geological events. *J. Geophys. Res.* **71**, 6025–6051.

Irving, E., and Couillard, G. W. (1973). Cretaceous normal polarity interval. *Nature (London), Phys. Sci.* **244**, 10–11.

Irving, E., and Major, A. (1964). Post-depositional detrital remanent magnetization in a synthetic sediment. *Sedimentology* **3**, 135–143.

Irving, E., and Monger, J. W. H. (1987). Preliminary paleomagnetic results from the Permian Asitka group. British Columbia. *Can. J. Earth Sci.* **24**, 1490–1497.

Irving, E., and Park, J. K. (1972). Hairpins and superintervals. *Can. J. Earth Sci.* **9**, 1318–1324.

Irving, E., and Parry, L. G. (1963). The magnetism of some Permian rocks from New South Wales. *Geophys. J. R. Astron. Soc.* **7**, 395–411.

Irving, E., and Pullaiah, G. (1976). Reversals of the geomagnetic field, magnetostratigraphy, and relative magnitude of paleosecular variation in the Phanerozoic. *Earth Sci. Rev.* **12**, 35–64.

Irving, E., and Strong, D. F. (1985). Paleomagnetism of rocks from Burin Peninsula, Newfoundland: Hypothesis of Late Paleozoic displacement of Acadia criticized. *J. Geophys. Res.* **90**, 1949–1962.

Izett, G. A. and Obradovich, J. D. (1991). Dating of the Matuyama-Brunhes boundary based on $^{40}Ar/^{39}Ar$ ages of the Bishops Tuff and the Cerro San Luis rhyolite. *Geol. Soc. Am. Abstr. Prog.* **23**, A106.

Izett, G. A., Obradovich, J. D., and Mehnert, H. H. (1988). The Bishop ash bed (Middle Pleistocene) and some older (Pliocene and Pleistocene) chemically and mineralogically similar ash beds in California, Nevada, and Utah. *Geol. Surv. Bull. (U.S.)* **1675**, 1–37.

Jackson, M. (1990). Diagenetic sources of stable remanence is remagnetized Paleozoic cratonic carbonates: A rock magnetic study. *J. Geophys. Res.* **95**, 2753–2761.

Jackson, M., Rochette, P., Fillion, G., Banerjee, S., and Marvin, J. (1993). Rock magnetism of remagnetized Paleozoic carbonates: Low-temperature behavior and susceptibility characteristics. *J. Geophys. Res.* **98**, 6217–6225.

Jacobs, J. A. (1994). "Reversals of the Earth's Magnetic Field." Cambridge Univ. Press, Cambridge, UK.

Johnson, E. A., Murphy, T., and Torrenson, O. W. (1948). Prehistory of the Earth's magnetic field. *Terr. Magn. Atmos. Electr.* **53**, 349–372.

Johnson, G. D., Johnson, N. M., Opdyke, N. D., and Tahirkheli, R. A. K. (1979). Magnetic reversal stratigraphy and sedimentary tectonic history of the upper Siwalik Group eastern Salt Range and southwestern Kashmir. *In* "Geodynamics of Pakistan" (A. Farah and K. A. DeJong, eds.), pp. 149–165. Geological Survey of Pakistan, Quetta.

Johnson, G. D., Opdyke, N. D., Tandon, S. K., and Nanda, A. C. (1983). The magnetic polarity stratigraphy of the Siwalik Group at Haritalyangar (India) and a new last appearance datum for Ramapithecus and Sivapithecus in Asia. *Palaeogeogr., Palaeoclimatol., Palaeoecol.* **44**, 223–249.

Johnson, H. P., Lowrie, W., and Kent, D. V. (1975). Stability of anhysteretic remanent magnetization in fine and coarse magnetite and maghemite particles. *Geophys. J. R. Astron. Soc.* **41**, 1–10.

Johnson, N. M., Opdyke, N. D., and Lindsay, E. H. (1975). Magnetic polarity stratigraphy of Pliocene Pliestocene terrestrial deposits and vertebrate faunas, San Pedro Valley, Arizona. *Geol. Soc. Am. Bull.* **86**, 5–12.

Johnson, N. M., Opdyke, N. D., Johnson, D. G., Lindsay, E. H., and Tahirkheli, R. A. K. (1982). Magnetic polarity stratigraphy and ages of Siwalik Group rocks of the Potwar Plateau, Pakistan. *Palaeogeogr., Palaeoclimatol., Palaeoecol.* **37**, 17–42.

Johnson, N. M., Officer, C. B., Opdyke, N. D., Woodard, G. D., Zeitler, P. K., and Lindsay, E. H. (1983). Rates of Cenozoic tectonism in the Vallecito–Fish Creek Basin, western Imperial Valley, California. *Geology* **11**, 664–667.

Johnson, N. M., Stix, J., Tauxe, L., Cerveny, P. F., and Tahirkeli, R. A. K. (1985). Paleomagnetic chronology, fluvial processes and tectonic implications of the Siwalik deposits near Chinji Village, Pakistan. *J. Geol.* **93**, 27–40.

Johnson, N. M., Sheikh, K. A., Dawson-Saunders, E., and McRae, L. E. (1988). The use of magnetic reversal time lines in stratigraphic analysis: A case study in measuring variability in sedimentation rates. *In* "New Perspectives in Basin Analysis" (K. L. Kleinspehn and C. Paola, eds.), pp. 189–200. Springer-Verlag, New York.

Johnson, R. G. (1982). Brunhes-Matuyama magnetic reversal dated at 790,000 yr B. P. by marine-astronomical correlations. *J. Quat. Res.* **17**, 135–147.

Jordan, T. E., Rutty, P. M., McRae, L. E., Beer, J. A., Tabbutt, K., and Damanti, J. F. (1990). Magnetic polarity stratigraphy of the Miocene Rio Azul section, Precordillera thrust belt, San Juan Province, Argentina. *J. Geol.* **98**, 519–539.

Juárez, M. T., Osete, M. L., Meléndez, G., Langereis, C. G., and Zijderveld, J. D. A. (1994). Oxfordian magnetostratigraphy of the Aquilón and Tosos sections (Iberian Range, Spain) and evidence of a pre-Oligocene overprint. *Phys. Earth Planet. Inter.* **85**, 195–211.

Juárez, M. T., Osete, M. L., Meléndez, G., and Lowrie, W. (1995). Oxfordian magnetostratigraphy in the Iberian Range. *Geophys. Res. Lett.* **22**, 2889–2892.

Kappelman, J., Simons, E. L., and Swisher, C. C., III (1992). New age determinations for the Eocene-Oligocene, boundary sediments in the Fayum Depression, Northern Egypt. *J. Geol.* **100**, 647–668.

Karlin, R. (1990a). Magnetic diagenesis in marine sediments from the Oregon continental margin. *J. Geophys. Res.* **95**, 4405–4419.

Karlin, R. (1990b). Magnetic mineral diagenesis in suboxic sediments at Bettis site W-N, NE Pacific Ocean. *J. Geophys. Res.* **95**, 4421–4436.

Karlin, R., and Levi, S. (1983). Diagenesis of magnetic minerals in recent hemipelagic sediments. *Nature (London)* **303**, 327–330.

Karlin, R., and Levi, S. (1985). Geochemical and sedimentological control on the magnetic properties of hemipelagic sediments. *J. Geophys. Res.* **90**, 10,373–10,392.

Karlin, R., Lyle, M., and Heath, G. R. (1987). Authigenic magnetite formation in suboxic marine sediments. *Nature (London)* **6**, 490–493.

Kawai, N. (1984). Paleomagnetic study of Lake Biwa sediments. *In* "Lake Biwa" (S. Horie, ed.), pp. 399–416. Dr. W. Junk Publ., Dordrecht, The Netherlands,

Keating, B. H., and Herrero-Bervera, E. (1984). Magnetostratigraphy of Cretaceous and Early Cenozoic sediments of Deep Sea Drilling Project, Site 530, Angola Basin. *In* Initial Reports of the Deep Sea Drilling Project" (W. W. Hay, J. C. Sibuet *et al.*, eds.), Vol. 75, pp. 1211–1218. U.S. Govt. Printing Office, Washington, DC.

Keating, B. H., Helsley, C. E., and Passagno, E. A. (1975). Late Cretaceous reversal sequence. *Geology* **2**, 75–79.

Keigwin, L. D., Jr. (1979). Late Cenozoic stable isotope stratigraphy and paleoceanography of Deep Sea Drilling Project sites from the east equatorial and central North Pacific Ocean. *Earth Planet. Sci. Lett.* **45**, 361–382.

Keller, M. H., Tahirkheli, R. A. K., Mirza, M. A., Johnson, G. D., Johnson, N. M., and Opdyke, N. D. (1977). Magnetic polarity stratigraphy of the upper Siwalik deposits, Pabbi Hills, Pakistan. *Earth Planet. Sci. Lett.* **36**, 187–201.

Kellogg, T. B., Duplessy, J. C., and Shackleton, N. J. (1978). Planktonic foraminiferal and oxygen isotopic stratigraphy and paleoclimatology of Norwegian Sea deep-sea cores. *Boreas* **7**, 61–73.

Kennedy, W. J., Cobban, W. A., and Scott, G. R. (1992). Ammonite correlation of the uppermost Campanian of western Europe, the U.S. Gulf Coast, Atlantic seaboard and Western Interior, the numerical age of the base of the Maastrichtian. *Geol. Mag.* **129**, 497–500.

Kennett, J. P., and Watkins, N. D. (1974). Late Miocene–Early Pliocene paleomagnetic stratigraphy, paleoclimatology, and biostratigraphy in New Zealand. *Geol. Soc. Am. Bull.* **85**, 1385–1398.

Kent, D. V. (1973). Post-depositional remanent magnetization in deep sea sediment. *Nature (London)* **246**, 32–34.

Kent, D. V. (1979). Paleomagnetism of the Devonian Onondaga limestone revisited. *J. Geophys. Res.* 3576–3588.

Kent, D. V. (1982). Apparent correlation of paleomagnetic intensity and climatic records in deep-sea sediments. *Nature (London)* **299**, 538–539.

Kent, D. W., and Gradstein, F. M. (1985). Cretaceous and Jurassic geochronology. *Geol. Soc. Am. Bull.* **91**, 1419–1427.

Kent D. V., and Lowrie, W. (1974). Origin of magnetic instability in sediment cores from the central North Pacific. *J. Geophys. Res.* **79**, 2987–3000.

Kent, D. V., and Opdyke, N. D. (1977). Paleomagnetic field intensity variation recorded in a Brunhes epoch deep sea sediment core. *Nature (London)* **266**, 156–159.

Kent, D. V., and Opdyke, N. D. (1985). Multicomponent magnetizations from the Mauch Chunk Formation of the central Appalachians and their tectonic implications. *J. Geophys. Res.* **90**, 5371–5384.

Kent, D. V., and Spariosu, D. J. (1983). High resolution magnetostratigraphy of Caribbean Plio-Pleistocene sediments, *Paleogeogr., Paleoclimatol., Paleoecol.* **42**, 47–64.

Kent, D. V., Olsen, P. E., and Witte, W. K. (1995). Late Triassic–Early Jurassic geomagnetic polarity sequence from drill cores in the Newark rift basin, eastern North America. *J. Geophys. Res.* **100**, 14,965–14,998.

Kent, J. T. (1982). The Fisher-Bingham distribution on a sphere. *J.R. Stat. Soc., Ser. B* **44**, 71–80.

Khan, M. J., Kent, D. V., and Miller, K. G. (1985). Magnetostratigraphy of Oliogocene to Pleistocene sediments of DSDP Leg 82. *In* "Initial Reports of the Deep Sea Drilling Project" 82, (H. Bougault, S. C. Cande *et al.*, eds.), Vol. 82, pp. 385–392. U.S. Govt. Printing Office, Washington, DC.

Khan, M. J., Opdyke, N. D., and Tahirkheli, R. A. K. (1988). Magnetic stratigraphy of the Siwalik Group, Bhitanni, Marwat, and Khasor Ranges, northwestern Pakistan and the timing of Neogene tectonics of the Trans Indus. *J. Geophys. Res.* **93**, 11773–11790.

Khramov, A. N. (1958). "Palaeomagnetism and Stratigraphic Correlation." Gostoptechjzdat, Leningrad (English translation by A. J. Lojkine. Published by Geophys. Dept., A.N.U., Canberra, 1960).

Khramov, A. N. (1974). "Paleomagnetism of the Paleozoic." NEDRA, Moscow (in Russian).

Khramov, A. N., and Rodionov, R. (1981). The geomagnetic field during Paleozoic time. *In* "Global Reconstruction and the Geomagnetic Field During the Paleozoic" (M. McElhinny *et al.,* eds.), Adv. Earth Planet. Sci., pp. 99–116. Center for Academic Publ., Tokyo.

King, J. W., and Channell, J. E. T. (1991). Sedimentary magnetism, environmental magnetism, and magnetistratigraphy. *In* "U.S. National Report to International Union of Geodesy and Geophysics," Rev. Geophys., Suppl., pp. 358–370.

King, J. W., Banerjee, S. K., Marvin, J., and Özdemir, O. (1982). A comparison of different magnetic methods for determining the relative grain size of magnetite in natural minerals, some results for lake sediments. *Earth Planet. Sci. Lett.* **59,** 404–419.

King, J. W., Banerjee, S. K., and Marvin, J. (1983). A new rock magnetic approach to selecting sediments for geomagnetic intensity studies: Application to paleointensity for the last 4000 years. *J. Geophys. Res.* **88,** 5911–5921.

Kirschvink, J. L. (1978). The Precambrian-Cambrian boundary problem: Magnetostratigraphy of the Amadeus basin, central Australia. *Geol. Mag.* **115**(2), 139–150.

Kirschvink, J. L. (1980). The least squares lines and plane analysis of paleomagnetic data. *Geophys. J. R. Astron. Soc.* **62,** 699–718.

Kirschvink, J. L., and Lowenstam, H. A. (1979). Mineralization and magnetization of chiton teeth: Paleomagnetic, sedimentologic, and biologic implications of organic magnetite. *Earth Planet. Sci. Lett.* **44,** 193–204.

Kirschvink, J. L., and Rozanov, A. Y. (1984). Magnetostratigraphy of lower Cambrian strata from the Siberian Platform: A paleomagnetic pole and a polarity time-scale. *Geol. Mag.* **121,** 189–203.

Kirschvink, J. L., Magaritz, M., Ripperdan, R. L., Zhuravlev, A. Y., and Rozanov, A. Y. (1991). The Pre-Cambrian/Cambrian boundary: Magnetostratigraphy and carbon isotopes resolve correlation problems between Siberia, Morocco and South China. *G.S.A. Today* **1**(4), 69–91.

Kligfield, R., and Channell, J. E. T. (1981). Widespread remagnetization of Helvetic limestones. *J. Geophys. Res.* **86,** 1888–1900.

Klitgord, K. D., Heustis, S. P., Mudie, J. D., and Parker, R. L. (1975). An analysis of near-bottom magnetic anomalies: Sea floor spreading and the magnetized layer. *Geophys. J. R. Astron. Soc.* **43,** 387–424.

Kobayashi, K., and Nomura, M. (1972). Iron sulfides in the sediment cores from the Sea of Japan and their geophysical implications. *Earth Planet. Sci. Lett.* **16,** 200–208.

Koenigsberger, J. G. (1938). Natural residual magnetism of eruptive rocks. Parts I and II. *Terr. Magn. Atmos. Electr.* **43,** 119–127, 299–320.

Kovacheva, M. (1983). Archeomagnetic data for Bulgaria and south eastern Yugoslavia. *In* "Geomagnetism of Baked Clays and Recent Sediments" (K. M. Creer, P. Tucholka, and C. E. Barton, eds.), pp. 106–109. Elsevier, Amsterdam.

Krijgsman, W., Hilgen, F. J., Langereis, C. G., and Zachariasse, W. J. (1994a). The age of the Tortonian/Messinian boundary. *Earth Planet. Sci. Lett.* **121,** 533–548.

Krijgsman, W., Langereis, C. G., Drams, R., and Van der Meulen, A. J. (1994b). Magnetostratigraphic dating of the middle Miocene climate change in the continental deposits of the Aragonian type area in the Calatayred-Ternel basin (central Spain). *Earth Planet. Sci. Lett.* **128,** 513–526.

Kristjansson, L., and Gudmundsson, A. (1980). Geomagnetic excursion in late glacial basalt outcrops in southwestern Iceland. *Geophys. Res. Lett.* **7,** 337–340.

Kroepnick, R. B., Denison, R. F., and Dahl, D. A. (1988). The Cenozoic sea water $^{87}Sr/^{86}Sr$ curve, data review and implications for correlation of marine strata. *Paleoceanography* **3,** 743–756.

Krumsick, K., and Hahn, G. G. (1989). Magnetostratigraphy near the Cretaceous: Tertiary boundary at Aix-en-Provence (southern France). *Cah. Reserve Geol. Haute-Provence Digne* **1**, 43–50.

Kukla, G., and An, Z. S. (1989). Loess stratigraphy in central China. *Palaeogeogr., Palaeoclimatol., Palaeoecol.* **72**, 2203–2225.

Kukla, G., Heller, F., Liu, X. M., Xu, T. C., Liu, T. S., and An, Z. S. (1988). Pleistocene climates in China dated by magnetic susceptibility. *Geology* **16**, 811–816.

LaBrecque, J. L., Kent, D. V., and Cande, S. C. (1977). Revised magnetic polarity time-scale for the Late Cretaceous and Cenozoic time. *Geology* **5**, 330–335.

LaBrecque, J. L., Hsu, K. J., Carmen, M. F., Jr., Karpoff, A. M., McKenzie, J. A., Percival, S. F., Jr., Petersen, N. P., Piscotto, K. A., Schreiber, E., Tauxe, L., Tucker, P., Weissert, H. J., and Wright, R. (1983). DSDP Leg 73: Contributions to Paleogene stratigraphy in nomenclature, chronology and sedimentation rates. *Paleogeogr., Paleoclimatol., Paleoecol.* **42**, 91–125.

Laj, C., Nordemann, D., and Pomeau, Y. (1979). Correlation function analysis of geomagnetic field reversals. *J. Geophys. Res.* **84**, 4511–4515.

Laj, C., Mazaud, A., Weeks, R., Fuller, M., and Herrero-Bervera, E. (1991). Geomagnetic reversal paths. *Nature (London)* **351**, 447.

Laj, C., Mazaud, A., Weeks, R., Fuller, M., and Herrero-Bervera, E. (1992). Statistical assessment of the preferred longitudinal bands for recent geomagnetic reversal records. *Geophys. Res. Lett.* **19**, 2003–2006.

Langereis, C. G., and Hilgen, F. J. (1991). The Rossello composite: A Mediterranean and global reference section for the Early to early Late Pliocene. *Earth Planet. Sci. Lett.* **104**, 211–225.

Langereis, C. G., Zachariasse, W. J., and Zijderveld, J. D. A. (1984). Late Miocene magnetobiostratigraphy of Crete. *Mar. Micropaleontol.* **8**, 265–281.

Langereis, C. G., Sen, S., Sumengen, M., and Unay, E. (1989). Preliminary magnetostratigraphic results of some Neogene mammal localities from Anatolia (Turkey). *In* "NATO Advanced Study Workshop in European Neogene Mammal Chronology" (E. Lindsay, P. Mein, and V. Fahlbush, eds.), NATO Ser. A, pp. 495–505. NATO, Dordrecht, The Netherlands.

Larson, E. E., Walker, T. R., Patterson, P. E., Hoblitt, R. P., and Rosenbaum, J. G. (1982). Paleomagnetism of the Moenkopi formation, Colorado Plateau: Basis for long-term model of acquisition of chemical remanent magnetization in red beds. *J. Geophys. Res.* **87**, 1081–1106.

Larson, R. L., and Chase, C. G. (1972). Late Mesozoic evolution of the western Pacific Ocean. *Geol. Soc. Am. Bull.* **83**, 3627–3644.

Larson, R. L., and Hilde, T. W. C. (1975). A revised time scale of magnetic reversals for the Early Cretaceous and Late Jurassic. *J. Geophys. Res.* **80**, 2586–2594.

Larson, R. L., and Olson, P. (1991). Mantle plumes control magnetic reversal frequency. *Earth Planet. Sci. Lett.* **107**, 437–447.

Larson, R. L., and Pitman, W. C., III (1972). World wide correlation of Mesozoic magnetic anomalies and its implications. *Geol. Soc. Am. Bull.* **83**, 3645–3662.

Larson, R. L., Fischer, A. G., Erba, E., and Premoli-Silva, I. (1993). "Apticore-Albicore," A Workshop Report on Global Events and Rhythms of the Mid-Cretaceous, 4–9 October 1992, Perugia, Italy.

Lazarenko, A. A., Bolikhovskaya, N. S., and Semenor, V. V. (1981). An attempt at a detailed stratigraphic subdivision of the loess association of the Tashkan region. *Int. Geol. Rev.* **23**, 1335–1360.

Ledbetter, M. (1983). Magnetostratigraphy of Middle-Upper Miocene and Upper Eocene sections in hole 512. *In* "Initial Reports of the Deep Sea Diving Project" (W. J. Ludwig, V. A. Krasheninnikov *et al.*, eds.), Vol. 71. U.S. Govt. Printing Office, Washington, DC.

Lehman, B., Laj, C., Kissel, C., Mazaud, A., Paterne, M., and Labeyrie, L. (1996). Relative changes of the geomagnetic field intensity during the last 280 kyear from piston cores in the Acores area. *Phys. Earth Planet. Inter.* **93**, 269–284.

Lerbekmo, J. F., Evans, M. E., and Baadsguard, H. (1979). Magnetostratigraphy, biostratigraphy and geochronology of Cretaceous-Tertiary boundary sediments, Red Deer Valley. *Nature (London)* **279**, 26–30.

Leslie, B. W., Hammond, D. E., Berelson, W. M., and Lund, S. P. (1990). Diagenesis in anoxic sediments from the California continental borderland and its influence on iron, sulfur and magnetite behavior. *J. Geophys. Res.* **95**, 4453–4470.

Leslie, B. W., Lund, S. P., and Hammond, D. E. (1990). Rock magnetic evidence for the dissolution and authigenic growth of magnetic minerals with anoxic marine sediments of the California continental borderland. *J. Geophys. Res.* **95**, 4437–4452.

Levi, S., and Karlin, R. (1989). A sixty thousand year paleomagnetic record from Gulf of California sediments: secular variation, late Quaternary excursions and geomagnetic implications. *Earth Planet. Sci. Lett.* **92**, 219–233.

Levi, S., Gudmunsson, H., Duncan, R. A., Kristjansson, L., Gillot, P. V., and Jacobsson, S. P. (1990). Late Pleistocene geomagnetic excursion in Icelandic lavas: Confirmation of the Laschamp excursion. *Earth Planet. Sci. Lett.* **96**, 443–457.

Levy, E. (1972). Kinematic reversal schemes for the geomagnetic dipole. *Astrophys. J.* **171**, 635–642.

Liddicoat, J. C. (1992). Mono Lake excursion in Mono Basin, California, and at Carson Sink and Pyramid Lake, Nevada. *Geophys. J. Int.* **108**, 442–452.

Liddicoat, J. C., and Coe, R. C. (1979). Mono Lake geomagnetic excursion. *J. Geophys. Res.* **84**, 261–271.

Liddicoat, J. C., Opdyke, N. D., and Smith, G. I. (1980). Paleomagnetic polarity in a 930m core from Searles Valley California. *Nature (London)* **286**, 22–25.

Lienert, B. R., and Helsley, C. E. (1980). Magnetostratigraphy of the Moenkopi Formation at Bears Ears, Utah. *J. Geophys. Res.* **85**, 1474–1480.

Lindsay, E. H. (1990). Sediments, geomorphology, magnetostratigraphy, and vertebrate paleontology in the San Pedro Valley, Arizona. *J. Geol.* **98**, 605–619.

Lindsay, E. H. (1995). Copemys and the Barstovian/Hemingfordian boundary. *J. Vertebr. Paleontol.* (in press).

Lindsay, E. H., Opdyke, N. D., and Johnson, N. M. (1980). Pliocene dispersal of the horse *Equus* and late Cenozoic mammalian dispersal events. *Nature (London)* **287**, 135–138.

Lindsay, E. H., Opdyke, N. D., and Johnson, N. M. (1984). Blancan-Hemphillian land mammal ages and late Cenozoic mammal dispersal events. *Annu. Rev. Earth Planet. Sci.* **12**, 445–488.

Linkova, T. I. (1965). "Some Results of Paleomagnetic Study of Arctic Ocean floor Sediments" (translated by E. R. Hope from "Natoyascheye i Proshloye Magnitnogo Polia Zemli"), pp. 279–281. Nauka, Moscow (Directorate of Scientific Information Services Publ., T463R, Canada, 1966).

Liu, T. S., An, Z. S., Yuan, B. Y., and Han, J. M. (1985). The loess-paleosol sequence in China and climatic history. *Episodes* **8**, 21–28.

Liu, X. M., Liu, T. S., Xu, T. C., Liu, C., and Cheng, M. Y. (1988). The primary study on magnetostratigraphy of a loess profile in Xifeng area, Gansu province. *Geophys. J. R. Astron. Soc.* **92**, 345–348.

Liu, X. M., Shaw, J., Liu, T. S., Heller, F., and Cheng, M. Y. (1992). Rock magnetic properties and paleoclimate of Chinese loess. *J. Geomagn. Geoelectr.* **45**, 117–124.

Longworth, G., Becker, L. W., Thompson, R., Oldfield, F., Dearing, J. A., and Rummery, T. A. (1979). Mossbauer and magnetic studies of secondary iron oxides in soils. *J. Soil Sci.* **30**, 93–100.

Loomis, D. P., and Burbank, D. W. (1988). The stratigraphic evolution of the El Paso Basin, southern California: Implications for the Miocene development of the Garlock fault and uplift on the Sierra Nevada. *Geol. Soc. Am. Bull.* **100**, 12–28.

Lovley, D. R. (1990). Magnetite formation during microbial dissimilatory iron reduction. *In* "Iron Biominerals" (R. B. Frankel and R. P. Blakemore, eds.), pp. 151–166. Plenum, New York.

Lovley, D. R., Stolz, J. F., Nord, G. L., Jr., and Phillips, E. J. P. (1987). Anaerobic production of magnetite by a dissimilatory iron-reducing microorganism. *Nature (London)* **330**, 252–254.

Lovlie, R. (1974). Post-depositional remanent magnetization in a re-deposited deep-sea sediment. *Earth Planet. Sci. Lett.* **21**, 315–320.

Lovlie, R., and Torsvik, T. (1984). Magnetic remanence and fabric properties of laboratory-deposited hematite bearing sandstone. *Geophys. Res. Lett.* **11**, 221–224.

Lovlie, R., Torsvik, T., Jelenska, M., and Levandowski, M. (1984). Evidence for detrital remanent magnetization carried by hematite in Devonian red beds from Spitsbergen: Palaeomagnetic implications. *Geophys. J. R. Astron. Soc.* **79**, 573–588.

Lowrie, W. (1989). Magnetostratigraphy and the geomagnetic polarity record. *Cuad. Geol. Iber.* **12**, 95–120.

Lowrie, W. (1990). Identification of ferromagnetic minerals in a rock by coercivity and unblocking temperature properties. *Geophys. Res. Lett.* **17**, 159–162.

Lowrie, W., and Alvarez, W. (1975). Paleomagnetic evidence for relation of the Italian peninsula. *J. Geophys. Res.* **80**, 1579–1592.

Lowrie, W., and Alvarez, W. (1977a). Upper Cretaceous to Paleocene magnetic stratigraphy at Gubbio, Italy, III. Upper Cretaceous magnetic stratigraphy. *Geol. Soc. Am. Bull.* **101**, 374–377.

Lowrie, W., and Alvarez, W. (1977b). Late Cretaceous geomagnetic polarity sequence: Detailed rock and palaeomagnetic studies of the Scaglia Rossa limestone at Gubbio, Italy. *Geophys. J. R. Astron. Soc.* **51**, 561–581.

Lowrie, W., and Alvarez, W. (1981). One hundred million years of geomagnetic polarity history. *Geology* **9**, 392–397.

Lowrie, W., and Alvarez, W. (1984). Lower Cretaceous magnetic stratigraphy in Umbrian pelagic limestone sections. *Earth Planet. Sci. Lett.* **71**, 315–328.

Lowrie, W., and Channell, J. E. T. (1984). Magnetostratigraphy of the Jurassic-Cretaceous boundary in the Maiolica limestone (Umbria, Italy). *Geology* **12**, 44–47.

Lowrie, W., and Fuller, M. (1971). On the alternating field demagnetization characteristics of multidomain thermoremanent magnetization in magnetite. *J. Geophys. Res.* **76**, 6339–6349.

Lowrie, W., and Heller, F. (1982). Magnetic properties of marine limestones. *Rev. Geophys. Space Phys.* **20**, 171–192.

Lowrie, W., and Kent, D. V. (1983). Geomagnetic reversal frequency since the late Cretaceous. *Earth Planet. Sci. Lett.* **62**, 305.

Lowrie, W., and Lanci, L. (1994). Magnetostratigraphy of Eocene-Oligocene boundary sections in Italy: No evidence for short subchrons within chrons 12R and 13R. *Earth Planet. Sci. Lett.* **126**, 247–258.

Lowrie, W., and Ogg, J. G. (1985/86). A magnetic polarity time scale for the Early Cretaceous and Late Jurassic. *Earth Planet. Sci. Lett.* **76**, 341–349.

Lowrie, W., Alvarez, W., Premoli Silva, I., and Monechi, W. (1980a). Lower Cretaceous magnetic stratigraphy in Umbrian pelagic carbonate rocks. *Geophys. J. R. Astron. Soc.* **60**, 263–281.

Lowrie, W., Channell, J. E. T., and Alvarez, W. (1980b). A review of magnetic stratigraphy investigations in Cretaceous pelagic carbonate rocks. *J. Geophys. Res.* **85**, 3597–3605.

Lowrie, W., Alvarez, W., Napoleone, G., Perch-Nielson, K., Premoli-Silva, I., and Toumarkine, M. (1982). Paleogene magnetic stratigraphy in Umbrian pelagic carbonate rocks: The Contessa sections, Gubbio. *Geol. Soc. Am. Bull.* **93**, 414–432.

Lund, S. P. (1989). Paleomagnetic secular variation. *In* "The Encyclopedia of Solid Earth Geophysics" (D. E. James, ed.), pp. 876–888. Van Nostrand-Reinhold, New York.

Lund, S. P. (1996). A comparison of Holocene paleomagnetic secular variation records from North America. *J. Geophys. Res.* **101**, 8007–8024.

Lund, S. P., and Banerjee, S. K. (1985). Late Quaternary paleomagnetic field secular variation from two Minnesotan lakes. *J. Geophys. Res.* **90**, 803–825.

Lund, S. P., and Keigwin, L. (1994). Measurement of the degree of smoothing in sediment paleomagnetic secular variation records: An example from late Quaternary deep-sea sediments of the Bermuda Rise, western North Atlantic Ocean. *Earth Planet. Sci. Lett.* **122**, 317–330.

Lund, S. P., Liddicoat, J. C., Lajoie, K. R., Henyey, T. L., and Robinson, S. W. (1988). Paleomagnetic evidence for longer term (10^4 year) memory and periodic behavior in the earth's core dynamic process. *Geophys. Res. Lett.* **15**, 1101–1104.

Lyle, M., Mayer, L., Pisias, N., Hagelburg, T., Dadey, K., Bloomer, S., and Shipboard Scientific Party of Leg 138 (1992). Downhole logging as a paleoceanographic tool on Ocean Drilling Program Leg 138: Interface between high-resolution stratigraphy and regional syntheses. *Paleoceanography* **7**, 691–700.

MacFadden, B. J. (1977). Magnetic polarity stratigraphy of the Chamita Formation stratotype (Mio-Pliocene) of north-central New Mexico. *Am. J. Sci.* **277**, 769–800.

MacFadden, B. J., Johnson, N. M., and Opdyke, N. D. (1979). Magnetic polarity stratigraphy of the Mio-Pliocene mammal-bearing Big Sandy Formation of western Arizona. *Earth Planet. Sci. Lett.* **44**, 349–364.

MacFadden, B. J., Siles, O., Zeitler, P., Johnson, N. M., and Campbell, K. E. (1983). Magnetic polarity stratigraphy of the middle Pleistocene (Ensenadan) Tarija Fm. of southern Bolivia. *Quat. Res.* **19**, 172–187.

MacFadden, B. J., Campbell, K. E., Cifelli, R. L., Siles, O., Johnson, N. M., Naeser, C. W., and Zeitler, P. K. (1985). Magnetic polarity stratigraphy and mammalian fauna of the Deseadan (late Oligocene–early Miocene) Salla beds of northern Bolivia. *J. Geol.* **93**, 223–250.

MacFadden, B. J., Anaya, F., Perez, H., Naeser, C. W., Zeitler, P. K., and Campbell, K. E., Jr. (1990a). Late Cenozoic paleomagnetism and chronology of Andean Basins of Bolivia: Evidence for possible oroclinal bending. *J. Geol.* **98**, 541–555.

MacFadden, B. J., Swisher, C. C., III, Opdyke, N. D., and Woodburn, M. O. (1990b). Paleomagnetism, geochronology, and possible tectonic rotation of the middle Miocene Barstow Formation, Mojave Desert, southern California. *Geol. Soc. Am. Bull.* **102**, 478–493.

MacFadden, B. J., Anaya, F., and Argollo, J. (1993). Magnetic polarity stratigraphy of Inchasi: A Pliocene mammal-bearing locality from the Bolivian Andes just before the Great American Interchange. *Earth Planet. Sci. Lett.* **114**, 229–241.

Mackereth, F. J. H. (1958). A portable core sampler for lake deposits. *Limnol. Oceanogr.* **3**, 181–91.

Mackereth, F. J. H. (1971). On the variation in direction of the horizontal component of remanent magnetization in lake sediments. *Earth Planet. Sci. Lett.* **12**, 332–338.

Madrid, V. M., Stuart, R. M., and Verosub, K. L. (1986). Magnetostratigraphy of the late Neogene Purisima formation, Santa Cruz County, California. *Earth Planet. Sci. Lett.* **79**, 431–440.

Maenaka, K. (1983). Magnetostratigraphic study of the Osaka Group, with special reference to the existence of pre and post-Jaramillo episodes in the Lake Matuyama polarity epoch. *Mem. Hanazono Univ.* **14**, 1–65.

Magnus, G., and Opdyke, N. D. (1991). A paleomagnetic investigation of the Minturn formation, Colorado: A study in establishing the timing of remanence acquisition. *Tectonophysics* **187**, 181–189.

Maher, B. A. (1986). Characterization of soils by mineral magnetic measurements. *Phys. Earth Planet. Inter.* **42**, 76–92.

Maher, B. A. (1988). Magnetic properties of some synthetic sub-micron magnetites. *Geophys. J. Int.* **94**, 83–96.

Maher, B. A., and Taylor, R. M. (1988). Formation of ultrafine-grained magnetite in soils. *Nature (London)* **336**, 368–370.

Maher, B. A., and Thompson, R. (1991). Mineral magnetic record of Chinese loess and paleosols. *Geology* **19**, 3–6.

Maher, B. A., and Thompson, R. (1992). Paleoclimatic significance of mineral magnetic record of the Chinese loess and paleosols. *Quat. Res.* **37**, 155–170.

Maher, B. A., Thompson, R., and Zhou, L. P. (1994). Spatial and temporal reconstructions of changes in the Asian palaeomonsoon: A new mineral magnetic approach. *Earth Planet. Sci. Lett.* **125**, 461–471.

Mahoney, J. J., Storey, M., Duncan, R., Spencer, K. J., and Pringle, M. (1993). Geochemistry and age of the Ontong-Java Plateau. *In* "Monograph on the Mesozoic Pacific" (M. S. Pringle, W. W. Sager, W. V. Sliter, and S. Stein, eds.), pp. 233–261. Am. Geophys. Union, Washington, DC.

Mankinen, E. A., and Dalrymple, G. B. (1979). Revised geomagnetic polarity time scale for the interval 0–5 m.y.b.p. *J. Geophys. Res.* **84**, 615–626.

Mankinen, E. A., and Grommé, C. S. (1982). Paleomagnetic data from the Cosa Range, California and current status of the Cobb Mountain normal geomagnetic polarity event. *Geophys. Res. Lett.* **9**, 1239–1282.

Mankinen, E. A., Donnelly, J. M., and Grommé, C. S. (1978). Geomagnetic polarity event recorded at 1.1 m.y.b.p. on Cobb Mountain Clear Lake volcanic field, California. *Geology* **6**, 653–656.

Marshall, L. G., Butler, R. F., Drake, R. E., Curtis, G. H., and Tedford, R. H. (1979). Calibration of the Great American Interchange. *Science* **204**, 272–279.

Marshall, L. G., Butler, R. F., Drake, R. E., and Curtis, G. H. (1982). Geochronology of Type Uguian (Late Cenozoic) land mammal age, Argentina. *Science* **216**, 986–989.

Marshall, L. G., Drake, R. E., Curtis, G. H., Butler, R. F., Flanagan, K. M., and Naeser, C. W. (1986). Geochronology of type Santacrucian (middle tertiary) land mammal age, Patagonia, Argentina. *J. Geol.* **94**, 449–457.

Martinson, D. G., Pisias, N. G., Hays, J. D., Imbrie, J., Moore, T. C., Jr., and Shackleton, N. J. (1987). Age dating and the orbital theory of the Ice Ages: Development of a high-resolution 0 to 300,000-year chronostratigraphy. *Quat. Res.* **27**, 1–29.

Márton, E. (1982). Late Jurassic/Early Cretaceous magnetic stratigraphy from the Sumeg section, Hungary. *Earth Planet. Sci. Lett.* **57**, 182–190.

Márton, E., Márton, P., and Heller, F. (1980). Remanent magnetization of a Pliensbachian limestone sequence at Bakenycsernye (Hungary). *Earth Planet. Sci. Lett.* **48**, 218–226.

Mary, C., Moreau, M., Orue-Etxebarria, X., Estibaliz, A., and Courtillot, V. (1991). Biostratigraphy and magnetostratigraphy of the Cretaceous/Tertiary Sopelana section (Basque country). *Earth Planet. Sci. Lett.* **106**, 133–150.

Mary, C., Iaccarino, S., Courtillot, V., Besse, J., and Aissaoui, D. M. (1993). Magnetostratigraphy of Pliocene sediments from the Stirone River, (P. Valley). *Geophys. J. Int.* **112**, 359–380.

Matuyama, M. (1929). On the direction of magnetization of basalts in Japan, Tyosen, and Manchuria. *Proc. Imp. Acad.* (*Tokyo*) **5**, 203.

Mauritsch, H. J., and Turner, P. (1975). The identification of magnetite in limestones using the low temperature transition. *Earth Planet. Sci. Lett.* **24**, 414–418.

Mazaud, A., Laj, C., and Bender, M. (1994). A geomagnetic chronology for antarctic ice accumulation. *Geophys. Res. Lett.* **21**, 337–340.

Mccabe, C., and Channell, J. E. T. (1994). Late Paleozoic remagnetization in limestones of the Craven Basin (northern England) and the rock magnetic fingerprint of remagnetization secondary carbonates. *J. Geophys. Res.* **99**, 4603–4612.

Mccabe, C., Van der Voo, R., Peacor, D. R., Scotese, C. R., and Freeman, R. (1983). Diagenetic magnetite carries ancient yet secondary remanence in some Paleozoic sedimentary carbonates. *Geology* **11**, 221–223.

McDougall, I., and Chamalaun, F. H. (1966). Geomagnetic polarity scale of time. *Nature* (*London*) **212**, 1415–1418.

McDougall, I., and Tarling, D. H. (1963a). Dating of reversals of the Earth's magnetic fields. *Nature* (*London*) **198**, 1012–1013.

McDougall, I., and Tarling, D. H. (1963b). Dating of polarity zones in the Hawaiian Islands. *Nature* (*London*) **200**, 54–56.

McDougall, I., and Tarling, D. H. (1964). Dating geomagnetic polarity zones. *Nature* (*London*) **202**, 171–172.

McDougall, I., and Wensink, J. (1966). Paleomagnetism and geochronology of the Pliocene-Pleistocene lavas in Iceland. *Earth Planet. Sci. Lett.* **1**, 232–236.

McDougall, I., Allsop, H. L., and Chamalaun, F. H. (1966). Isotopic dating of the New Volcanic series of Victoria, Australia and geomagnetic polarity epochs. *J. Geophys. Res.* **71**, 6107–6118.

McDougall, I., Saemundson, K., Johannesson, H., Watkins, N. D., and Kristjansson, L. (1977). Extension of the geomagnetic polarity time scale to 6.5 my.: K-Ar dating, geological and paleomagnetic study of a 3500 M lava succession in western Iceland. *Bull. Geol. Soc. Am.* **88**, 1–15.

McDougall, I., Kristjanson, L., and Saemundson, K. (1984). Magnetostratigraphy and geochronology of northwest Iceland. *J. Geophys. Res.* **82**, 7029–7060.

McDougall, I., Brown, F. H., Cerling, T. E., and Hillhouse, J. W. (1992). A reappraisal of the geomagnetic polarity time scale to 4 Ma using data from Turkana Basin, East Africa. *Geophys. Res. Lett.* **19**, 2349–2352.

McElhinny, M. W. (1964). Statistical significance of the fold test in paleomagnetism. *Geophys. J. R. Astron. Soc.* **8**, 338–340.

McElhinny, M. W. (1973). "Paleomagnetism and Plate Tectonics." Cambridge Univ. Press, Cambridge, UK.

McElhinny, M. W., and Opdyke, N. D. (1973). The remagnetization hypothesis discounted, a paleomagnetic study of the Trenton limestone (New York State). *Geol. Soc. Am. Bull.* **84**, 3697–3708.

McElhinny, M. W., and Senanayake, W. E. (1982). Variations in the geomagnetic dipole. 1. The past 50,000 years. *J. Geomagn. Geoelectr.* **34**, 39–51.

McFadden, P. L. (1989). Geomagnetic reversal sequence: Statistical structure. *In* "The Encyclopedia of Solid Earth Geophysics" (D. E. James, ed.), pp. 556–560. Van Nostrand-Reinhold, New York.

McFadden, P. L. (1990). A new fold test for paleomagnetic studies. *Geophys. J. Int.* **103**, 163–169.

McFadden, P. L., and Jones, D. L. (1981). The fold test in palaeomagnetism. *Geophys. J. R. Astron. Soc.* **67**, 53–58.

McFadden, P. L., and McElhinny, M. W. (1988). The combined analysis of remagnetization circles and direct observations in paleomagnetism. *Earth Planet. Sci. Lett.* **87**, 161–172.

McFadden, P. L., and McElhinny, M. W. (1990). Classification of the reversal test in paleomagnetism. *Geophys. J. Int.* **103**, 725–729.

McFadden, P. L., and Merrill, R. T. (1984). Lower mantle convection and geomagnetism. *J. Geophys. Res.* **89**, 3354–3362.

McFadden, P. L., Ma, X. H., McElhinny, M. W., and Zhang, Z. K. (1988). Permo-Triassic magnetostratigraphy in China: Northern Tarim. *Earth Planet. Sci. Lett.* **87**, 152–160.

McIntosh, W. C., Hargraves, R. B., and West, C. L. (1985). Paleomagnetism and oxide mineralogy of Upper Triassic to Lower Jurassic red beds and basalts in the Newark basin. *Geol. Soc. Am. Bull.* **96**, 463–480.

McKenna, M. C. (1975). Fossil mammals and early Eocene Atlantic land continuity. *Ann. Mo. Bot. Gard.* **62**, 335–353.

McMahon, B. E., and Strangway, D. W. (1968). Stratigraphic implications of paleomagnetic data from upper Palaeozoic–lower Triassic redbeds of Colorado. *Geol. Soc. Am. Bull.* **79**, 417–428.

McNeill, D. F. (1990). Biogenic magnetite from surface Holocene carbonate sediments. Great Bahama Bank. *J. Geophys. Res.* **95**, 4363–4371.

McNeill, D. F., and Kirschvink, J. L. (1993). Early dolomitization of platform carbonates and the preservation of magnetic polarity. *J. Geophys. Res.* **98**, 7977–7986.

McNeill, D. F., Ginsburg, R. N., Chang, S. B. R., and Kirschvink, J. L. (1988). Magnetostratigraphic dating of shallow-water carbonates from San Salvador, the Bahamas. *Geology* **16**, 8–12.

McRae, L. E. (1990a). Paleomagnetic isochrons, unsteadiness, and non-uniformity of sedimentation in Miocene fluvial strata of the Siwalik group, northern Pakistan. *J. Geol.* **98**, 433–456.

McRae, L. E. (1990b). Paleomagnetic isochrons, unsteadiness, and uniformity of sedimentation in Miocene Itermotane Basin sediments at Salla, eastern Andean Cordillera, Bolivia. *J. Geol.* **98**, 479–500.

Mead, G. A. (1996). Correlation of Cenozoic–Late Cretaceous geomagnetic polarity time scales: An internet archive. *J. Geophys. Res.* **101**, 8107–8109.

Mead, G. A., and Hodell, D. A. (1995). Controls on the $^{87}Sr/^{86}Sr$ composition of seawater from the middle Eocene to Oligocene: Hole 689B, Maud Rise, Antarctica. *Paleoceanography* **10**, 327–346.

Mead, G. A., Tauxe, L., and LaBrecque, J. L. (1986). Oligocene paleoceanography of the South Atlantic: Paleoclimatic implications of sediments accumulation rates and magnetic susceptibility. *Paleoceanography* **1**, 272–284.

Menning, M., Katzung, G., and Lutzner, H. (1988). Magnetostratigraphic investigation in the Rotliegendes (300–252 Ma) of Central Europe. 2. *Geol. Wiss. Berlin* **16**(11/12), 1045–1063.

Mercanton, P. L. (1926). Inversion de l'inclinaison aux âges géologique. *Terr. Magn. Atmos. Electr.* **31**, 187–190.

Merrill, R. T., and McElhinny, M. W. (1977). Anomalies in the time-averaged paleomagnetic field and their implications for the lower mantle. *Rev. Geophys. Space Phys.* **15**, 309–323.

Merrill, R. T., and McElhinny, M. W. (1983). "The Earth's Magnetic Field." Academic Press, London.

Merrill, R. T., McElhinny, M. W. and Stevenson, D. J. (1979). Evidence for long-term asymmetries in the Earth's magnetic field and possible implications for dynamo theories. *Phys. Earth Planet. Inter.* **20**, 75–82.

Metallova, V. V., Iosifidi, A. G., and Mostrukov, V. P. (1984). Paleomagnetic stratigraphy from the early Ordovician, River Lena Sibera. *Iv. Akad. Nauk. SSSR, Ser. Geofiz.* **10**, 66–70.

Meynadier, L., Valet, J. P., Weeks, R., Shackleton, N. J., and Hagee, V. L. (1992). Relative geomagnetic intensity of the field during the last 140 ka. *Earth Planet. Sci. Lett.* **114**, 39–57.

Meynadier, L., Valet, J.-P., Bassinot, F. C., Shackleton, N. J., and Guyodo, Y. (1994). Asymmetrical saw-tooth pattern of the geomagnetic field intensity from equatorial sediments in the Pacific and Indian oceans. *Earth Planet. Sci. Lett.* **126**, 109–128.

Mienert, J., and Bloemendal, J. (1989). A comparison of acoustic and rock magnetic properties of equatorial Atlantic deep-sea sediments: Paleoclimatic implications. *Earth Planet. Sci. Lett.* **94**, 291–300.

Milankovitch, M. (1930). Mathematische Klimalehre und Astronomische Theore der Klimaschwankungen. *In* "Handbook der Klimatologic" (W. Koppen and R. Geiger, eds.), Vol. I, Part A, pp. 1–176. Borntraeger, Berlin.

Miller, J. D., and Kent, D. V. (1989). Paleomagnetism of the Upper Ordovician Juniata fm. of the central Appalachians revisited again. *J. Geophys. Res.* **94**, 1843–1849.

Miller, J. D., and Opdyke, N. D. (1985). The magnetostratigraphy of the Red Sandstone Creek section, Vail, Colorado. *Geophys. Res. Lett.* **12**, 133–136.

Miller, K. G., Aubry, M. P., Kahn, M. J., Melillo, A. J., Kent, D. V., and Berggren, W. A. (1985). Oligocene-Miocene biostratigraphy, magnetostratigraphy, and isotopic stratigraphy of the western North Atlantic. *Geology* **13**, 257–261.

Miller, K. G., Fairbanks, R. G., and Mountian, G. S. (1987). Testing oxygen isotope synthesis of sea level history and continental margin erosion. *Paleoceanography* **2**, 1–19.

Miller, K. G., Feigensen, M. D., Kent, D. V., and Olsen, R. K. (1988). Upper Eocene to Oligocene isotope ($^{87}Sr/^{86}Sr$, $\delta^{18}O$, $\delta^{13}C$) standard section DSDP site 522. *Paleoceanography* **3**, 223–233.

Miller, K. G., Wright, J. D., and Brower, A. N. (1989). Oligocene to Middle Miocene stable isotope stratigraphy and planktonic foraminiferal biostratigraphy of the Sierra Leone Rise (Sites 366–667). *In* "Proceedings of the Ocean Drilling Program, Scientific Results" W. Ruddiman, M. Sarnthein *et al.*, eds.), Vol. 108, pp. 279–291. ODP, College Station, TX.

Miller, K. G., Kent, D. V., Brower, A. N., Bybell, L. M., Feigenson, M. D., Olson, R. K., and Poore, R. Z. (1990). Eocene-Oligocene sea-level changes on the New Jersey coastal plain linked to the deep-sea record. *Geol. Soc. Am. Bull.* **102**, 331–339.

Miller, K. G., Wright, J. D., and Fairbanks, R. G. (1991a). Unlocking the ice house: Oligocene-Miocene oxygen, eustasy and margin erosion. *J. Geophys. Res.* **96**, 6829–6848.

Miller, K. G., Feigenson, M. D., Wright, J. D., and Clement, B. M. (1991b). Miocene isotopic reference section, Deep Sea Drilling Project Site 608: An evaluation of isotope and biostratigraphic resolution. *Paleoceanography* **6**, 33–52.

Miller, K. G., Thomason, P. R., and Kent, D. V. (1993). Integrated Late Eocene–Oligocene stratigraphy of the Alabama coastal plain: Correlation of hiatuses and stratal surfaces to glacioeustatic lowerings. *Paleoceanography* **8**, 313–331.

Miller, K. G., Wright, J. D., Van Fossen, M. C., and Kent, D. V. (1994). Miocene stable isotope stratigraphy and magnetostratigraphy of Buff Bay, Jamaica. *Geol. Soc. Am. Bull.* **106**, 1605–1620.

Molina-Garza, R. S., Van der Voo, R., and Geissman, J. W. (1989). Paleomagnetism of the Dewey Lake Formation northwest Texas: end of the Kiaman superchron in North America. *J. Geophys. Res.* **94**, 17,881–17,888.

Molina-Garza, R. S., Geissman, J. W., Van der Voo, R., Lucas, S. G., and Hayden, S. H. (1991). Paleomagnetism of the Moenkopi and Chinle Fm. in central New Mexico. Implications for the North American apparent polar wander path and Triassic magnetostratigraphy. *J. Geophys. Res.* **96**, 14239–14262.

Molostovsky, E. A. (1992). Paleomagnetic stratigraphy of the Permian system. *In. Geol. Rev.* **34**, 1001–1007.

Molyneux, L., Thompson, R., Oldfield, F., and McCallan, M. E. (1972). Rapid measurement of the remanent magnetization of long cores of sediment. *Nature (London)* **237**, 42–43.

Monechi, S., and Thierstein, H. R. (1985). Late Cretaceous–Eocene nannofossil and magneto-stratigraphic correlations near Gubbio, Italy. *Mar. Micropaleontol.* **9**, 419–440.

Monechi, S., Bleil, U., and Backman, J. (1985). Magnetobiochronology of Late Cretaceous, Paleogene and Late Cenozoic pelagic sedimentary sequences from the northwest Pacific. DSDP Leg 86, Site 577. *In* "Initial Reports of the Deep Sea Drilling Project" (G. R. Heath, L. H. Burkle *et al.*, eds.), Vol. 86, pp. 787–797. U.S. Govt. Printing Office, Washington, DC.

Montanari, A., Deino, A. L., Drake, R. E., Turrin, B. D., DePaolo, D. J., Odin, G. S., Curtis, G. H., Alvarez, W., and Bice, D. (1988). Radioisotopic dating of the Eocene-Oligocene boundary in the pelagic sequences of the northeastern Apennines. *In* "The Eocene-Oligocene Boundary in the Marche-Umbria Basin (Italy)" (I. Premoli-Silva, R. Cocioni, and A. Montanari, eds.), Intl. Subcomm. Paleogene Stratigraphy Report, Eocene/Oligo-cene Boundary Meet., Ancona, pp. 195–208.

Morner, N. A., Lanser, J. P., and Hospers, J. (1971). Late Weichselian paleomagnetic reversal. *Nature (London)* **234**, 173–174.

Moskowitz, B. M., Frankel, R. B., Flanders, P. J., Blakemore, R. P., and Schwartz, B. B. (1988). Magnetic properties of magnetotactic bacteria. *J. Magn. Magn. Mater.* **73**, 273–288.

Moskowitz, B. M., Frankel, R. B., and Bazylinski, D. A. (1994). Rock magnetic criteria for the detection of biogenic magnetite. *Earth Planet. Sci. Lett.* **120**, 283–300.

Mothersill, J. S. (1979). The paleomagnetic record of the late Quaternary sediments of Thunder Bay. *Can. J. Earth Sci.* **16**, 1016–1023.

Mothersill, J. S. (1981). Late Quaternary paleomagnetic record of Goderich Basin, Lake Huron. *Can. J. Earth Sci.* **18**, 448–456.

Mullender, T. A. T., van Velzen, A. J. and Dekkers, M. J. (1993). Continuous drift correction and separate identification of the ferrimagnetic and paramagnetic contributions in ther-momagnetic runs. *Geophys. J. Int.* **114**, 663–672.

Mullins, C. E. (1977). Magnetic susceptibility of the soil and its significance in soil science—a review. *J. Soil Sci.* **28**, 233–246.

Muttoni, G., and Kent, D. V. (1994). Paleomagnetism of Latest Anisian (Middle Triassic) sections of the Prezzo Limestone and the Buchenstein formation, Southern Alps, Italy. *Earth Planet. Sci. Lett.* **122**, 1–18.

Muttoni, G., Channell, J. E. T., Nicora, A., and Rettori, R. (1994). Magnetostratigraphy and biostratigraphy of an Anisian-Ladinian (Middle Triassic boundary section from Hydra, Greece). *Paleogeogr., Palaeoclimatol., Palaeoecol.*, **111**, 249–262.

Muttoni, G., Kent, D. V., and Gaetani, M. (1995). Magnetostratigraphy of a candidate global stratotype section and point for the base of the Anisian from Chios (Greece). Implications for the Early to Middle Triassic geomagnetic polarity sequence. *Phys. Earth Planet. Inter.* **92**, 245–260.

Nabel, P. E. A., and Valencio, D. A. (1981). La magnetostratigraphia del ensenadense de la ciudad de Buenos Aires, su significado geologicas. *Asoc. Geol. Argent. Rev.* **36**(1), 70–78.

Nagata, T. (1952). Reverse thermal-remanent magnetism. *Nature (London)* **169**, 704.

Nagata, T. (1961). "Rock Magnetism." Maruzen, Tokyo.

Nagata, T., Uyeda, S., and Ozima, M. (1957). Magnetic interaction between ferromagnetic minerals contained in rocks. *Philos. Mag., Suppl. Adv. Phys.* **6**, 264–287.

Nagata, T., Arai, Y., and Momose, K. (1963). Secular variation of the geomagnetic total force during the last 5000 years. *J. Geophys. Res.* **68**, 5277–5282.

Nagata, T., Kobayashi, K., and Fuller, M. (1964). Identification of magnetite and hematite in rocks by magnetic observation at low temperature. *J. Geophys. Res.* **69**, 2111–2120.

Nagy, E. A., and Valet, J. P. (1993). New advances for paleomagnetic studies of sediment cores using U-channels. *Geophys. Res. Lett.* **20**, 671–674.

Nakanishi, M., Tamaki, K., and Kobayashi, K. (1989). Mesozoic magnetic anomaly lineations and seafloor spreading history of the northwestern Pacific. *J. Geophys. Res.* **94**, 15437–15462.

Nakanishi, M., Tamaki, K., and Kobayashi, K. (1992). Magnetic anomaly lineations from Late Jurassic to Early Cretaceous in the west-central Pacific Ocean. *Geophys. J. Int.* **109**, 701–719.

Napoleone, G., and Ripepe, M. (1990). Cyclic geomagnetic changes in Mid-Cretaceous rhythmites, Italy. *Terra Nova* **1**, 437–442.

Napoleone, G., Premoli-Silva, F., Heller, F., Cheli, P., Corezzi, S., and Fischer, A. G. (1983). Eocene magnetic stratigraphy at Gubbio, Italy, and its implications for Paleogene geochronology. *Geol. Soc. Am. Bull.* **94**, 181–191.

Needham, J. (1962). "Science and Civilization in China," Vol. 4, Part 1. Cambridge Univ., Cambridge, UK.

Neel, L. (1951). L'inversion de l'aimentation permanente des roches. *Ann. Geophys.* **7**, 90–102.

Negrini, R. M., Davis, J. O., and Verosub, K. L. (1984). Mono Lake geomagnetic excursion found at Summer Lake, Oregon. *Geology* **12**, 643–646.

Negrini, R. M., Verosub, K. L., and Davis, J. O. (1987). Long-term nongeocentric axial dipole directions and a geomagnetic excursion from the middle Pleistocene sediments of the Humboldt River canyon, Pershing County, Nevada. *J. Geophys. Res.* **92**, 10,617–10,627.

Negrini, R. M., Erbes, D. B., Roberts, A. P., Verosub, K. L., Sarna-Wojcicki, A. M., and Meyer, C. E. (1994). Repeating waveform initiated by a 180–190 ka geomagnetic excursion in western North America: Implications for field behavior during polarity transitions and subsequent secular variation. *J. Geophys. Res.* **99**, 24,105–24,119.

Ness, G., Levi, S., and Couch, R. (1981). Marine magnetic anomaly time scale for the Cenozoic and Late Cretaceous: A précis, critique, and synthesis. *Rev. Geophys. Space Phys.* **18**, 753–770.

Neville, C., Opdyke, N. D., Lindsay, E. H., and Johnson, M. N. (1979). Magnetic stratigraphy of Pliocene deposits of the Glenns Ferry formation, Idaho, and its implications for North American mammalian biostratigraphy. *Am. J. Sci.* **279**, 503–526.

Nick, K., Xia, K., and Elmore, R. D. (1991). Paleomagnetic and petrographic evidence for early magnetizations in successive terra rosa paleosols, Lower Pennsylvania Black Price Limestone, Arizona. *J. Geophys. Res.* **96**, 9873–9885.

Ninkovich, D., Opdyke, N. D., Heezen, B. C., and Foster, J. H. (1966). Paleomagnetic stratigraphy, rates of deposition and tephrachronology in North Pacific deep-sea sediments. *Earth Planet. Sci. Lett.* **1**, 476–492.

Nocchi, M., Parisi, G., Monaco, P., Monechi, S., Mandile, M., Napoleone, G., Ripepe, M., Orlando, M., Premoli-Silva, I., and Brice, D. M. (1986). The Eocene-Oligocene boundary in the Umbrian pelagic regression. *In* "Terminal Eocene Events, Developments in

Palaeontology and Stratigraphy" (C. Pomerol and I. Premoli-Silva, eds.), Vol. 9, pp. 25–40. Elsevier, Amsterdam.

Obradovich, J. D. (1993). A Cretaceous time scale. *Spec. Pap.–Geol. Assoc. Can.* **39,** 379–396.

Obradovich, J. D., and Cobban, W. A. (1975). A time-scale for the Late Cretaceous of the western interior of North America. *Spec. Pap.—Geol. Assoc. Can.* **13,** 31–54.

Obradovich, J. D., Sutter, J. F., and Kunk, M. J. (1986). Magnetic polarity chron tie points for the Cretaceous and Early Tertiary. *Terra Cognita* **6,** 140.

Oda, H., and Shibuya, H. (1996). Deconvolution of long-core paleomagnetic data of the Ocean Drilling Program by Akaike's Bayesian Information Criterion Minimization. *J. Geophys. Res.* **101,** 2815–2834.

Odin, G. S., Curry, D. S., Gale, N. H., and Kennedy, W. J. (1982). The Phanerozic time scale in 1981. *In* "Numerical Dating in Stratigraphy" (G. S. Odin, ed.), pp. 957–960. Wiley, New York.

Odin, G. S., Montanari, A., Deino, A., Drake, R., Guise, P. G., Kreuzer, H., and Rex, D. C. (1991). Reliability of volcanic-sedimentary biotite ages across the Eocene-Oligocene boundary (Apennines, Italy). *Chem. Geol., Isot. Geosci. Sect.* **86,** 203–224.

Ogg, J. G. (1988). Early Cretaceous and Tithonian magnetostratigraphy of the Galicia Margin (ODP Leg 103). *In* "Proceedings of the Ocean Drilling Program, Scientific Results" (G. Boillot, E. L. Winterer et al., eds.), Vol. 103, pp. 659–681. ODP, College Station, TX.

Ogg, J. G., and Lowrie, W. (1986). Magnetostratigraphy of the Jurassic-Cretaceous boundary. *Geology* **4,** 547–550.

Ogg, J. G., and Steiner, M. B. (1991). Early Triassic magnetic polarity time scale—integration of magnetostratigraphy, ammonite zonations and sequence stratigraphy from stratotype sections (Canadian Arctic Archipelago). *Earth Planet. Sci. Lett.* **107,** 69–89.

Ogg, J. G., Steiner, M. B., Oloriz, F., and Tavera, J. M. (1984). Jurassic magnetostratigraphy. 1. Kimmeridgian-Tithonian of Sierra Gorda and Carcabuey, southern Spain. *Earth Planet. Sci. Lett.* **71,** 147–162.

Ogg, J. G., Steiner, M. B., Company, M., and Tavera, J. M. (1988). Magnetostratigraphy across the Berriasian-Valanginian stage boundary (Early Cretaceous), at Cehegin (Murcia Province, southern Spain). *Earth Planet. Sci. Lett.* **87,** 205–215.

Ogg, J. G., Steiner, M. B., Wiczorck, J., and Hoffman, M. (1991a). Jurassic magnetostratigraphy. 4. Early Callovian through Middle Oxfordian of the Krakow Uplands (Poland). *Earth Planet. Sci. Lett.* **104,** 488–504.

Ogg, J. G., Kodama, K., and Wallick, B. P. (1991b). Lower Cretaceous magnetostratigraphy and paleolatitude off northwest Australia, ODP Site 765 and DSDP Site 261, Argo abyssal plain, and ODP Site 766, Gascoyne abyssal plain. *In* "Proceedings of the Ocean Drilling Project, Scientific Results" (F. M. Gradstein, J. N. Ludden et al., eds.), Vol. 123, pp. 523–554. ODP, College Station, TX.

Ogg, J. G., Hasenyager, R. W., Wimbledon, W. A., Channell, J. E. T., and Bralower, T. J. (1991c). Magnetostratigraphy of the Jurassic-Cretaceous boundary interval—Tethyan and English faunal realms. *Cretaceous Res.* **12,** 455–482.

Oldfield, F., Dearing, J. A., and Battarbee, R. W. (1983). New approaches to recent environmental change. *Geogr. J.* **149,** 167–181.

Olsen, P. E. (1980). The latest Triassic and Early Jurassic formations of the Neward basin of eastern North America, (Newark Supergroup): Stratigraphy, structure and sedimentation. *N.J. Acad. Sci. Bull.* **25,** 25–51.

Olson, P. (1983). Geomagnetic polarity reversals in a turbulent core. *Phys. Earth Planet. Inter.* **33,** 260–274.

Olson, R. K., and Poore, R. Z. (1990). Eocene-Oligocene sea-level changes on the New Jersey coastal plain linked to the deep-sea record. *Geol. Soc. Am. Bull.* **102,** 331–339.

Omarzi, S. K., Coe, R. S. and Barron, J. A. (1993). Magnetostratigraphy: A powerful tool for high resolution age-dating and correlation in the Miocene Monterey Formation of California: Results from Shell Beach Section, Pismo Basin. *In* "Applications of Paleomagnetism to Sedimentary Geology" (D. J. Aissaoni, D. F. McNeill, and N. F. Hurley, eds.), SEPM Society for Sedimentary Geology Spec. Publ. No. 49, pp. 95–111.

Onstott, T. C. (1980). Application of the Bingham distribution function in paleomagnetic studies. *J. Geophys. Res.* **85**, 1500–1510.

Opdyke, N. D. (1968). The paleomagnetism of oceanic cores. *In* "History of the Earth's Crust" (R. A. Phinney, ed.), pp. 67–72. Princeton Univ. Press, Princeton, NJ.

Opdyke, N. D. (1972). Paleomagnetism of deep sea cores. *Rev. Geophys. Space Phys.* **10**, 213–249.

Opdyke, N. D. (1990). Magnetic stratigraphy of Cenozoic terrestrial sediments and mammalian dispersal. *J. Geol.* **98**, 621–637.

Opdyke, N. D., and DiVenere, V. J. (1994). Magnetic polarity stratigraphy of the Mauch Chunk Formation (Pennsylvania). *EOS, Trans. Am. Geophys. Union, Spring Meet.* **75**, 130.

Opdyke, N. D., and Foster, J. H. (1970). The paleomagnetism of cores from the North Pacific, geological investigation of the North Pacific. *Mem.—Geol. Soc. Am.* **126**, 83–119.

Opdyke, N. D., and Glass, B. (1969). The paleomagnetism of sediment cores from the Indian Ocean. *Deep-Sea Res.* **16**, 249–261.

Opdyke, N. D., and Henry, K. W. (1969). A test of the dipole hypothesis. *Earth Planet. Sci. Lett.* **6**, 139–151.

Opdyke, N. D., and Runcorn, S. K. (1956). New evidence for reversal of the geomagnetic field near the Plio-Pleistocene boundary. *Science* **123**, 1126–1127.

Opdyke, N. D., Claoue-Long, J., Roberts, J., Irving, E. (1995). Preliminary Magnetostratigraphic results from the Carboniferous of Eastern Australia. EOS, Abstracts, Fall Meeting, 171.

Opdyke, N. D., Glass, B., Hays, J. P., and Foster, J. (1966). Paleomagnetic study of Antarctica deep-sea cores. *Science* **154**, 349–357.

Opdyke, N. D., Ninkovich, D., Lowrie, W., and Hayes, J. D. (1972). The palaeomagnetism of two Aegean deep-sea cores. *Earth Planet. Sci. Lett.* **14**, 145–149.

Opdyke, N. D., Kent, D. V., and Lowrie, W. (1973). Details of magnetic polarity transitions recorded in a high deposition rate deep sea core. *Earth Planet. Sci. Lett.* **20**, 315–324.

Opdyke, N. D., Burkle, L. H., and Todd, A. (1974). The extension of the magnetic time scale in sediments of the central Pacific Ocean. *Earth Planet. Sci. Lett.* **22**, 300–306.

Opdyke, N. D., Lindsay, E. H., Johnson, N. M., and Downs, T. (1977). The paleomagnetism and magnetic polarity stratigraphy of the mammal-bearing section of Anza-Borrego State Park, California. *Quat. Res.* **7**, 316–326.

Opdyke, N. D., Lindsay, E., Johnson, G. D., Johnson, N., Tahirkheli, R. A. K., and Mirza, M. A. (1979). Magnetic polarity stratigraphy and vertebrate paleontology of the upper Siwalik subgroup of northern Pakistan. *Palaeogeogr., Palaeoclimatol., Palaeoecol.* **27**, 1–34.

Opdyke, N. D., Johnson, N. M., Johnson, G., Lindsay, E., and Tahirkheli, R. A. K. (1982). Paleomagnetism of the middle Siwalik formations of northern Pakistan and rotation of the Salt Range Décollement. *Palaeogeogr., Palaeoclimatol., Palaeoecol.* **37**, 1–15.

Opdyke, N. D., Mein, P., Moissenet, E., Perez-Gonzalez, A., Lindsay, E., and Petko, M. (1989). The magnetic stratigraphy of the late Miocene sediments of the Cabriel basin, Spain. *In* "European Neogene Mammal Chronology" (E. Lindsay *et al.* (eds.), pp. 507–514. Plenum, New York.

Opdyke, N. D., Khramov, A. W., Gurevitch, E., Iosifidi, A. G., and Makarov, I. A. (1993). A paleomagnetic study of the middle Carboniferous of the Donetz Basin, Ukraine. *EOS, Trans. Am. Geophys. Union, Spring Meet.*, p. 118.

Opdyke, N. D., Mein, P., Lindsay, E., Perez-Gonazales, A., Moissenet, E., and Norton, V. L. (1996). Continental Deposits, Magnetostratigraphy and Vertebrate Paleontology, Eastern Spain. *Palaeogeogr. Palaeoclimatol., Palaeocol.,* submitted for publication.

O'Reilly, W. (1984). "Rock and Mineral Magnetism." Blackie, Glasgow and London.

Orguira, M. J. (1990). Paleomagnetism of late Cenozoic fossiliferous sediments from Barranca de los Lobos (Buenos Aires Province, Argentina), The magnetic ages of the South American land-mammal ages. *Phys. Earth Planet. Inter.* **64,** 121–132.

Orguira, M. J., and Valencio, D. A. (1984). Estudio paleomagnetico de los sedimentos asignado al cenzoico tardio aflorantes en la barranca de los Lobos, Provincia de Buenos Aires. *Actas Congr. Geol. Argent.* [N.S.] **4,** 162–173.

O'Sullivan, P. E., Oldfield, R., and Battarbee, R. W. (1972). Preliminary studies of Lough Neagh sediments. I. Stratigraphy, chronology and pollen analysis. *In* "Quaternary Planet Ecology" (H. J. B. Birks and R. G. West, eds.). Blackwell, Oxford.

Özdemir, Ö., and Banerjee, S. K. (1982). A preliminary magnetic study of soil samples from west-central Minnesota. *Earth Planet. Sci. Lett.* **59,** 393–403.

Özdemir, Ö., Dunlop, D. J., and Moskowitz, B. M. (1993). The effect of oxidation on the Verwey transition in magnetite. *Geophys. Res. Lett.* **20,** 1671–1674.

Palmer, D. F., Hanyey, T. L., and Dodson, R. E. (1979). Paleomagnetic and sedimentological studies at Lake Tahoe, California-Nevada. *Earth Planet. Sci. Lett.* **46,** 125–137.

Palmer, J. A., Perry, S. P. G., and Tarling, D. H. (1985). Carboniferous magnetostratigraphy. *J. Geol. Soc. London* **142,** 945–955.

Parker, E. N. (1969). The occasional reversal of the geomagnetic field. *Astrophys. J.* **158,** 815–827.

Parry, L. G. (1980). Shape-related factors in the magnetization of immobilized magnetite particles. *Phys. Earth Planet. Inter.* **22,** 144–154.

Parry, L. G. (1982). Magnetization of immobilized particles dispersions with two distinct particle sizes. *Phys. Earth Planet. Inter.* **28,** 230–241.

Payne, M. E., Wolberg, D. L., and Hunt, A. (1983). Magnetostratigraphy of a core from the Raton Basin, New Mexico, implications for synchroneity of Cretaceous-Tertiary boundary events, New Mexico. *Geology* **5,** 41–44.

Peck, J. A., King, J. W., Colman, S. M., and Kravchinsky, V. A. (1994). A rock-magnetic record from Lake Baikal, Siberia: Evidence for Late Quaternary climate change. *Earth Planet. Sci. Lett.* **122,** 221–238.

Peck, J. A., King, J. W., Colman, S. M., and Kravinsky, V. A. (1996). An 84-kyr paleomagnetic record from sediments of Lake Baikal, Siberia. *J. Geophys. Res.* 101, 11,365–11,385.

Peng, L., and King, J. W. (1992). A Late Quaternary geomagnetic secular variation record from Lake Waiau, Hawaii, and the question of the Pacific non-dipole low. *J. Geophys. Res.* **97,** 4407–4424.

Pessagno, E. A., and Blome, C. D. (1990). Implications of new Jurassic stratigraphic, geochrononometric and paleolatitudinal data from the western Klamath terrane (Smith River and Rogue Valley subterranes). *Geology* **18,** 665–668.

Petersen, N., Heller, F., and Lowrie, W. (1984). Magnetostratigraphy of the Cretaceous/ Tertiary, geological boundary. *In* "Initial Reports of the Deep Sea Drilling Project" (K. J. Hsu, J. L. LaBrecque *et al.,* eds.), Vol. 73, 657–661. U.S. Govt. Printing Office, Washington, DC.

Petersen, N., von Dobeneck, T., and Vali, H. (1986). Fossil bacterial magnetite in deep-sea sediments from the South Atlantic Ocean. *Nature (London)* **320,** 611–615.

Pevzner, M. A., Vangengeium, E. A., Zazhigan, V. S., and Liskun, I. G. (1983). Correlation of the Upper Neogene sediments of Central Asia and Europe on the basis of paleomagnetic and biostratigraphic data. *Int. Geol. Rev.* **25,** 1075–1088.

Phillips, J. D. (1977). Time variation and asymmetry in the statistics of geomagnetic reversal sequences. *J. Geophys. Res.* **82**, 835–843.

Picard, M. D. (1964). Paleomagnetic correlation of units within Chugwater (Triassic) formation, west-central Wyoming. *Am. Assoc. Pet. Geol. Bull.* **48**, 269–291.

Pick, T., and Tauxe, L. (1993a). Holocene paleointensities: Tellier experiments on submarine basaltic glass from the East Pacific Rise. *J. Geophys. Res.* **98**, 17,946–17,949.

Pick, T., and Tauxe, L. (1993b). Geomagnetic paleointensities during the Cretaceous normal superchron measured using submarine basaltic glass. *Nature (London)* **366**, 238–242.

Piper, J. D. A. (1987). "Palaeomagnetism and the Continental Crust." Wiley, New York.

Pitman, W. C., III, (1978). Relationship between eustacy and stratigraphic sequences of passive margins. *Geol. Soc. Am. Bull.* **89**, 1389–1403.

Pitman, W. C., III, and Heirtzler, J. R. (1966). Magnetic anomalies over the Pacific-Antarctic ridge. *Science* **154**, 1164–1171.

Pluhar, C. J., Kirschvinck, J. L., and Adams, R. W. (1991). Stratigraphy and intrabasin correlation of the Mojave River formation, central Mojave Desert, California. *San Bernardino Co. Mus. Assoc. Q.* **38**, 31–42.

Pluhar, C. J., Holt, J. W., and Kirschvink, J. L. (1992). Magnetostratigraphy of Plio-Pleistocene lake sediments in the Confidence Hills of southern Death Valley California. *San Bernardino Co. Mus. Assoc. Q.* **39**, 12–19.

Poore, R. Z., Tauxe, L., Percival, S. F., LaBrecque, J. L., Wright, R., Petersen, N. P., Smith, C. A., Tucker, P., and Hsü, K. J. (1984). Late Cretaceous–Cenozoic magnetostratigraphy and biostratigraphy correlations of the South Atlantic Ocean, DSDP leg 73. *In* "Initial Reports of the Deep Sea Drilling Project" (K. J. Hsu, J. L. LaBrecque *et al.*, eds.), Vol. 73, pp. 645–655. U.S. Govt. Printing Office, Washington, DC.

Preisinger, A., Aslanian, S., Stoykova, K., Grass, F., Mauritsch, H. J., and Scholgar, R. (1993). Cretaceous/Tertiary boundary sections of the Black Sea near Bjala (Bulgaria). *Palaeogeogr., Palaeoclimatol., Palaeoecol.* **104**, 219–228.

Preisinger, A., Zobetz, E., Gratz, A. J., Lahodynsky, R., Becke, M., Mauritsch, H. J., Eder, G., Grass, F., Rogl, F., Stradner, H., and Surenian, R. (1986). The Cretaceous-Tertiary boundary in the Gosau Basin, Austria. *Nature (London)* **322**, 794–799.

Premoli-Silva, I. (1977). Upper Cretaceous–Paleocene magnetic stratigraphy at Gubbio, Italy. II. Biostratigraphy. *Geol. Soc. Am. Bull.* **88**, 371–374.

Premoli-Silva, I., Orlando, M., Monechi, S., Madile, M., Napoleone, G., and Ripepe, M. (1988). Calcareous plankton biostratigraphy and magnetostratigraphy at the Eocene-Oligocene transition in the Gubbio area. *Int. Subcomm. Paleogene Stratgr., (E/O Meet.,* Anacona, *1987,* Spec. Publ. II, Vol. 6, pp. 137–160.

Prentice, M. L., and Matthews, R. K. (1991). Tertiary ice sheet dynamics, the snow gun hypothesis. *J. Geophys. Res.* **96**, 6811–6827.

Prévot, M., and Camps, P. (1993). Absence of preferred longitude sectors for poles from volcanic records of geomagnetic polarity reversals. *Nature (London)* **366**, 53–56.

Prévot, M., Mankinen, E. A., and Grommé, S. (1985). The Steens Mountain (Oregon) geomagnetic polarity transition. 2. Field intensity variations and discussion of reversal models. *J. Geophys. Res.* **90**, 10,417–10,448.

Prévot, M., Derder, M. E., McWilliams, M., and Thompson, J. (1990). Intensity of the Earth's magnetic field: Evidence for a Mesozoic dipole low. *Earth Planet. Sci. Lett.* **97**, 129–139.

Prince, R. A., Heath, G. R., and Kominz, M. (1980). Palaeomagnetic studies of central North Pacific sediment cores: Stratigraphy, sedimentation rates, and the origin of magnetic instability. *Geol. Soc. Am. Bull., Part 2* **91**(8), 1789–1835.

Pringle, M. S., Staudigel, H., Duncan, R. A., Christie, D. M., ODP Leg 143 and 144 Scientific Staffs (1993). ^{40}Ar/^{39}Ar ages of basement lavas at Resolution, MIT, and Wodejebato

Guyots compared with magneto- and bio-stratigraphic results from ODP Legs 143/144. EOS, *Trans. Am. Geophys. Union* **73**(43), 353.

Prothero, D. R. (1985). Chadronian (early Oligocene) magnetostratigraphy of eastern Wyoming: Implications for the age of the Eocene-Oligocene boundary. *J. Geol.* **93**, 555–565.

Prothero, D. R. (1991). Magnetic stratigraphy of Eocene-Oligocene mammal localities in southern San Diego County. In "Eocene Geologic History of the San Diego Region" (P. L. Abbott and J. A. May, eds.), Pac. Sect. SEPM Society for Sedimentary Geology, No. 68, pp. 125–130.

Prothero, D. R., and Armentrout, J. M. (1985). Magnetostratigraphic correlation of the Lincoln Creek Formation, Washington: Implications for the age of the Eocene-Oligocene boundary. *Geology* **13**, 103–211.

Prothero, D. R., and Swisher, C. C. (1992). Magnetostratigraphy and geochronology of the terrestrial Eocene-Oligocene transition in North America. In "Eocene-Oligocene Climatic and Biotic Evolution" (D. R. Prothero and W. Berggren, eds.), pp. 46–73. Princeton Univ. Press, Princeton, NJ.

Prothero, D. R., Denham, C. R., and Farmer, H. G. (1983). Magnetostratigraphy of the White River Group and its implications for Oligocene geochronology. *Palaeogeogr., Palaeoclimatol., Palaeoecol.* **42**, 151–166.

Pullaiah, G., Irving, E., Buchan, K. L., and Dunlop, D. J. (1975). Magnetization changes caused by burial and uplift. *Earth Planet. Sci. Lett.* **28**, 133–143.

Purucker, M. E., Elston, D. P., and Shoemaker, E. M. (1980). Early acquisition of characteristic magnetization in red beds of the Moenkopi Formation (Triassic), Gray Mountains, Arizona. *J. Geophys. Res.* **95**, 997–1012.

Quade, J., Cerling, F. E., and Bowman, J. R. (1989). Development of Asian monsoon revealed by marked ecological shift during the latest Miocene in northern Pakistan. *Nature (London)* **342**, 163–166.

Quidelleur, X., and Valet, J.-P. (1994). Paleomagnetic records of excursions and reversals: Possible biases caused by magnetization artifacts. *Phys. Earth Planet. Inter.* **82**, 27–48.

Radhakrishnamurty, C., Likhite, S. D., Amin, B. S., and Somayajulu, B. L. K. (1968). Magnetic susceptibility stratigraphy in ocean sediment cores. *Earth Planet. Sci. Lett.* **4**, 464–468.

Raff, A. D. (1966). Boundaries of an area of very long magnetic anomalies in the north east Pacific. *J. Geophys. Res.* **71**, 2631–3636.

Rapp, S. D., MacFadden, B. J., and Schiebout, J. A. (1983). Magnetic polarity stratigraphy of the early Tertiary Black Peaks Formation, Big Bend National Park, Texas. *J. Geol.* **91**, 555–572.

Raymo, M. E., Ruddiman, W. E., and Clement, B. M. (1986). Pliocene-Pleistocene paleoceanography of the North Atlantic at Deep Sea Drilling Project Site 609. In "Initial Reports of the Deep Sea Drilling Project" (W. F. Ruddiman, R. B. Kidd, E. Thomas, *et al.*, eds.), Vol. 94, pp. 895–901. U.S. Govt. Printing Office, Washington, DC.

Raymo, M. E., Ruddiman, W. F., Backman, J., Clement, B. M., and Martinson, D. G. (1989). Late Pliocene variation in Northern Hemisphere ice sheets and North Atlantic deep water circulation. *Paleoceanography* **4**, 413–446.

Raynolds, R. G. H., and Johnson, G. D. (1985). Rates of Neogene depositional and deformational processes, northwest Himalayan fore deep margin, Pakistan. In "The Chronology of the Geological Record" (N. J. Snelling, ed.), pp. 297–311. Blackwell, Oxford.

Reeve, S. C., and Helsley, C. E. (1972). Magnetic reversal sequence in the upper portion of the Chinle Formation, Montoya, New Mexico. *Geol. Soc. Am. Bull.* **83**, 3795–3812.

Reeve, S. C., Leythaeuser, D., Helsley, C. E., and Bay, K. W. (1974). Paleomagnetic results from the Upper Triassic of East Greenland. *J. Geophys. Res.* **79**, 3302–3307.

Renné, P. R., Fulford, M. M., and Busby-Spera, C. (1991). High resolution ^{40}Ar/^{39}Ar chrono-stratigraphy of the Late Cretaceous El Gallo Formation, Baja California Del Norte, Mexico. *Geophys. Res. Lett.* **18,** 459–462.

Renné, P. R., Walter, R., Verosub, K., Sweitzer, M., and Aronson, J. (1993). New data from Hadar (Ethiopia) support orbitally tuned time scale to 3.3 Ma. *Geophys. Res. Lett.* **20,** 1067–1070.

Renné, P. R., Deino, P. L., Walter, R. C., Turrin, B. D., Swisher, C. C., III, Becker, T. A., Curtis, G. H., Sharp, W. D., and Abdur-Rahim, J. (1994). Intercalibration of astronomical and radioisotopic time. *Geology* **22,** 783–786.

Reynolds, R. L., Tuttle, M. L., Rice, C. A., Fishman, N. S., Karachewski, J. A., and Sherman, D. M. (1994). Magnetization and geochemistry of greigite-bearing Cretaceous strata, north slope basin, Alaska. *Am. J. Sci.* **294,** 485–528.

Rieck, H. J., Sarna-Wojcick, A. M., Meyer, C. E., and Adam, D. P. (1992). Magnetostratigraphy and tephrochronology of an upper Pliocene to Holocene record in lake sediments at Tulelake, northern California. *Geol. Soc. Am. Bull.* **104,** 409–428.

Rio, D., Sprovieri, R., and Channell, J. E. T. (1990). Pliocene–early Pleistocene chronostratig-raphy and the Tyrrhenian deep-sea record from site 653. *In* "Proceedings of the Ocean Drilling Program, Scientific Results" (K. A. Kastens, J. Muscle *et al.*, eds.), Vol. 107, pp. 705–715. ODP, College Station, TX.

Rio, D., Sprovieri, R., and Thunell, R. (1991). Pliocene-lower Pleistocene chronostratigraphy: A re-evaluation of the Mediterranean type sections. *Geol. Soc. Am. Bull.* **103,** 1049–1058.

Ripperdan, R. L., and Kirschvink, J. L. (1992). Paleomagnetic results from the Cambrian Ordovician boundary section at Black Mountain, Georgina Basin, Western Queensland, Australia. *In* "Global Perspectives on Ordovician Geology" (B. D. Webby and J. R. Laurie, eds.), pp. 93–103. Balkema, Rotterdam.

Roberts, A. P., and Turner, G. M. (1993). Diagenetic formation of ferrimagnetic iron sulphide minerals in rapidly deposited marine sediments, South Island, New Zealand. *Earth Planet. Sci. Lett.* **115,** 257–273.

Roberts, A. P., Turner, G. M., and Vella, P. P. (1994a). Magnetostratigraphic chronology of late Miocene to early Pliocene biostratigraphic and oceanographic events in New Zealand. *Geol. Soc. Am. Bull.* **106,** 665–683.

Roberts, A. P., Verosub, K. L., and Negrini, R. M. (1994b). Middle/Late Pleistocene relative palaeointensity of the geomagnetic field from lacustrine sediments, lake Chewaucan, western United States. *Geophys. J. Int.* **118,** 101–110.

Robinson, S. G. (1986). The Late Pleistocene palaeoclimatic record of North Atlantic deep-sea sediments revealed by mineral-magnetic measurements. *Phys. Earth Planet. Inter.* **42,** 22–46.

Robinson, S. G. (1990). Applications of whole-core magnetic susceptibility measurements of deep-sea sediments. *In* "Proceedings of the Ocean Drilling Program, Scientific Results" (R. A. Duncan, J. Backman, L. C. Peterson, *et al.*, eds.), Vol. 115, pp. 737–771. ODP, College Station, TX.

Robinson, S. G., Maslin, M. A. and McCave, I. N. (1995). Magnetic susceptibility variations in Upper Pleistocene deep-sea sediments of the NE Atlantic: Implications for ice rafting and paleocirculation at the last glacial maximum. *Paleoceanography* **10,** 221–250.

Roche, A. (1950). Sur les caractères magnétiques du système éruptif de géogovie. *C. R. Hebd. Seances Acad. Sci.* **230,** 113–115.

Roche, A. (1951). Sue les inversions de l'aimentation remanente des roches volcaniques dans les mont d'Auvergne. *C. R. Hebd. Seances Acad. Sci.* **223,** 1132–1134.

Roche, A. (1956). Sur la date de la dernière inversion du champ magnétique terrestre. *C. R. Hebd. Seances Acad. Sci.* **243,** 812–814.

Rochette, P., and Fillion, G. (1989). Field and temperature behavior of remanence in synthetic goethite: Paleomagnetic implications. *Geophys. Res. Lett.* **16,** 851–854.

Rodda, P., McDougall, I., Cassil, R. A., Falvey, D. A., Todd, R., and Wilcoxon, J. A. (1985). Isotopic ages, magnetostratigraphy, and biostratigraphy of the Early Pliocene Suva Marl, Fiji. *Bull. Geol. Soc. Am.* **96,** 529–538.

Rodionov, V. P. (1966). Dipole character of the geomagnetic field in the late Cambrian and Ordovician in the south of the Siberian Platform. *Geol. Geofiz.* **1,** 94–101.

Roggenthen, W. A. (1976). Magnetic stratigraphy of the Paleocene. A comparison between Spain and Italy. *Mem. Geol. Soc. Ital.* **15,** 73–82.

Roggenthen, W. A., and Napoleone, G. (1977). Upper Cretaceous–Paleocene magnetic stratigraphy at Gubbio, Italy. IV. Upper Maastrichtian–Paleocene magnetic stratigraphy. *Geol. Soc. Am. Bull.* **88,** 378–382.

Rowley, D. B., and Lottes, A. L. (1988). Plate-kinematic reconstructions of the North Atlantic and Arctic late Jurassic to present. *Tectonophysics* **155,** 73–120.

Roy, J. L., and Morris, W. A. (1983). A review of paleomagnetic results from the carboniferous of North America: The concept of carboniferous geomagnetic field horizon markers. *Earth Planet. Sci. Lett.* **65,** 167–181.

Ruddiman, W. F., McIntyre, A., and Raymo, M. (1986). Matuyama 41,000 year cycles: North Atlantic Ocean and northern hemisphere ice sheets. *Earth Planet. Sci. Lett.* **80,** 117–129.

Ruddiman, W. F., Raymo, M. E., Martinson, D. G., Clement, B. M., and Backman, J. (1989). Pleistocene evolution: Northern hemisphere ice sheet and north Atlantic Ocean. *Paleoceanography* **4,** 353–412.

Rummery, T. A. (1983). The use of magnetic measurements in interpreting the fire histories of lake drainage basins. *Hydrobiologia* **103,** 53–58.

Rummery, T. A., Bloemendal, J., Dearing, J., Oldfield, F., and Thompson, R. (1979). The persistence of fire-induced magnetic oxides in soils and lake sediments. *Ann. Geophys.* **35,** 103–107.

Runcorn, S. K. (1959). On the theory of the geomagnetic secular variation. *Ann. Geophys.* **15,** 87–92.

Runcorn, S. K. (1992). Polar paths in geomagnetic reversals. *Nature (London)* **356,** 654–656.

Ruocco, M. (1989). A 3 Ma paleomagnetic record of coastal continental deposits in Argentina. *Palaeogeogr., Palaeoclimatol., Palaeoecol.* **72,** 105–113.

Rutten, M. G., and Wensink, H. (1960). Paleomagnetic dating, glaciations and chronology of the Plio-Pleistocene in Iceland. *Int. Geol. Congr., Rep. Sess., Norden, 21st,* Vol. 4, pp. 62–70.

Rutter, N., Zhongli, D., Evans, M. E., and Yuchum, W. (1990). Magnetostratigraphy of the Baojo loess–paleosol section in the North Central China Loess Plateau. *Quat. Int.* **7/8,** 97–102.

Ryan, W. B. F., and Flood, J. D. (1972). Preliminary paleomagnetic measurements on sediments from the Ionian (site 125) and Tyrrhenian (site 132) basins of the Mediterranean Sea. *In* "Initial Reports of the Deep Sea Drilling Project" (W. B. F. Ryan, K. J. Hsu *et al.,* eds.), Vol. 13, pp. 599–603. U.S. Govt. Printing Office, Washington, DC.

Sadler, P. M. (1981). Sediment accumulation rates and the completeness of stratigraphic sections. *J. Geol.* **89,** 569–584.

Saka, H., and Keating, B. (1991). Paleomagnetism of Leg 119, Holes 737A, 738C, 742A, 745B, and 746A. *In* "Proceedings of the Ocean Drilling Program, Scientific Results" (J. Barran, B. Larsen *et al.,* eds.), Vol. 119, pp. 751–770. ODP, College Station, TX.

Salloway, J. C. (1983). Paleomagnetism of sediments from Deep Sea Drilling Project Leg 71. *In* "Initial Reports of the Deep Sea Drilling Project" (W. J. Ludwig, A. Krasheninnikov *et al.,* eds.), Vol. 71, pp. 1073–1091. U.S. Govt. Printing Office, Washington, DC.

Sayre, W. O. (1981). Preliminary report on the paleomagnetism of Aptian and Albian limestone and trachytes from the mid-Pacific mountains and Hess Rise. *In* "Initial Reports of the Deep Sea Drilling Project" (J. Theide, T. L. Vallier, and C. G. Adelseck, eds.), Vol. 62, pp. 983–993. U.S. Govt. Printing Office, Washington, DC.

Schneider, D. A. (1993). An estimate of late Pleistocene geomagnetic intensity variations from the Sulu sea sediments. *Earth Planet. Sci. Lett.* **120**, 301–310.

Schneider, D. A. (1995). Paleomagnetism of some Ocean Drilling Program Leg 138, sediments: Detailing Miocene magnetostratigraphy. *In* "Proceedings of the Ocean Drilling Program, Scientific Results" (N. G. Pisias, L. A. Mayer, T. R. Janecek *et al.,* eds.), Vol. 138, pp. 59–72. College Station, TX.

Schneider, D. A., and Kent, D. V. (1988). Inclination anomalies from Indian Ocean sediments and the possibility of a standing non-dipole field. *J. Geophys. Res.* **93**, 11,621–11,630.

Schneider, D. A., and Kent, D. V. (1990a). The time average paleomagnetic field. *Rev. Geophys.* **28**, 71–96.

Schneider, D. A., and Kent, D. V. (1990b). Paleomagnetism of Leg 115 sediments: Implication for Neogene magnetostratigraphy and paleolatitude of the Reunion hotspot. *In* "Proceedings of the Ocean Drilling Project, Scientific Results" (R. A. Duncan, J. Blackman, L. C. Peterson *et al.,* eds.), Vol. 115, pp. 717–736. ODP, College Station, TX.

Schnepp, E., and Hradetzky, H. (1994). Combined paleointensity and $^{40}Ar/^{39}Ar$ age spectrum data from volcanic rocks of the West Eifel field (Germany): Evidence for an early Brunhes geomagnetic excursion. *J. Geophys. Res.* **99**, 9061–9076.

Schult, A. (1976). Self-reversal above room temperature due to N-type magnetization in basalt. *J. Geophys.* **42**, 81–84.

Schulze, D. G., and Dixon, J. B. (1979). High gradient magnetic separation of iron oxides and other magnetic minerals from soil clays. *Soil Sci. Soc. Am. J.* **43**, 793–799.

Schwarzacher, W., and Fischer, A. G. (1982). Limestone-shale bedding and perturbations in the earth's orbit. *In* "Cyclic Event Stratification" (F. Einsele and A. Seilacher, eds.), pp. 72–95. Springer-Verlag, New York.

Schweickert, R. A., Bogen, A., Girty, N. L., Hanson, R. E., and Nerguerian, C. (1984). Timing and structural expression of the Nevadan orogeny, Sierra Nevada, California. *Geol. Soc. Am. Bull.* **95**, 967–979.

Sclater, J. G., and Fisher, R. L. (1974). Evolution of the Central Indian Ocean with emphasis on the tectonic setting of the ninetyeast ridge. *Geol. Soc. Am. Bull.* **85**, 683–702.

Scotese, C. R., Van der Voo, R., and McCabe, C. (1982). Paleomagnetism of the Upper Silurian and Lower Devonian carbonates of New York State: Evidence for secondary magnetizations residing in magnetite. *Phys. Earth Planet. Inter.* **30**, 385–395.

Sen, S. (1989). Hipparion datum and its chronologic evidence in the Mediterranean area. *In* "NATO Advanced Study Workshop in European Neogene Mammal Chronology" (E. Lindsay, P. Mein, and V. Fahlbush, eds.), NATO Ser. A, pp. 495–505. Plenum, New York.

Sen, S., Valet, J., and Ioakim, C. (1986). Magnetostratigraphy and biostratigraphy of the Neogene deposits of Kastellios Hill (central Crete). *Palaeogeogr., Palaeoclimatol., Palaeoecol.* **53**, 321–334.

Shackleton, N. J. (1977). Carbon-13 in Uvigerina: Tropical rainforest history and the equatorial Pacific carbonate dissolution cycles. *In* "The Fate of Fossil Fuel CO_2 in the Oceans" (N. R. Anderson and A. Melahoff, eds.), pp. 401–427. Plenum, New York.

Shackleton, N. J., and Opdyke, N. D. (1973). Oxygen isotope and paleomagnetic stratigraphy of equatorial pacific core V28-238: Oxygen isotope temperatures and ice volumes on a 10^5 and 10^6 year scale. *J. Quat. Res.* **3**(1), 39–55.

Shackleton, N. J., and Opdyke, N. D. (1976). Oxygen-isotope and paleomagnetic stratigraphy of Pacific core V28-239 Late Pliocene to Latest Pleistocene. *Mem.—Geol. Soc. Am.* **145**, 449–464.

Shackleton, N. J., and Opdyke, N. D. (1977). Oxygen isotope and palaeomagnetic evidence for early Northern Hemisphere glaciation. *Nature (London)* **270**, 216–219.

Shackleton, N. J., Backman, J., Zimmerman, H., Kent, D. V., Hall, M. A., Roberts, D. G., Schnitker, D., Baldant, J. G., Desprairics, A., Homrighausen, R., Huddlcgton, P., Keene, J. B., Kaltenback, A. J., Krumsick, K. A. O., Morton, A. C., Murray, J. W., and Westlong-Smith, J. (1984). Oxygen isotope calibration of the onset of ice-rafting and history of glaciation in the North Atlantic region. *Nature (London)* **307**, 620–623.

Shackleton, N. J., Berger, A., and Peltier, W. R. (1990). An alternative astronomical calibration of the lower Pleistocene timescale based on ODP Site 677. *Trans. R. Soc. Edinburgh* **81**, 251–261.

Shackleton, N. J., and the Shipboard Scientific Party (1992). Sedimentation rates: Toward a grape stratigraphy for Leg 138 carbonate sections. In "Proceedings of the Ocean Drilling Program, Initial Reports" (N. G. Pisias, L. A. Mayer, T. R. Janecek *et al.*, eds.), Vol. 138, pp. 87–92. ODP, College Station, TX.

Shackleton, N. J., Crowhurst, S., Hagelberg, T., Pisias, N. G., and Schneider, D. A. (1995a). A new Late Neogene time scale: Application to Leg 138 sites. In "Proceedings of the Ocean Drilling Program, Scientific Results." Vol. 138, pp. 73–104, College Station, TX.

Shackleton, N. J., Hall, M. A., and Pate, D. (1995b). Pliocene stable isotope stratigraphy at Site 846. In "Proceedings of the Ocean Drilling Program, Scientific Results" (N. G. Pisias, L. A. Mayer, T. R. Janecek *et al.*, eds.), Vol. 138, pp. 337–353, College Station, TX.

Shane, P., Black, T. and Westgate, J. (1994). Isothermal plateau fission-track age for a paleo-magnetic excursion in the Mamaku Ignimbrite, New Zealand, and implications for late Quaternary stratigraphy. *Geophys. Res. Lett.* **21**, 1695–1698.

Shi, N. (1994). The late Cenozoic stratigraphy chronology, palynology and environmental development in the Yushe Basin, North China. *Striae* **36**, 1–90.

Shipboard Party (1989). Site 758. In "Proceedings of the Ocean Drilling Program, Initial Reports" (J. Pierce, J. Weissel *et al.*, eds.), Vol. 121, pp. 359–453. ODP, College Station, TX.

Shive, P. N., Steiner, M. B., and Huycke, D. T. (1984). Magnetostratigraphy of the Triassic Chugwater Formation of Wyoming. *J. Geophys. Res.* **89**, 1801–1815.

Singh, G., Opdyke, N. D., and Bowler, J. M. (1981). Late Cainozoic stratigraphy, palaeomagnetic chronology and vegetational history from Lake George, N. S. W. *J. Geol. Soc. Aust.* **28**, 435–452.

Sinito, A. M., Valencio, D. A., and Vilas, J. F. (1979). Paleomagnetism of a sequence of Upper Paleozoic–Lower Mesozoic rec beds from Argentina. *Geophys. J. R. Astron. Soc.* **58**, 237–247.

Sliter, W. V. (1992). Biostratigraphic zonation for Cretaceous planktonic foraminifers examined in thin section. *J. Foraminiferal Res.* **19**, 1–19.

Smith, G., and Creer, K. M. (1986). Analysis of geomagnetic secular variations 10,000 to 30,000 years B.P., Lac du Bouchet, France. *Phys. Earth Planet. Inter.* **4**, 1–14.

Smith, J. D., and Foster, J. H. (1969). Geomagnetic reversal in the Brunhes normal polarity epoch. *Science* **163**, 565–567.

Smith, P. J. (1968). Pre-Gilbertian conceptions of terrestrial magnetism. *Tectonophysics* **6**, 499.

Smith, P. J. (1970). Petrus Peregrinus Epistola—the beginning of experimental studies of magnetism in Europe, Atlas. *Earth Sci. Rev., New Suppl.* **6**, A11.

Snowball, I. (1991). Magnetic hysteresis properties of greigite (Fe_3S_4) and a new occurrence in Holocene sediments from Swedish Lapland. *Phys. Earth Planet. Inter.* **68**, 32–40.

Snowball, I. (1994). Bacterial magnetite and the magnetic properties of sediments in a Swedish lake. *Earth Planet. Sci. Lett.* **126**, 129–142.

Snowball, I., and Thompson, R. (1988). The occurrence of greigite in sediments from Loch Lomond. *J. Quat. Sci.* **3**, 121–125.

Snowball, I., and Thompson, R. (1990). A stable chemical remanence in Holocene sediments. *J. Geophys. Res.* **95**, 4471–4479.

Soffel, H. C. (1977). Pseudo-single domain effects and single-domain multidomain transition in natural pyrrhotite deduced from domain structure observations. *J. Geophys.* **42**, 351–359.

Soffel, H. C. (1981). Domain structure of natural fine-grained pyrrhotite in a rock matrix (diabase). *Phys. Earth Planet. Inter.* **26**, 98–106.

Sparks, N. H. C., Mann, S., Bazylinski, D. A., Lovley, D. R., Jannasch, H. W., and Frankel, R. B. (1990). Structure and morphology of magnetite anaerobically-produced by a marine magnetotactic bacterium and a dissimilatory iron-reducing bacterium. *Earth Planet. Sci. Lett.* **98**, 14–22.

Speiss, V. (1990). Cenozoic magnetostratigraphy of Leg 113 drill sites, Maud Rise, Weddell Sea, Antarctica. *In* "Proceedings of the Ocean Drilling Program, Scientific Results" 113, (P. F. Barker, J. P. Kennett *et al.*, eds.), Vol. 113, pp. 261–290. College Station, TX.

Spell, T. L., and McDougall, I. (1992). Revisions to the age of the Brunhes/Matuyama boundary and the Pleistocene geomagnetic polarity timescale. *Geophys. Res. Lett.* **19**, 1182–1184.

Sprovieri, R. (1992). Mediterranean Pliocene biochronology: A high resolution record based on quantitative planktonic foraminifera distribution. *Riv. Ital. Paleontol. Strat.* **98**, 61–100.

Sprowl, D. R., and Banerjee, S. K. (1989). The Holocene paleosecular variation record from Elk Lake, Minnesota. *J. Geophys. Res.* **94**, 9369–9388.

Stamatakos, J., and Kodama, K. (1991). The effects of grain-scale deformation on the Bloomsburg formation pole. *J. Geophys. Res.* **96**, 1991.

Stearwald, B. A., Clark, D. L., and Andrew, J. A. (1968). Magnetic stratigraphy and faunal patterns in Arctic Ocean sediments. *Earth Planet. Sci. Lett.* **5**, 79–85.

Steiner, M. B. (1978). Magnetic polarity during the Middle Jurassic as recorded in the Summerville and Curtis formations. *Earth Planet. Sci. Lett.* **38**, 331–345.

Steiner, M. B. (1983). Detrital remanent magnetization in hematite. *J. Geophys. Res.* **88**, 6523–6539.

Steiner, M. (1988). Paleomagnetism of the late Pennsylvanian and Permian. *J. Geophys. Res.* **93**, 2201–2215.

Steiner, M., and Helsley, C. (1974). Magnetic polarity sequence of the Kayenta formation. *Geology* **2**, 191–194.

Steiner, M. B., and Helsley, C. E. (1975a). Late Jurassic magnetic polarity sequence. *Earth Planet. Sci. Lett.* **27**, 108–112.

Steiner, M. B., and Helsley, C. E. (1975b). Reversal pattern and apparent polar wander for the late Jurassic. *Geol. Soc. Am. Bull.* **86**, 1537–1574.

Steiner, M. B., and Lucas, S. (1992). A Middle Triassic paleomagnetic pole for North America. *Geol. Soc. Am. Bull.* **104**, 993–998.

Steiner, M. B., and Ogg, J. G. (1988). Early and Middle Jurassic magnetic polarity time scale (R. B. Rocha and A. F. Soares, eds.), pp. 1097–1111. 2nd International symposium on Jurassic stratigraphy, Lisbon.

Steiner, M. B., and Wallick, B. P. (1992). Jurassic to Paleocene paleolatitudes of the Pacific plate derived from the paleomagnetism of the sedimentary sequences at sites 800, 801, and 802. *In* "Proceedings of the Ocean Drilling Program, Scientific Results" (R. L. Larson, Y. Lancelot *et al.*, eds.), Vol. 129, pp. 431–446. ODP, College Station, TX.

Steiner, M. B., Ogg, J. G., Melendez, G., and Sequeiros, L. (1985/86). Jurassic magnetostratigraphy. 2. Middle-Late Oxfordian of Aquilon, Iberian Cordillera, northern Spain. *Earth Planet. Sci. Lett.* **76**, 151–166.

Steiner, M. B., Ogg, J. G., and Sandoval, J. (1987). Jurassic magnetostratigraphy, 3. Bathonian-Bajocian of Carcabuey Sierra Harana and Campillo de Arenas (Subbectic Cordillera, southern Spain). *Earth Planet. Sci. Lett.* **82**, 357–372.

Steiner, M. B., Ogg, J., Zhang, A., and Sun, S. (1989). The Late Permian/Early Triassic magnetic polarity timescale and plate motions of South China. *J. Geophys. Res.* **84**, 7343–7363.

Steiner, M. B., Morales, M., and Shoemaker, E. M. (1993). Magnetostratigraphic, biostratigraphic, and lithologic correlations in Triassic strata of the western United States. *In* "Applications of Palaeomagnetism to Sedimentary Geology" (D. M. Aissaoui, D. F. McNeill, and N. F. Hurley, eds.), SEPM Society for Sedimentary Geology Spec. Publ. No. 49, pp. 41–57.

Steiner, M. B., Lucas, S. G., and Shoemaker, E. M. (1994). Correlation and age of the Late Jurassic Morrison formation from magnetostratigraphy analysis. *In* "Mesozoic Systems of the Rocky Mountain Region, USA, Rocky Mountain Section", (M. V. Caputo, J. A. Peterson, and K. J. Franczyk, eds.), SEPM Spec. Publ., Denver, CO. pp. 315–330.

Stephenson, A. (1971). Single domain grain distributions. 1. A method for the determination of single domain size distributions. *Phys. Earth Planet. Inter.* **4**, 353–360, 361–369.

Stober, J. C., and Thompson, R. (1977). Palaeomagnetic secular variation studies of Finnish lake sediment and the carriers of remanence. *Earth Planet. Sci. Lett.* **37**, 139–149.

Stober, J. C., and Thompson, R. (1979). Magnetic remanence acquisition in Finnish lake sediments. *Geophys. J. R. Astron. Soc.* **57**, 727–739.

Stolz, J. F., Lovley, D. R., and Haggerty, S. E. (1990). Biogenic magnetite and the magnetization of sediments. *J. Geophys. Res.* **95**, 4355–4361.

Stoner, J. S. Channell, J. E. T., Hillaire-Marcel, C., and Mareschal, J. C. (1994). High resolution rock magnetic study of a Late Pleistocene core from the Labrador Sea. *Can. J. Earth Sci.* **31**, 104–114.

Stoner, J. S., Channell, J. E. T., and Hillaire-Marcel, C. (1995a). Magnetic properties of deep-sea sediments off southwest Greenland: Evidence for major differences between the last two deglaciations. *Geology* **23**, 241–244.

Stoner, J. S., Channell, J. E. T., and Hillaire-Marcel, C. (1995b). Late Pleistocene relative geomagnetic paleointensity from the deep Labrador Sea: Regional and global correlations. *Earth Planet. Sci. Lett.* **134**, 237–252.

Stoner, J. S., Channell, J. E. T., and Hillaire-Marcel, C. (1996). The magnetic signature of rapidly deposited detrital layers from the deep Labrador Sea: Relationship to North Atlantic Heinrich layers. *Paleoceanography* **11**, 309–325.

Suk, D., Peacor, R., and Van der Voo, R. (1990). Replacement of pyrite framboids by magnetite in limestone and implications for palaeomagnetism. *Nature (London)* **345**, 611–613.

Suk, D., Van der Voo, R., and Peacor, D. (1993). Origin of magnetite for remagnetization of Early Paleozoic limestones of New York State. *J. Geophys. Res.* **98**, 419–434.

Sun, W., and Jackson, M. (1994). Scanning electron microscopy and rock magnetic carriers in remagnetized Early Paleozoic carbonates from Missouri. *J. Geophys. Res.* **99**, 2935–2942.

Sutter, J. F. (1988). Innovative approaches to the dating of igneous events in the early Mesozoic basins of the eastern United States. *Geol. Surv. Bull. (U.S.)* **1776**, 194–200.

Swisher, C. C., III, and Knox, R. W. O. (1991). The age of the Paleocene/Eocene boundary: [40]Ar/[39]Ar dating of the lower part of NP10, North Sea Basin and Denmark. *Rep. Int. Correl. Proj.*, p. 308.

Swisher, C. C., III, and Prothero, D. R. (1990). Single crystal [40]Ar/[39]Ar dating of the Eocene-Oligocene transition in North America. *Science* **240**, 760–762.

Swisher, C. C., III, Grajales-Nishimura, J. M., Montanari, A., Margolis, S. V., Claeys, P., Alvarez, W., Renné, P. R., Cedillo-Pardo, E., Maurrasse, F. J. M. R., Curtis, G., Smit, S., and McWilliams, M. O. (1992). Coeval ^{40}Ar/^{39}Ar ages of 65.0 million years ago from Chicxulub crater melt rock and Cretaceous-Tertiary boundary tektites. *Science* **257**, 954–958.

Symonds, D. T. A. (1990). Early Permian pole: Evidence from the Pictou red beds, Prince Edward Island, Canada. *Geology* **18**, 234–237.

Talling, P. J., Burbank, D. W., Lawston, T. F., Hobbs, R. S., and Lund, S. P. (1994). Magneto-stratigraphic chronology of Cretaceous to Eocene thrust belt evolution, central Utah, U.S.A. *J. Geol.* **102**, 181–196.

Tamaki, K., and Larson, R. L. (1988). The Mesozoic tectonic history of the Magellan microplate in the western central Pacific. *J. Geophys. Res.* **93**, 2857–2874.

Tarduno, J. A. (1990). Brief reversed polarity interval during the Cretaceous normal polarity superchron. *Geology* **18**, 683–686.

Tarduno, J. A., Sliter, W. V., Bralower, T. J., McWilliams, M., Premoli-Silva, I., and Ogg, J. G. (1989). M-sequence reversals recorded in DSDP sediment cores from the western Mid-Pacific Mountains and Magellan Rise. *Geol. Soc. Am. Bull.* **101**, 1306–1316.

Tarduno, J. A., Sliter, M. V., Kroenke, L., Leckie, M., Mayer, H., Mahoney, J. J., Musgrave, R., Storey, M., and Winterer, E. L. (1991). Rapid formation of the Ontong Java Plateau by Aptian mantle plume volcanism. *Science* **254**, 399–403.

Tarduno, J. A., Lowrie, W., Sliter, W. V., Bralower, T. J., and Heller, F. (1992). Reversed polarity characteristics magnetizations in the Albian Contessa section, Umbrian Appen-ines, Italy: Implications for the existence of a Mid-Cretaceous mixed polarity interval. *J. Geophys. Res.* **97**, 241–271.

Tarling, D. H. (1983). "Paleomagnetism." Chapman & Hall, London.

Tarling, D. H., and Mitchell, J. G. (1976). Revised Cenozoic polarity time scale. *Geology* **4**, 133–136.

Tauxe, L. (1993). Sedimentary records of relative paleointensity of the geomagnetic field: Theory and practice. *Rev. Geophys.* **31**, 319–354.

Tauxe, L., and Butler, R. (1987). Magnetostratigraphy: In pursuit of missing links. *Rev. Geophys.* **25**, 939–950.

Tauxe, L., and Clark, D. R. (1987). New paleomagnetic results from the Eureka Sound Group: Implications for the age of the early Tertiary Arctic biota. *Geol. Soc. Am. Bull.* **99**, 739–747.

Tauxe, L., and Kent, D. V. (1984). Properties of a detrital remanence carried by hematite from study of modern river deposits and laboratory redeposition experiments. *Geophys. J. R. Astron. Soc.* **77**, 543–561.

Tauxe, L., and Opdyke, N. D. (1982). A time framework based on magnetostratigraphy for the Siwalik sediments of the Khaur area, northern Pakistan. *Palaeogeogr., Palaeoclimatol., Palaeoecol.* **37**, 43–61.

Tauxe, L., and Shackleton, N. J. (1994). Relative paleointensity records from the Ontong-Java plateau. *Geophys. J. Int.* **117**, 769–782.

Tauxe, L., and Valet, J. P. (1989). Relative paleointensity of the Earth's magnetic field from marine sedimentary cores: A global perspective. *Phys. Earth Planet. Inter.* **56**, 59–68.

Tauxe, L., and Watson, G. S. (1994). The fold test: An eigen analysis approach. *Earth Planet. Sci. Lett.* **122**, 331–342.

Tauxe, L., and Wu, G. (1990). Normalized remanence in sediments from western equatorial Pacific: Relative paleointensity of the goemagnetic field. *J. Geophys. Res.* **95**, 12337–12350.

Tauxe, L., Kent, D. V., and Opdyke, N. D. (1980). Magnetic components contributing to the NRM of Middle Sidwalik red beds. *Earth Planet. Sci. Lett.* **47**, 279–284.

Tauxe, L., Opdyke, N. D., Pasini, G., and Elmi, C. (1983a). Age of the Plio-Pleistocene boundary in the Vrica section, southern Italy. *Nature (London)* **304,** 125–129.

Tauxe, L., LaBrecque, J. L., Dodson, R., and Fuller, M. (1983b). U-channels—a new technique for paleomagnetic analysis of hydraulic piston cores. *EOS, Trans. Am. Geophys. Union* **64,** 219.

Tauxe, L., Tucker, P., Peterson, N. P., and LaBrecque, J. L. (1983c). The magnetostratigraphy of leg 73 sediments. *Palaeogeogr., Palaeoclimatol., Palaeoecol.* **42,** 65–90.

Tauxe, L., Tucker, P., Peterson, N. P., and LaBrecque, J. L. (1984). Magnetostratigraphy of leg 73 sediments. *In* "Initial Reports of the Deep Sea Drilling Project" (K. J. Hsu, J. L. LaBrecque *et al.,* eds.), Vol. 73, pp. 609–621. U.S. Govt. Printing Office, Washington, DC.

Tauxe, L., Monaghan, M., Drake, R., Curtis, G., and Staudigel, H. (1985). Paleomagnetism of Miocene East African rift sediments and the calibration of the geomagnetic reversal time scale. *J. Geophys. Res.* **90,** 4639–4646.

Tauxe, L., Valet, J., and Bloemendal, J. (1989). Magnetostratigraphy of leg 108 advanced piston cores. *In* "Proceedings of the Ocean Drilling Project, Scientific Results" (W. Ruddiman, M. Sarthein *et al.,* eds.), Vol. 108, pp. 429, 440. ODP, College Station, TX.

Tauxe, L., Kylstra, N., and Constable, C. (1991). Bootstrap statistics for paleomagnetic data. *J. Geophys. Res.* **96,** 11,723–11,740.

Tauxe, L., Deino, A. L., Behrensmeyer, A. K., and Potts, R. (1992). Pinning down the Brunhes/Matuyama and upper Jaramillo boundaries; a reconciliation of orbital and isotopic time scales. *Earth Planet. Sci. Lett.* **190,** 561–572.

Tauxe, L., Gee, J., Gallet, Y., Pick, T., and Brown, J. (1994). Magnetostratigraphy of the Willwood Formation, Bighorn Basin, Wyoming: New constraints on the location of Paleocene/Eocene boundary. *Earth Planet. Sci. Lett.* **125,** 159–172.

Tedford, R. H., Skinner, M. F., Field, R. W., Rensberger, D. P., Whistler, D. P., Galusha, T., Taylor, B. E., MacDonald, J. R., and Webb, S. D. (1987). Biochronology of the Arikarean through Hemphillian interval (Late Oligocene through earliest Pliocene epochs). *In* "Cenozoic Mammals of North America" (M. O. Woodburne, ed.), pp. 153–358. Univ. of California Press, Berkeley.

Tedford, R. H., Flynn, L. J., Zhanxiang, Q., Opdyke, N. D., and Down, W. R. (1991). Yushe Basin, China: Paleomagnetically calibrated mammalian biostratigraphic standard for the late Neogene of Eastern Asia. *J. Vertebr. Paleontol.* **1194,** 519–526.

Testamarte, M. M., and Gosé, W. A. (1979). Magnetostratigraphy of the Eocene-Oligocene Vieja Group, Trans-Pecos Texas. *Bur. Econ. Geol., Univ. Tex. Austin Guideb.* **19,** 55–66.

Thellier, E., and Thellier, O. (1959). Sur l'intensité du champ magnétique terrestre dans le passé historique et géologique. *Ann. Geophys.* **15,** 285–376.

Theyer, F., and Hammond, S. R. (1974). Cenozoic magnetic time scale in deep sea cores: Completion of the Neogene. *Geology* **2,** 487–492.

Thierstein, H. R. (1973). Lower Cretaceous calcareous nannoplankton biostratigraphy. *Abh. Geol. Bundesanst. (Austria)* **29,** 1–52.

Thierstein, H. R. (1976). Mesozoic calcareous nannoplankton biostratigraphy of marine sediments. *Mar. Micropaleontol.* **1,** 325–362.

Thomas, C., and Briden, J. C. (1976). Anomalous geomagnetic field during the Late Ordovician. *Nature (London)* **259,** 380–382.

Thomas, E., Barrera, E., Hamilton, N., Huber, B. T., Kennett, J. P., O'Connell, S. B., Pospichal, J. J., Spiess, V., Stott, L. O., Wei, W., and Wise, S. W. (1990). Upper Cretaceous–Paleocene stratigraphy of sites 689 and 690, Maud Rise, Antarctica. *In* "'Proceedings of the Ocean Drilling Project, Scientific Results" (P. F. Barker, J. P. Kennett *et al.,* eds.), Vol. 113, pp. 901–910. ODP, College Station, TX.

Thompson, R. (1973). Palaeolimnology and palaeomagnetism. *Nature (London)* **242,** 182–184.

Thompson, R. (1986). Modeling magnetization data using SIMPLEX. *Phys. Earth Planet. Inter.* **42**, 113–127.

Thompson, R., and Barraclough, D. R. (1982). Geomagnetic secular variation based on spherical harmonic and cross validation analysis of historical and archaeomagnetic data. *J. Geomagn. Geoelectr.* **34**, 245–263.

Thompson, R., and Berglund, B. (1976). Late Weichselian geomagnetic "reversal" as a possible example of the reinforcement syndrome. *Nature (London)* **264**, 490–491.

Thompson, R., and Clark, R. M. (1989). Sequence slotting for stratigraphic correlation between cores: Theory and practice. *J. Paleolimnol.* **2**, 173–184.

Thompson, R., and Morton, D. J. (1979). Magnetic susceptibility and particle-size distribution in recent sediments of the Loch Lomond drainage basin, Scotland. *J. Sediment. Petrol.* **49**(3), 801–812.

Thompson, R., and Oldfield, F. (1986). "Environmental Magnetism." Allen & Unwin, New York.

Thompson, R., Battarbee, R. W., O'Sullivan, P. E., and Oldfield, F. (1975). Magnetic susceptibility of lake sediments. *Limnol. Oceanogr.* **20**, 687–698.

Thompson, R., Turner, G. M., Stiller, M., and Kaufman, A. (1985). Near East palaeomagnetic secular variation recorded in sediments from the sea of Galilee (Lake Kinneret). *Quat. Res.* **23**, 175–188.

Thouveny, N. (1987). Variations of the relative paleointensity of the geomagnetic field in western Europe in the interval 25–10 kyr BP as deduced from analyses of lake sediments. *Geophys. J. R. Astron. Soc.* **91**, 123–142.

Thouveny, N., and Creer, K. M. (1992). Geomagnetic excursions in the last 60 km: Ephemeral secular variation features. *Geology* **20**, 399–402.

Thouveny, N., and Servant, M. (1989). Paleomagnetic stratigraphy of Pliocene continental deposits of the Bolivian Altiplano. *Palaeogeogr., Palaeoclimatol., Palaeoecol.* **70**, 331–334.

Thouveny, N., Creer, K. M., and Blunk, I. (1990). Extension of the Lac du Bouchet paleomagnetic record over the last 120,000 years. *Earth Planet. Sci. Lett.* **97**, 140–161.

Thouveny, N., de Beaulieu, J.-L., Bonifay, E., Creer, K. M., Gulot, J., Icole, M., Johnsen, S., Jouzel, J., Reille, M., Williams, T., and Williamson, D. (1994). Climate variations in Europe over the past 140 kyr deduced from rock magnetism. *Nature (London)* **371**, 503–506.

Tomida, Y., and Butler, R. F. (1980). Dragonian mammals and paleogene magnetic polarity stratigraphy North Horn Formation, Central Utah. *Am. J. Sci.* **280**, 787–811.

Tonni, E. P., Laberdi, M. T., Prado, J. L., Bargo, M. S., and Cionne, A. L. (1992). Changes of the mammal assemblages in the Pampean region (Argentina) and their relation with the Plio-Pleistocene boundary. *Palaeogeogr., Palaeoclimatol., Palaeoecol.* **95**, 179–194.

Torii, M., Yue, L., Hayashida, A., Maenaka, K., Yokoyama, T., Wang, Y., and Sasajima, S. (1984). Natural remanent magnetization of loess/paleosol deposits in Luochuan area. *In* "The Recent Research of Loess in China" (S. Sasajima and Yongyan, eds.), pp. 32–41. Kyoto University, Kyoto/Northwest Univ.

Torsvik, T. H., and Trench, A. (1991). Ordovician magnetostratigraphy: Llanvirn-Caradoc limestone of the Baltic Platform. *Geophys. J. Int.* **107**, 171–184.

Torsvik, T. H., Lyse, O., Attaras, G., and Bluck, B. J. (1989). Paleozoic paleomagnetic results from Scotland and their bearing on the British apparent polar wandering path. *Phys. Earth Planet. Inter.* **55**, 93–105.

Townsend, H. A. (1985). The paleomagnetism of sediments acquired from the Goban spur on Deep Sea Drilling Project Leg 80. *In* "Initial Reports of the Deep Sea Drilling Project" (P. R. Graciansky, C. W. Pong *et al.*, eds.), Vol. 80, pp. 389–414. U.S. Govt. Printing Office, Washington, DC.

Townsend, H. A., and Hailwood, E. (1985). Magnetostratigraphic correlations of Palaeogene sediments in the Hampshire and London basins, southern U.K. *J. Geol. Soc., London* **142**, 957–982.

Trench, A., McKerrow, W. S., and Torsvik, T. H. (1991). Ordovician magnetostratigraphy: A correlation of global data. *J. Geol. Soc., London* **148**, 949–958.

Trench, A., McKerrow, W. S., Torsvik, T. H., Li, Z. X., and McCraken, S. R. (1993). The polarity of the Silurian magnetic field: Indications from a global data compilation. *J. Geol. Soc., London* **150**, 823–831.

Tric, E., Laj, C., Jéhanno, C., Valet, J.-P., Kissel, C., Mazaud, A., and Iaccarino, S. (1991a). High-resolution record of the Olduvai transition from Po valley (Italy) sediments: support for dipolar transition geometry. *Phys. Earth Planet. Inter.* **65**, 319–336.

Tric, E., Laj, C., Valet, J.-P., Tucholka, P., Paterne, M., and Guichard, F. (1991b). The Blake geomagnetic event: Transition geometry, dynamical characteristics and geomagnetic significance. *Earth Planet. Sci. Lett.* **102**, 1–13.

Tric, E., Valet, J. P., Tucholka, P., Paterne, M., Labeyrie, L., Guichard, F., Tauxe, L., and Fontugne, M. (1992). Paleointensity of the geomagnetic field for the last 80,000 years. *J. Geophys. Res.* **97**, 9337–9351.

Tsuchi, R., Takayanagi, Y., and Shibata, K. (1981). Neogene bioevents in the Japanese islands. *In* "Neogene of Japan: Its Biostratigraphy and Chronology" (T. Tsuchi, ed.), pp. 15–32. Kurofune Printing Co., Shizuoka.

Tucholka, P., Fontugue, M., Guichard, F., and Paterne, M. (1987). The Blake polarity episode in cores from the Mediterranean Sea. *Earth Planet. Sci. Lett.* **86**, 320–326.

Tucker, P. (1980). Stirred remanent magnetization: A laboratory analogue of post-depositional realignment. *J. Geophys.* **48**, 153–157.

Tucker, P. (1981). Paleointensities from sediments: Normalization by laboratory redeposition. *Earth Planet. Sci. Lett.* **56**, 398–404.

Tucker, P., and Tauxe, L. (1984). The downhole variation of the rock-magnetic properties of Leg 73 sediments. *In* "Initial Reports of the Deep Sea Drilling Project" (K. J. Hsu, J. L. LaBrecque *et al.*, eds.), Vol. 73, pp. 673–685. U.S. Govt. Printing Office, Washington, DC.

Turner, G. M. (1987). A 5000 year geomagnetic paleosecular variation record from western Canada. *Geophys. J. R. Astron. Soc.* **91**, 103–121.

Turner, G. M., and Thompson, R. (1979). Behaviour of the Earth's magnetic field as recorded in the sediments of Loch Lomond. *Earth Planet. Sci. Lett.* **42**, 412–426.

Turner, G. M., and Thompson, R. (1981). Lake sediment record of the geomagnetic secular variation in Britain during Holocene times. *Geophys. J. R. Astron. Soc.* **65**, 703–725.

Turner, G. M., and Thompson, R. (1982). Detransformation of the British geomagnetic secular variation record for Holocene times. *Geophys. J. R. Astron. Soc.* **70**, 789–792.

Turner, P. (1975). Depositional magnetization of the Carboniferous limestones from the Craven Basin of northern England. *Sedimentology* **22**, 563–581.

Turner, P., Metcalfe, I., and Tarling, D. H. (1979). Paleomagnetic studies of some Dinantian limestones from the Craven Basin and a contribution to Aspian magnetostratigraphy. *Proc. Yorks. Geol. Soc.* **42**, 371–396.

Turner, P., Turner, A., Ramos, A., and Sopeña, A. (1989). Palaeomagnetism of Permo-Triassic rocks in the Iberian Cordillera, Spain: Acquisition of secondary and characteristic remanence. *J. Geol. Soc., London* **146**, 61–76.

Turrin, B. D., Donnelly-Nolan, J. M., and Hearn, B. C., Jr. (1994). $^{40}Ar/^{39}Ar$ ages from the rhyolite of Alder Creek, California: Age of the Cobb Mountain normal polarity subchron revisited. *Geology* **22**, 251–254.

Urrutia-Fucugauchi, J., Bohnel, H. M., and Valencio, D. A. (1990). Magnetostratigraphy of a Middle Jurassic red bed sequence from southern Mexico. *Phys. Earth Planet. Inter.* **64**, 237–246.

Uyeda, S. (1958). Thermoremanent magnetism as a medium of palaeomagnetism with special reference to reverse thermoremanent magnetism. *Jpn. J. Geophys.* **2**, 1–123.

Vail, P. R., and Hardenbol, J. (1979). Sea level changes during the Tertiary. *Oceanus* **22**, 71–79.

Vail, P. R., Mitchum, R. M., Jr., and Thompson, S., III (1977). Global cycles of relative changes of sea level. *In* "Seismic Stratigraphy: Applications to Hydrocarbon Exploration" (C. E. Payton, ed.), SEPM (Society for Sedimentary Geology Spec. Publ. No. 45, pp. 83–97.

Valencio, D. A. (1981a). Paleomagnetism of Lower Ordovician and Upper Pre-Cambrian rocks from Argentina. *In* "Global Reconstruction and the Geomagnetic Field During the Paleozoic" (M. McElhinny *et al.*, eds.), Adv. Earth Planet. Sci., Vol. 10, pp. 71–75. Center for Academic Publ., Tokyo.

Valencio, D. A. (1981b). Reversals and excursions of the geomagnetic field as defined by paleomagnetic data from upper Paleozoic–lower Mesozoic sediments and igneous rocks. *In* "Global Reconstruction and the Geomagnetic Field During the Paleozoic" (M. McElhinny et al., eds.), Adv. Earth Planet. Sci., Vol. 10, pp. 137–142. Center for Academic Publ., Tokyo.

Valencio, D. A., and Orgeira, M. J. (1983). La magnetostratigrafia del ensenadense y bonarense de la ciudad de Buenos Aires. Parte II. *Asoc. Geol. Argent. Rev.* **38**(1), 24–33.

Valencio, D. A., Vilas, J. F., and Mendia, J. E. (1977). Paleomagnetism of a sequence of red beds of the Middle and Upper sections of Paganzo Group (Argentina) and the correlation of Upper Paleozoic–Lower Mesozoic rocks. *Geophys. J. R. Astron. Soc.* **62**, 27–39.

Valencio, D. A., Vilas, J. F., and Pacca, I. G. (1983). The significance of the palaeomagnetism of Jurassic-Cretaceous rocks from South America: Pre-drift movements, hairpins and magnetostratigraphy. *Geophys. J. R. Astron. Soc.* **73**, 135–151.

Valet, J. P., and Meynadier, L. (1993). Geomagnetic field intensity and reversals during the past four million years. *Nature (London)* **366**, 234–238.

Valet, J. P., Tauxe, L., and Clement, B. M. (1989). Equatorial and mid-latitude records of the last geomagnetic reversal from the Atlantic Ocean. *Earth Planet. Sci. Lett.* **94**, 371–384.

Valet, J. P., Tucholka, P., Courtillot, V., and Meynadier, L. (1992). Paleomagnetic constraints on the geometry of the geomagnetic field during reversals. *Nature (London)* **356**, 400–407.

Valet, J. P., Meynadier, L., Bassinot, F. C., and Garnier, F. (1994). Relative paleointensity across the last geomagnetic reversal from sediments of the Atlantic, Indian and Pacific Oceans. *Geophys. Res. Lett.* **21**, 485–488.

Vali, H., Förster, O., Amarantidis, G., and Petersen, N. (1987). Magnetotactic bacteria and their magnetofossils in sediments. *Earth Planet. Sci. Lett.* **86**, 389–400.

Vandamme, D., Courtillot, V., Besse, J., and Montigny, R. (1991). Paleomagnetism and age determinations of the Deccan Traps (India): Results of a Naupur-Bombay traverse, and review of earlier work. *Rev. Geophys.* **29**, 159–190.

Vandenberg, J., and Wonders, A. A. H. (1980). Paleomagnetism of late Mesozoic pelagic limestones from the southern Alps. *J. Geophys. Res.* **85**, 3623–3637.

Van der Voo, R. (1990). Phanerozoic paleomagnetic poles from Europe and North America and comparisons with continental reconstructions. *Rev. Geophys.* **28**, 167–206.

Van der Voo, R. (1993). "Paleomagnetism of the Atlantic, Tethys and Iapetus Oceans." Cambridge Univ. Press, Cambridge, UK.

Verosub, K. L. (1977a). Depositional and post-depositional processes in the magnetization of sediments. *Rev. Geophys. Space Phys.* **15**(2), 129–143.

Verosub, K. L. (1977b). The absence of the Mono Lake geomagnetic excursion from the paleomagnetic record of Clear Lake, California. *Earth Planet. Sci. Lett.* **36**, 219–230.

Verosub, K. L., and Banerjee, S. K. (1977). Geomagnetic excursions and their paleomagnetic record. *Rev. Geophys.* **15**, 145–155.

Verosub, K. L., Davis, J. O., and Valastro, S. (1980). A paleomagnetic record from Pyramid Lake, Nevada, and its implications for proposed geomagnetic excursions. *Earth Planet. Sci. Lett.* **49**, 141–148.

Verosub, K. L., Mehringer, P. J., Jr., and Waterstraat, P. (1986). Holocene secular variation in western North America: Paleomagnetic record from Fish Lake, Harney County, Oregon. *J. Geophys. Res.* **91**, 3609–3623.

Verosub, K. L., Haggart, J. W., and Ward, P. D. (1989). Magnetostratigraphy of Upper Cretaceous strata of the Sacramento Valley, California. *Geol. Soc. Am. Bull.* **101**, 521–533.

Vincent, E., and Berger, W. H. (1985). Carbon dioxide and polar cooling in the Miocene: The Monterey hypothesis. *In* "The Carbon Cycle and Atmospheric CO_2: Natural Variations Archean to Present" (E. T. Sundquist and W. S. Brecker, eds.), Geophys. Monogr., Vol. 32, pp. 455–468. Am. Geophys. Union, Washington, DC.

Vincent, E., Killingsley, J. and Berger, W. H. (1985). Miocene oxygen and carbon isotope stratigraphy of the tropical Indian Ocean. *Mem. Geol. Soc. Am.* **163**, 103–130.

Vine, F. J. (1966). Spreading of the ocean floor: New evidence. *Science* **154**, 1405–1415.

Vine, F. J., and Matthews, D. H. (1963). Magnetic anomalies over oceanic ridges. *Nature (London)* **199**, 947–949.

Vine, F. J., and Wilson, J. T. (1965). Magnetic anomalies over a young ocean ridge off Vancouver Island. *Science* **150**, 485–489.

Vogt, P. R., Anderson, C. N., and Bracey, D. R. (1971). Mesozoic magnetic anomalies, sea-floor spreading, and geomagnetic reversals in the southwestern North Atlantic. *J. Geophys. Res.* **76**, 4792–4823.

Walker, T. R., Larson, E. E., and Hoblitt, R. P. (1981). Nature and origin of hematite in the Moenkopi Formation (Triassic), Colorado plateau: A contribution to the origin of magnetism in red beds. *J. Geophys. Res.* **86**, 317–334.

Walter, R. C. (1994). Age of Lucy and First Family: Single crystal $^{40}Ar/^{39}Ar$ dating of the Denen Dora and lower Kada Hadar members of the Hadar Formation, Ethiopia. *Geology* **22**, 6–10.

Walter, R. C., and Aronson, J. L. (1993). Age and source of the Sidi Hakoma tuff, Hadar Formation, Ethiopia. *J. Hum. Evol.* **25**, 229–240.

Walter, R. C., Manega, P. C., Hay, R. L., Drake, R. E., and Curtis, G. H. (1991). Laser fusion $^{40}Ar/^{39}Ar$ dating of Bed 1, Olduvai Gorge, Tanzania. *Nature (London)* **354**, 145–149.

Walton, A. H. (1992). Magnetostratigraphy of the Lower and Middle members of the Devils Graveyard Formation (Middle Eocene) Trans-Pecos, Texas. *In* "Eocene-Oligocene Climatic and Biotic Evolution" (D. R. Prothero and W. Berggren, eds.), pp. 75–86. Princeton Univ. Press, Princeton, NJ.

Watkins, N. D. (1968). Short period geomagnetic polarity events in deep sea sedimentary cores. *Earth Planet. Sci. Lett.* **4**, 341–349.

Watkins, N. D. (1972). Review of the development of the geomagnetic polarity timescale and discussion of prospects for its fine definition. *Geol. Soc. Am. Bull.* **83**, 551–574.

Watkins, N. D., and Walker, G. P. L. (1977). Magnetostratigraphy of eastern Iceland. *Am. J. Sci.* **277**, 513–584.

Watkins, N. D., McDougall, I., and Kristjansson, L. (1975). A detailed paleomagnetic survey of the type location for the Gilsa geomagnetic polarity event. *Earth Planet. Sci. Lett.* **27**, 436–444.

Watkins, N. D., McDougall, I., and Kristjansson, L. (1977). Upper Miocene and Pliocene geomagnetic secular variation in the Borgarfjordar Area of western Iceland. *Geophys. J. R. Astron. Soc.* **49**, 609–632.

Watson, G. S. (1956a). Analysis of dispersion on a sphere. *Mon. Not. R. Astron. Soc. Geophys., Suppl.* **7**, 153–159.

Watson, G. S. (1956b). A test for randomness of directions. *Mon. Not. R. Astron. Soc. Geophys.,* *Suppl.* **7,** 160–161.

Watson, G. S., and Enkin, R. J. (1993). The fold test in paleomagnetism as a parameter estimation problem. *Geophys. Res. Lett.* **20,** 2135–2137.

Weaver, P. P. E., and Clement, B. M. (1986). Synchroneity of Pliocene plankton foraminiferal datums in the North Atlantic. *Mar. Micropaleontol.* **10,** 295–307.

Webb, S. D., and Opdyke, N. D. (1995). Global climatic influence on Cenozoic land mammal faunas. *In* "Effects of Past Global Changes on Life, Studies in Geophysics" *Spec. Publ. Natl. Acad. Sci.* pp. 247–274.

Weeks, R., Laj, C., Endignoux, L., Fuller, M., Roberts, A., Manganne, R., Blanchard, E., and Goree, W. S. (1993). Improvements in long-core measurement techniques: Applications in palaeomagnetism and palaeoceanography. *Geophys. J. Int.* **114,** 651–662.

Weeks, R. J., Laj, C., Endignoux, L., Mazaud, A., Labeyrie, L., Roberts, A. P., Kissel, C., and Blanchard, E. (1995). Normalised NRM intensity during the last 240,000 years in piston cores from Central North Atlantic Ocean: geomagnetic field intensity or environmental signal? *Phys. Earth Planet. Inter.* **87,** 213–229.

Wei, W. (1995). Revised age calibration points for the geomagnetic polarity time scale. *Geophys. Res. Lett.* **22,** 957–960.

Weinrich, N., and Theyer, F. (1985). Paleomagnetism of Deep Sea Drilling Project Leg 85 sediments: Neogene magnetostratigraphy and tectonic history of the Central Equatorial Pacific. *In* "Initial Reports of the Deep Sea Drilling Project" (L. Mayer, and F. Theyer, eds.), Vol. 85, pp. 849–901. U.S. Govt. Printing Office, Washington, DC.

Weissert, H., McKenzie, J. A., and Channell, J. E. T. (1985). Natural variations in the carbon cycle during the Early Cretaceous. *In* "The Carbon Cycle and Atmospheric CO_2: Natural Variations Archean to Present" (E. T. Sundquist and W. S. Broecker, eds.), Geophys. Monogr. Vol. 32, pp. 531–545. Am. Geophys. Union, Washington, DC.

Wensink, H. (1964). Paleomagnetic stratigraphy of younger basalts and intercalated Plio-Pleistocene tillites in Iceland. *Geol. Rundsch.* **54,** 364–384.

Westgate, J. A., Stemper, B. A., and Péwé, J. L. (1990). A 3 MY record of Pliocene-Pleistocene loess in interior Alaska. *Geology* **18**(9), 858–861.

Westphal, M., and Durand, J. P. (1990). Magnetostratigraphic des séries continental flurio-lacustres du Cretace supérieu dans synclinal de l'Arc (région d'Aien Provence, France). *Bull. Soc. Geol. Fr.* **8**(4, Part 1), 609–620.

Whistler, D. P., and Burbank, D. W. (1992). Miocene biostratigraphy and biochronology of the Dove Spring Formation, Mojave Desert, California, and characterization of the Clarendonian mammal age (late Miocene) in California. *Geol. Soc. Am. Bull.* **104,** 644–658.

Whitelaw, M. J. (1989). Magnetic polarity stratigraphy and mammalian fauna of the Late Pliocene (Early Matuyama) section at Batesford (Victoria), Australia. *J. Geol.* **97,** 624–631.

Whitelaw, M. J. (1990). Magnetic polarity stratigraphy of the Pliocene Hamilton and Forsyth's Bank Local sections, Hamilton (Victoria) Australia. *J. Geol.* **99,** 310–315.

Whitelaw, M. J. (1991a). Magnetic polarity stratigraphy of Pliocene and Pleistocene fossil vertebrate localities in southeastern Australia. *Geol. Soc. Am. Bull.* **103,** 1493–1503.

Whitelaw, M. J. (1991b). Magnetic polarity stratigraphy of the Fisherman's Cliff and Bone Bulch vertebrate fossil faunas from the Murray Basin New South Wales, Australia. *Earth Planet. Sci. Lett.* **104,** 417–423.

Whitelaw, M. J. (1992). Magnetic polarity stratigraphy of three Pliocene sections and inferences for the ages of vertebrate fossil sites near Bacchus Marsh Victoria, Australia. *Aust. J. Earth Sci.* **39,** 521–528.

Whitelaw, M. J. (1993). Age constraints on the Duck Ponds and Limeburner's Point mammalian faunas based on magnetic polarity stratigraphy in the Geelong area (Victoria), Australia. *Quat. Res.* **39,** 120–124.

Wiedenmayer, F. (1980). Die Ammoniten der mediterranean Provinz im Pliensbachian und im unteren Toarcian aufgrund neuer Unterschungen im Generosos-Becken (Lombardische Alpen). *Denkschr. Schweiz. Naturforsch. Ges.* **93,** 1–260.

Wignall, P. B., and Hallam, A. (1993). Greisbachian (earliest Triassic) palaeoenvironmental changes in the Salt Range, Pakistan and southeast China and their bearing on the Permo-Triassic mass extinction. *Palaeogeogr., Palaeoclimatol., Palaeoecol.* **102,** 215–237.

Wilson, D. S. (1993). Confirmation of the astronomical calibration of the magnetic polarity timescale from sea-floor spreading rates. *Nature (London)* **364,** 788–790.

Wilson, R. L. (1961). Palaeomagnetism in Northern Ireland. Pt. 1. The thermal demagnetization of natural magnetic movements of rocks. *Geophys. J. Roy. Astron. Soc.* **5,** 45–69.

Wilson, R. L. (1962). The paleomagnetism of baked contact rocks and reversals of the earth's magnetic field. *Geophys. J. R. Astron. Soc.* **7,** 194–202.

Wilson, R. L. (1972). Paleomagnetic differences between normal and reversed field sources and the problem of far-sided and right-handed pole positions. *Geophys. J. R. Astron. Soc.* **28,** 295–304.

Wilson, R. L., and Lomax, R. (1972). Magnetic remanence related to slow rotation of ferromagnetic material in alternating magnetic fields. *Geophys. J. R. Astron. Soc.* **30,** 295–304.

Wing, S. L., Bown, T. M., and Obradovich, J. D. (1991). Early Eocene biotic and cliamtic change in interior western North America. *Geology* **19,** 1189–1192.

Witte, W. K., and Kent, D. V. (1989). A middle Carnian to early Norian (~225 Ma) paleopole from sediments of the Neward Basin Pennsylvania. *Geol. Soc. Am. Bull.* **101,** 1175–1181.

Witte, W. K., Kent, D. V., and Olsen, P. E. (1991). Magnetostratigraphy and paleomagnetic poles from the Late Triassic–earliest Jurassic strata of the Neward basin. *Geol. Soc. Am. Bull.* **103,** 1648–1662.

Wollin, G., Ericson, D. B., Ryan, W. B. F., and Foster, J. H. (1971). Magnetism of the Earth and climatic changes. *Earth Planet. Sci. Lett.* **12,** 175–183.

Wollin, G., Ryan, W. B. F., Ericson, D. B., and Foster, J. H. (1977). Palaeoclimate, palaeomagnetism and the eccentricity of the earth's orbit. *Geophys. Res. Lett.* **44,** 267.

Wood, H. E., Chanely, R. W., Clary, J., Colbert, E. H., Jepson, G. L., Reesile, J. B., Jr., and Stock, C. (1941). Nomenclature and correlation of the North American continental Tertiary. *Geol. Soc. Am. Bull.* **52,** 1–48.

Woodburne, M. D. (1987). Principles, classifications and recommendations. *In* "Cenozoic Mammals of North America" (M. O. Woodburne, ed.), pp. 9–17. Univ. of California Press, Berkeley.

Woodruff, F., and Savin, S. M. (1991). Mid Miocene isotopic stratigraphy in the deep-sea: High resolution correlations, Paleoclimatic cycles in sediment preservation. *Paleoceanography* **6,** 755–806.

Wright, J. D., and Miller, K. G. (1992). Miocene stable isotope stratigraphy, site 747, Kerguelen Plateau *Proc. Ocean Drill Program* **120,** Sci. Results, part B., 855–866.

Wu, F., Van der Voo, R., and Liang, Q. Z. (1989). Reconnaissance magnetostratigraphy of the Pre Cambrian–Cambrian boundary section at Meishueun, south west China. *Cuad. Geol. Iber.* **12,** 205–222.

Wynne, P. J., Irving, E., and Osadetz, K. G. (1983). Paleomagnetism of the Esayoo Formation (Permian) of northern Ellesmere Island: Possible clue for the solution of the Nares Strait dilemma. *Tectonophysics* **100,** 241–256.

Wynne, P. J., Irving, E., and Osudetz, K. G. (1988). Paleomagnetism of Cretaceous volcanic rocks of the Sverdrup Basin—magnetostratigraphy, paleolatitudes, and rotations. *Can. J. Earth Sci.* **25,** 1220–1239.

Yamazaki, T., and Ioka, N. (1994). Long-term secular variation of the geomagnetic field during the last 200 kyr recorded in sediment cores from the western equatorial Pacific. *Earth Planet. Sci. Lett.* **128**, 527–544.

Yamazaki, T., and Katsura, T. (1990). Magnetic grain size and viscous remanent magnetization of pelagic clay. *J. Geophys. Res.* **95**, 4373–4382.

Yamazaki, T., Katsura, I., and Marumo, K. (1991). Origin of stable remanent magnetization of siliceous sediments in the central equatorial Pacific. *Earth Planet. Sci. Lett.* **105**, 81–93.

Yang, Z., Moreau, M.-G., Bucher, H., Dommergues, J.-L., and Trouiller, A. (1996). Hettangian and Sinemurian magnetostratigraphy from Paris Basin. *J. Geophys. Res.* **101**, 8025–8042.

Yaskawa, K., Nakajima, T., Kawai, N., Torii, M., Natsuhara, N., and Horie, S. (1973). Paleomagnetism of a core from lake Biwa (I). *J. Geomagn. Geoelectr.* **25**, 447–474.

Yue, L., Nishida, J., Sun, W., Miao, J., Wang, Y., and Sasajima, S. (1984). The magnetostratigraphic study of Chinese loess section from Shaan Xian and Xifeng. *In* "The Recent Research of Loess in China" (S. Sasajima and W. Yonyan, eds.), pp. 49–62. Kyoto University, Kyoto/Northwest Univ.

Zachariasse, W. J., Zijderveld, J. D. A., Langereis, C. G., Hilgen, F. J., and Verhallen, P. J. J. M. (1989). Early Late Pliocene biochronology and surface water temperature variations in the Mediterranean. *Mar. Micropaleontol.* **14**, 339–355.

Zachariasse, W. J., Gudjonsson, L., Hilgen, F. J., Langereis, C. G., Lourens, L. J., Verhallen, P. J. J. M., and Zijderveld, J. D. A. (1990). Late Gauss to early Matuyama invasions of *Neogloboquadrina atlantica* in the Mediterranean and associated records of climate change. *Paleoceanography* **5**, 239–252.

Zheng, H., An, Z., and Shaw, J. (1992). New contributions to Chinese Plio-Pleistocene magnetostratigraphy. *Phys. Earth Planet. Inter.* **70**, 146–133.

Zhou, L. P., Oldfield, F., Wintle, A. G., Robinson, S. G., and Wang, J. T. (1990). Partly pedogenic origin of magnetic variations in Chinese loess. *Nature (London)* **346**, 737–739.

Zhu, R. X., Zhou, L. P., Laj, C., Mazaud, A., and Ding, D. L. (1994). The Blake geomagnetic polarity episode recorded in Chinese loess. *Geophys. Res. Lett.* **21**, 697–700.

Zijderveld, J. D. A. (1967). AC demagnetization of rocks: Analysis of results. *In* "Methods in Paleomagnetism" (D. W. Collinson, K. M. Creer, and S. K. Runcorn, eds.), pp. 254–286. Elsevier, Amsterdam.

Zijderveld, J. D. A., Zachariasse, W. J., Verhallen, P. J. J. M., and Hilgen, F. J. (1986). The age of the Miocene-Pliocene boundary. *Newsl. Stratigr.* **16**, 169–181.

Zijderveld, J. D. A., Langereis, C. G., Hilgen, F. J., Verhallen, P. J. J. M., and Zachariasse, W. J. (1991). Integrated magnetostratigraphy and biostratigraphy of the Upper Pliocene–lower Pleistocene from the Monte Singa and Crotone areas in southern Calabria (Italy). *Earth Planet Sci. Lett.* **107**, 697–714.

Index

International Geophysics Series

EDITED BY

RENATA DMOWSKA
Division of Applied Science
Harvard University
Cambridge, Massachusetts

JAMES R. HOLTON
Department of Atmospheric Sciences
University of Washington
Seattle, Washington

* Out of Print

ISBN 0-12-527470-X